T0130939

MALDI MASS SPECTROMETRY FOR SYNTHETIC POLYMER ANALYSIS

CHEMICAL ANALYSIS

A SERIES OF MONOGRAPHS ON ANALYTICAL CHEMISTRY AND ITS APPLICATIONS

Series Editor
J. D. WINEFORDNER

Volume 175

A complete list of the titles in this series appears at the end of this volume.

MALDI MASS SPECTROMETRY FOR SYNTHETIC POLYMER ANALYSIS

EDITED BY
LIANG LI

WILEY

A JOHN WILEY & SONS, INC., PUBLICATION

For general information on our other products and services or for technical support, please contact our Customer Care Department within the United States at (800) 762-2974, outside the United States at (317) 572-3993 or fax (317) 572-4002.

Wiley also publishes its books in a variety of electronic formats. Some content that appears in print may not be available in electronic formats. For more information about Wiley products, visit our web site at www.wiley.com.

Library of Congress Cataloging-in-Publication Data
MALDI mass spectrometry for synthetic polymer analysis / edited by Liang Li.
    p. cm.
   Includes index.
   ISBN 978-0-471-77579-9 (cloth)
  1. Polymers–Spectra.   2. Matrix-assisted laser desorption-ionization.   I. Li, Liang.
   QD139.P6M35 2010
   547′.70154365–dc22

                                                 2009025918

Printed in the United States of America.

10 9 8 7 6 5 4 3 2 1

CONTENTS

v

PREFACE

WITH THE INCREASE in structural and compositional complexity of synthetic polymers, detailed characterization of a newly developed polymeric material is often a challenging task. Mass spectrometry (MS) has played an increasingly important role in polymer analysis. In particular, the technique of matrix-assisted laser desorption ionization (MALDI) MS has transformed the practice of polymer characterization.

Compared to other mass spectrometric techniques, MALDI MS provides several unique attributes that, taken together, make it a powerful tool for analyzing a wide range of polymers. However, because of the diversity in polymer chemistry, there is no universal approach in MALDI MS for the analysis of polymers. The major challenge in applying MALDI MS to characterize a particular polymeric system lies in developing a suitable analytical protocol tailored to this polymer. This usually involves screening and selecting a suitable matrix from a list of known matrices used for the analysis of similar polymers. In some cases, searching for new matrices is required. Once an appropriate matrix is found, paying attention to the many details in the MALDI MS analysis procedure is still required to ensure the final mass spectrometric results reflect the true chemical nature of the polymeric system.

There are a number of pitfalls during the MALDI MS analysis of a polymer sample. It is often easy to collect some mass spectral signals, but is not trivial to generate a correct mass spectrum that truly represents the chemical composition of the sample. Many experimental and instrumental factors can contribute to measurement errors. The intent of this book is to provide a tutorial on each important subject area related to MALDI MS. It is written for those who wish to gain a better understanding of the processes involved in MALDI MS polymer analysis. Through this book, we hope the reader will appreciate the subtlety of each step of the analysis that can influence the results and be cautious in data interpretation. We also hope the technical discussion often accompanied with useful hints and comments may inspire new ideas for developing suitable protocols tailored to a specific application. We note that many aspects of the MALDI MS technique are not fully or well understood at present and thus, a few similar topics (e.g., ionization mechanism and the role of the matrix) are discussed in different chapters from different angles or viewpoints.

I would like to thank all the contributors for their enthusiastic support and their patience. They have written the chapters with a high educational content and provided many expert views on the subjects. They have treated the topics with professionalism and with a balanced view of the pros and cons of each subject.

Because of their excellent work, I believe this book will serve well researchers entering the field as well as graduate students who wish to learn polymer MS. I would also like to thank the anonymous reviewers of each chapter for their critical comments and suggestions.

I wish to express my appreciation to Michael Leventhal of John Wiley for his assistance and patience. I would like to take this opportunity to acknowledge my students and postdoctoral fellows, Dr. Randy M. Whittal, Dr. David C. Schriemer, Dr. Talat Yalcin, Dr. Yuqin Dai, Dr. Wojciech Gabryelski, Ms. Honghui Zhu, Ms. Xinlei Yu, Dr. Nan Zhang, and Dr. Rui Chen, for their hard work on polymer MS. I thank many collaborators in the polymer industry, government institutions, and universities for their guidance, inspirational discussion, and friendship. Finally, I would like to thank my wife, Monica Li, for her love and support and our lovely children, Matthew and Stephanie, for their understanding and joyful smiles.

Liang Li
Edmonton, Alberta, Canada
May 2009

CONTRIBUTORS

Sabine Borgmann, Department of Chemistry and Biochemistry, University of Arkansas; Email: sabine-borgmann@web.de

Jana Falkenhagen, Federal Institute for Materials Research and Testing (BAM), Department I, Analytical Chemistry, Reference Materials, Richard-Willstaetter-Strasse 11, Berlin, Germany; Email: Jana.Falkenhagen@bam.de

Charles Guttman, Polymers Division, National Institute of Standards and Technology, Gaithersburg, MD; Email: charles.guttman@nist.gov

Scott D. Hanton, Hanton, Air Products & Chemicals, Inc., Allentown, PA; Email: hantonsd@airproducts.com

Anthony T. Jackson, Expert Capability Group Leader, Measurement and Analytical Science, Analytics and Physics (CAP), AkzoNobel RD&I, Deventer, The Netherlands; Email: tony.jackson@akzonobel.com

Anthony J. Kearsley, Mathematical and Computational Sciences Division, National Institute of Standards and Technology, Gaithersburg, MD; Email: ajk@nist.gov

Liang Li, Department of Chemistry, University of Alberta, Edmonton, Alberta, Canada; Email: liang.li@ualberta.ca

Kevin G. Owens, Department of Chemistry, Drexel University, Philadelphia, PA; Email: kevin.owens@drexel.edu

Michael J. Polce, Lubrizol Advanced Materials, Cleveland, OH; Email: michael.polce@lubrizol.com

Sarah Trimpin, Department of Chemistry, Wayne State University, Detroit, MI; Email: strimpin@chem.wayne.edu

William E. Wallace, Polymers Division, National Institute of Standards and Technology, Gaithersburg, MD; Email: william.wallace@nist.gov

Steffen M. Weidner, Federal Institute for Materials Research and Testing (BAM), Department I, Analytical Chemistry, Reference Materials, Richard-Willstaetter-Strasse 11, Berlin, Germany; Email: steffen.weidner@bam.de

Chrys Wesdemiotis, Department of Chemistry, The University of Akron, Knight Chemical Laboratory, Akron, OH; Email: wesdemiotis@uakron.edu

Randy M. Whittal, Department of Chemistry, University of Alberta, Edmonton, Alberta, Canada; Email: randy.whittal@ualberta.ca

Charles L. Wilkins, Department of Chemistry and Biochemistry, University of Arkansas, Fayetteville, AR; Email: cwilkins@mail.uark.edu

Renato Zenobi, Department of Chemistry and Applied Biosciences ETH Zurich, Switzerland; Email: ZENOBI@org.chem.ethz.ch

CHAPTER 1

OVERVIEW OF MS AND MALDI MS FOR POLYMER ANALYSIS

Liang Li

Department of Chemistry, University of Alberta, Edmonton, Alberta, Canada

POLYMER ANALYSIS involves many different activities ranging from average molecular mass determination to detailed characterization of chemical structures or compositions. It is a key step in understanding the relations of chemophysical properties of a polymeric material and its functions. It is essential for quality control of polymeric products as well as in troubleshooting of a polymer manufacturing process. Currently, there is a high demand for developing specialty materials in many applications such as safe water treatment, recyclable materials in high-tech products, and drug delivery systems with improved therapeutic efficacies. Many of the new materials are based on synthetic polymers. One can expect that the field of polymer analysis will continue to become increasingly important in polymer industry and polymer science.

Polymer analysis is often a challenging task, particularly when a polymer is made from new polymer chemistry, catalysis, or formulation process. Fortunately, a great number of analytical techniques are now available for polymer analysis. They include separation techniques [1–7], spectroscopic methods [8–11], NMR [12–14], X-ray [15–17], microscopy [18–21] and other surface characterization tools [22–24], and mass spectrometry (MS) [25–35]. For a given polymeric system, these techniques each having advantages and limitations over the others are often combined to provide a detailed characterization of the system [33–41]. Among them, MS has become an indispensable tool for polymer analysis and has widely been used to study polymer structure, polymer composition, molecular mass and molecular mass distribution, bulk and surface properties, impurity contents, and so on.

MS offers several important attributes for polymer analysis [25–35]. MS, based on accurate mass measurement and/or tandem MS (MS/MS) analysis, can generate rich chemical information that is highly specific for polymer structural analysis. MS is also very sensitive, allowing the detection and identification of minor polymer components or impurities in a composed polymeric material and any by-

MALDI Mass Spectrometry for Synthetic Polymer Analysis, Edited by Liang Li

product of polymerization reaction of a desired polymer. Rapid MS analysis can be done for many polymer samples where no prior sample treatment or extensive separation is needed. MS can potentially provide quantitative information required for determination of the average molecular mass and molecular mass distribution of a polymer or characterization of relative amounts of different components of a polymer mixture. Some forms of MS such as secondary ion mass spectrometry (SIMS) can also be used to characterize polymer surfaces [35].

Almost all mass spectrometric techniques developed so far have been tried for polymer analysis, with varying degrees of success; some have been more widely used than the others. Traditional techniques such as electron impact ionization (EI), chemical ionization (CI), and gas chromatography/mass spectrometry (GC/MS) are still used for generating structural or compositional information on polymers, albeit applicable to only low-mass polymers or their precursors (monomers, dimers, etc.). Pyrolysis MS or GC/MS uses a pyrolysis process to thermally degrade polymers, including very high-molecular-mass polymers, to fragment products that are sub-jected to MS or GC/MS analysis [25]. This technique is still useful for generating chemical information on monomer structures, copolymer repeat units, end groups, and impurities or additives, particularly for those polymers not amendable to modern MS techniques [25]. In applying this technique, one needs to be cautious in interpret-ing the mass spectral results, as thermal degradation often alters the chemical struc-tures of the polymeric materials. One interesting recent development in pyrolysis MS is to carry out a controlled degradation of polymer so that low-mass polymers or oligomers, instead of very low-mass products suitable for EI or CI, are generated. The low-mass polymers are analyzed using modern techniques such as matrix-assisted laser desorption ionization (MALDI) MS [42–44].

In the 1970s and earlier 1980s, active researches were being pursued to develop new ionization sources to overcome the limitations of EI and CI to handle thermally unstable and/or large molecules. Desorption techniques such as SIMS, fast atom bombardment (FAB), and laser desorption/ionization (LDI) were rapidly developed, and many researchers demonstrated that these techniques were useful for not only biopolymer characterization, but also synthetic polymer analysis. In the late 1980s, the introduction of MALDI and electrospray ionization (ESI) opened a new era in mass spectrometric analysis of biomolecules and synthetic polymers. Before MALDI and ESI, MS was limited to the analysis of relatively low-molecular-mass polymers of less than 3000 Da. Only in some favorable polymeric systems, such as poly(ethylene glycol) (PEG), MS analysis of up to 10,000 Da could be done [45]. Even for these polymers, the analysis was not routine, and was usually done by an experienced researcher.

ESI can be particularly useful for analyzing high-mass molecules that are easy to form multiply charged gas phase ions. PEG was one of the chemical samples used to illustrate the power of ESI for high-mass analysis in the early development of the ESI technique [46]. PEG can readily form multiply charged ions by attaching mul-tiple sodium or potassium ions to a polymer chain and they are stable in the gas phase. However, a majority of synthetic polymers do not form multiply charged ions easily, and hence ESI has a limited applicability in synthetic polymer analysis, especially for high-mass polymers. Many polymers require the use of metal ions to form adduct ions. These adduct ions are not stable in the gas phase and do not survive during the translation from atmosphere to the mass analyzer via an ESI interface. Even if multiply charged ions are formed, to deduce molecular mass or molecular mass distribution information, oligomer ions must be resolved to determine the charge states. This is not trivial as many peaks from oligomers with different numbers of repeat units and different charge states may overlap. In addition, the oligomer distribution from the ESI ions is very sensitive to the experimental conditions such as the skimmer voltages, making accurate determination of the average mass and mass distribution difficult [47–50].

Despite these limitations, ESI can be a very powerful technique for characterizing ESI-able polymers such as low-mass PEGs or some water-soluble polymers with diverse end-group structures [29, 51–53]. This is because ESI offers several important attributes. First of all, ESI can be readily interfaced with solution-based separation techniques such as high-performance liquid chromatography (HPLC) for separating complex mixtures [54–56]. More recently, ion mobility separation in the gas phase can be used to further separate closely related polymeric ions [57–60]. Second, many different types of ESI tandem mass spectrometers, including those capable of carrying out multiple-stage dissociation of an ion, are available for generating structural and compositional information on a polymer sample [61]. Third, multiply charged ions produced by ESI are relatively easier to dissociate than a singly charged ion often found in MALDI [62–66]. Finally, high-performance ESI mass spectrometers such as high-resolution time-of-flight (TOF), Orbitrap, and high-field Fourier transform ion cyclotron resonance (FT-ICR) MS are now available, offering excellent mass resolving power and mass measurement accuracy. Overall, the combination of new separation tools with high-performance ESI MS will likely increase its popularity in polymer analysis, particularly for detailed characterization of chemical structures and compositions of a polymeric system. But, the major obstacle still remains: many polymers simply do not ionize well by ESI.

MALDI MS transforms the practice of polymer characterization. It has now become a widely used technique for analyzing a great variety of polymers. In this book, many aspects of the MALDI MS technique will be discussed in great detail. There are several unique attributes of MALDI MS, together making it a powerful technique for polymer characterization. In MALDI MS, molecular mass and molecular mass distribution information can be obtained for polymers of narrow polydispersity with high precision and speed. The accuracy, although difficult to determine due to the lack of well-characterized standards, appears to be good as well [67, 68]. The MALDI analysis of polymers does not require the use of polymer standards for mass calibration. Furthermore, this technique uses a minimum amount of solvents and other consumables, which translates into low operational costs. MALDI MS can also provide structural information, if the instrumental resolution is sufficient to resolve oligomers. In this case, monomer and end-group masses can be deduced from the accurate measurement of the mass of individual oligomers. This is particularly true when a high-resolution instrument such as FT-ICR MS is used for polymer analysis. With the use of MALDI MS/MS, structural characterization can be facilitated. Finally, impurities, by-products, and subtle changes in polymer distributions can often be detected even for relatively complex polymeric systems such as copolymers.

Because of these attributes, in many labs, MALDI MS has become a routine tool for polymer characterization. This is evident from an increasing number of publications in polymer science literature (e.g., *Macromolecules*), which indicate the use of MALDI MS as a tool for characterizing newly grafted or synthesized polymers. In the industry dealing with polymeric materials, MALDI MS is often combined with other analytical techniques to provide detailed analyses of a polymeric system. In some cases, MALDI MS is the only technique that can provide the information required to solve a practical problem. One example is in the area of product failure analysis involving four copolymer samples [69].

However, current MALDI techniques have several limitations for polymer analysis. Some narrow polydispersity polymers cannot be readily analyzed. There is a need to develop sample preparation protocols to analyze important polymers such as polyethylene, perfluoropolymer, and polycationic polymers. The analysis of polystyrenes with molecular mass up to 1.5 million has been demonstrated. But it remains to be seen how the technique can apply to other high-molecular-mass polymers (MW > 500,000). Searching for sensitive sample preparation methods and improving the overall detection sensitivity of the current MALDI instrument for high-mass analysis are required. The MALDI technique has not yet generated reliable results in direct analysis of broad polydispersity polymers. The instrumental and chemistry problems associated with the analysis of these broad polydispersity polymers have been studied by a number of groups, as discussed in several chapters in this book. We now have a better understanding of the issues involved, and we will see more development in this area in addressing these issues in the near future. Alternatively, gel permeation chromatography or size exclusion chromatography can be combined with MALDI MS for polymers of broad polydispersity. Finally, it is difficult, at present, to deduce quantitative information on polymer mixtures from the MALDI spectra. The overall detection sensitivity for different polymers is not the same. There is no direct correlation between the relative peak areas in the spectra and the relative amounts in the mixture. However, information on relative changes in polymer composition can be obtained, if several polymeric systems containing the same polymer mixture, but different relative amounts, are available for interrogation.

In conclusion, many different mass spectrometric techniques are currently available for detailed characterization of a synthetic polymer. Among them, MALDI MS has been widely used for analyzing a great variety of polymers. It offers some unique attributes while some limitations remain. However, future advances in fundamental studies of the MALDI process, sample preparation methods, ion detection techniques, and data processing issues will undoubtedly enhance its role in polymer science in general. Many of these topics will be the subject of discussion throughout this book.

REFERENCES

1. Chang, T.Y., Recent advances in liquid chromatography analysis of synthetic polymers, *Advanced Polymer Science*, **2003**, 163, 1–60.
2. Colfen, H.; Antonietti, M., Field-flow fractionation techniques for polymer and colloid analysis, *New Developments in Polymer Analytics I*, **2000**, 150, 67–187.

3. Berek, D., Coupled liquid chromatographic techniques for the separation of complex polymers, *Progress in Polymer Science*, **2000**, 25(7), 873–908.
4. Pasch, H., Hyphenated techniques in liquid chromatography of polymers, *New Developments in Polymer Analytics I*, **2000**, 150, 1–66.
5. Philipsen, H.J.A., Determination of chemical composition distributions in synthetic polymers, *Journal of Chromatography A*, **2004**, 1037(1–2), 329–350.
6. Messaud, F.A.; Sanderson, R.D.; Runyon, J.R.; Otte, T.; Pasch, H.; Williams, S.K.R., An overview on field-flow fractionation techniques and their applications in the separation and characterization of polymers, *Progress in Polymer Science*, **2009**, 34(4), 351–368.
7. Mori, S.; Barth, H.G. (editors), *Size Exclusion Chromatography*. Berlin: Springer, **1999**.
8. Heigl, N.; Petter, C.H.; Rainer, M.; Najam-Ul-Haq, M.; Vallant, R.M.; Bakry, R.; Bonn, G.K.; Huck, C.W., Near infrared spectroscopy for polymer research, quality control and reaction monitoring, *Journal of Near Infrared Spectroscopy*, **2007**, 15, 269–282.
9. Young, R.J.; Eichhorn, S.J., Deformation mechanisms in polymer fibres and nanocomposites, *Polymer*, **2007**, 48(1), 2–18.
10. Ngamna, O.; Morrin, A.; Killard, A.J.; Moulton, S.E.; Smyth, M.R.; Wallace, G.G., Inkjet printable polyaniline nanoformulations, *Langmuir*, **2007**, 23(16), 8569–8574.
11. Baessler, H. (editor), *Optical Techniques to Characterize Polymer Systems*. Amsterdam: Elsevier, **1989**.
12. Bruch, M.D. (editor), *NMR Spectroscopy Techniques, Vol. 21; Practical Spectroscopy Series*. 2nd edition. New York: Marcel Dekker, **1996**.
13. Brandolini, A.J., Chemical and physical characterization of polymer systems by NMR spectroscopy. In *NMR Spectroscopy Techniques*, Vol. 21; Practical Spectroscopy Series. 2nd edition (Bruch, M.D., editor). New York: Marcel Dekker, **1996**, pp. 525–555.
14. Kitayama, T.; Hatada, K., *NMR Spectroscopy of Polymers*. New York: Springer Laboratory, **2004**.
15. Medhioub, H.; Zerrouki, C.; Fourati, N.; Smaoui, H.; Guermazi, H.; Bonnet, J.J., Towards a structural characterization of an epoxy-based polymer using small-angle X-ray scattering, *Journal of Applied Physiology*, **2007**, 101(4), 6.
16. Elmoutaouakkil, A.; Fuchs, G.; Bergounhon, P.; Peres, R.; Peyrin, F., Three-dimensional quantitative analysis of polymer foams from synchrotron radiation x-ray microtomography, *Journal of Physics D: Applied Physics*, **2003**, A37–A43.
17. Ha, C.S.; Gardella, J.A., X-ray photoelectron spectroscopy studies on the surface segregation in poly(dimethylsiloxane) containing block copolymers, *Journal of Macromolecular Science-Polymer Reviews*, **2005**, C45(1), 1–18.
18. Brostow, W.; Gorman, B.P.; Olea-Mejia, O., Focused ion beam milling and scanning electron microscopy characterization of polymer plus metal hybrids, *Materials Letters*, **2007**, 61(6), 1333–1336.
19. Wanakule, N.S.; Nedoma, A.J.; Robertson, M.L.; Fang, Z.; Jackson, A.; Garetz, B.A.; Balsara, N.P., Characterization of micron-sized periodic structures in multicomponent polymer blends by ultra-small-angle neutron scattering and optical microscopy, *Macromolecules*, **2008**, 41(2), 471–477.
20. Opdahl, A.; Koffas, T.S.; Amitay-Sadovsky, E.; Kim, J.; Somorjai, G.A., Characterization of polymer surface structure and surface mechanical behaviour by sum frequency generation surface vibrational spectroscopy and atomic force microscopy, *Journal of Physics-Condensed Matter*, **2004**, 16(21), R659–R677.
21. Gracias, D.H.; Chen, Z.; Shen, Y.R.; Somorjai, G.A., Molecular characterization of polymer and polymer blend surfaces. Combined sum frequency generation surface vibrational spectroscopy and scanning force microscopy studies, *Accounts of Chemical Research*, **1999**, 32(11), 930–940.
22. Chan, C.M.; Weng, L.T., Applications of X-ray photoelectron spectroscopy and static secondary ion mass spectrometry in surface characterization of copolymers and polymers blends, *Reviews in Chemical Engineering*, **2000**, 16(4), 341–408.
23. Sarac, A.S.; Geyik, H.; Parlak, E.A.; Serantoni, M., Electrochemical composite formation of thiophene and N-methylpyrrole polymers on carbon fiber microelectrodes: morphology, characterization by surface spectroscopy, and electrochemical impedance spectroscopy, *Progress in Organic Coatings*, **2007**, 59(1), 28–36.

24. Werner, C.; Jacobasch, H.J., Surface characterization of polymers for medical devices, *International Journal of Artificial Organs*, **1999**, 22(3), 160–176.

25. Sobeih, K.L.; Baron, M.; Gonzalez-Rodriguez, J., Recent trends and developments in pyrolysis-gas chromatography, *Journal of Chromatography A*, **2008**, 1186(1–2), 51–66.

26. Montaudo, G.; Samperi, F.; Montaudo, M.S., Characterization of synthetic polymers by MALDI-MS, *Progress in Polymer Science*, **2006**, 31(3), 277–357.

27. Klee, J.E., Mass spectrometry of step-growth polymers, *European Journal of Mass Spectrometry*, **2005**, 11(6), 591–610.

28. Murgasova, R.; Hercules, D.M., Polymer characterization by combining liquid chromatography with MALDI and ESI mass spectrometry, *Analytical and Bioanalytical Chemistry*, **2002**, 373(6), 481–489.

29. Hanton, S.D., Mass spectrometry of polymers and polymer surfaces, *Chemical Reviews*, **2001**, 101(2), 527–569.

30. Adriaens, A.; Van Vaeck, L.; Adams, F., Static secondary ion mass spectrometry (S-SIMS) Part 2: material science applications, *Mass Spectrometry Reviews*, **1999**, 18(1), 48–81.

31. Nielen, M.W.F., MALDI time-of-flight mass spectrometry of synthetic polymers, *Mass Spectrometry Reviews*, **1999**, 18(5), 309–344.

32. Wang, F.C.Y., Polymer analysis by pyrolysis gas chromatography, *Journal of Chromatography A*, **1999**, 843(1–2), 413–423.

33. Pasch, H.; Schrepp, W., *MALDI-TOF Mass Spectrometry of Synthetic Polymers*. New York: Springer, **2003**.

34. Montaudo, G.; Lattimer, R.P. (editors), *Mass Spectrometry of Polymers*. New York: CRC, **2001**.

35. Briggs, D., *Surface Analysis of Polymers by XPS and Static SIMS*. New York: Cambridge University Press, **2005**.

36. Brady, R.F. (editor), *Comprehensive Desk Reference of Polymer Characterization and Analysis*. Washington DC: American Chemical Society, **2003**.

37. Pethrick, R.A.; Dawkins, J.V. (editors), *Modern Techniques for Polymer Characterization*. New York: Wiley, **1999**.

38. Simon, G.P. (editor), *Polymer Characterization Techniques and Their Application to Blends*. Washington DC: American Chemical Society, **2003**.

39. Crompton, T.R., *Polymer Reference Book*. New York: Rapra Technology, **2006**.

40. Schroeder, E.; Mueller, G.; Arndt, K.F., *Polymer Characterization*. Cincinnati: Hanser, **1990**.

41. Kroschwitz, J.I. (editor), *Polymers: Polymer Characterization and Analysis*. New York: Wiley, **1990**.

42. Barton, Z.; Kemp, T.J.; Buzy, A.; Jennings, K.R., Mass spectral characterization of the thermal degradation of poly(propylene oxide) by electrospray and matrix-assisted laser desorption ionization, *Polymer*, **1995**, 36(26), 4927–4933.

43. Puglisi, C.; Samperi, F.; Carroccio, S.; Montaudo, G., MALDI-TOF investigation of polymer degradation: pyrolysis of poly(bisphenol A carbonate), *Macromolecules*, **1999**, 32(26), 8821–8828.

44. Montaudo, G.; Carroccio, S.; Puglisi, C., Thermal and thermoxidative degradation processes in poly(bisphenol A carbonate), *Journal of Analytical and Applied Pyrolysis*, **2002**, 64(2), 229–247.

45. Lubman, D.M. (editor), *Lasers and Mass Spectrometry*. New York: Oxford University Press, **1990**.

46. Fenn, J.B.; Mann, M.; Meng, C.K.; Wong, S.F.; Whitehouse, C.M., Electrospray ionization for mass-spectrometry of large biomolecules, *Science*, **1989**, 246(4926), 64–71.

47. Latourte, L.; Blais, J.C.; Tabet, J.C.; Cole, R.B., Desorption behavior and distributions of fluorinated polymers in MALDI and electrospray ionization mass spectrometry. *Analytical Chemistry*, **1997**, 69(14), 2742–2750.

48. Parees, D.M.; Hanton, S.D.; Clark, P.A.C.; Willcox, D.A., Comparison of mass spectrometric techniques for generating molecular weight information on a class of ethoxylated oligomers. *Journal of the American Society for Mass Spectrometry*, **1998**, 9(4), 282–291.

49. Yan, W.Y.; Ammon, D.M.; Gardella, J.A.; Maziarz, E.P.; Hawkridge, A.M.; Grobe, G.L.; Wood, T.D., Quantitative mass spectrometry of technical polymers: a comparison of several ionization methods, *European Mass Spectrometry*, **1998**, 4(6), 467–474.

50. Maziarz, E.P.; Baker, G.A.; Lorenz, S.A.; Wood, T.D., External ion accumulation of low molecular weight poly(ethylene glycol) by electrospray ionization Fourier transform mass spectrometry, *Journal of the American Society for Mass Spectrometry*, **1999**, 10(12), 1298–1304.

51. Guittard, J.; Tessier, M.; Blais, J.C.; Bolbach, G.; Rozes, L.; Marechal, E.; Tabet, J.C., Electrospray and matrix-assisted laser desorption/ionization mass spectrometry for the characterization of polyesters, *Journal of Mass Spectrometry*, **1996**, 31(12), 1409–1421.

52. Hart-Smith, G.; Lammens, M.; Du Prez, F.E.; Guilhaus, M.; Barner-Kowollik, C., ATRP poly(acrylate) star formation: a comparative study between MALDI and ESI mass spectrometry, *Polymer*, **2009**, 50(9), 1986–2000.

53. Benomar, S.H.; Clench, M.R.; Allen, D.W., The analysis of alkylphenol ethoxysulphonate surfactants by high-performance liquid chromatography, liquid chromatography-electro spray ionisation-mass spectrometry and matrix-assisted laser desorption ionisation-mass spectrometry, *Analytica Chimica Acta*, **2001**, 445(2), 255–267.

54. Feldermann, A.; Toy, A.A.; Davis, T.P.; Stenzel, M.H.; Bamer-Kowollik, C., An in-depth analytical approach to the mechanism of the RAFT process in acrylate free radical polymerizations via coupled size exclusion chromatography-electrospray ionization mass spectrometry (SEC-ESI-MS), *Polymer*, **2005**, 45(19), 8448–8457.

55. Hart-Smith, G.; Chaffey-Millar, H.; Barner-Kowollik, C., Living star polymer formation: detailed assessment of poly(acrylate) radical reaction pathways via ESI-MS, *Macromolecules*, **2008**, 41(9), 3023–3041.

56. Gruendling, T.; Guilhaus, M.; Barner-Kowollik, C., Quantitative LC-MS of polymers: determining accurate molecular weight distributions by combined size exclusion chromatography and electrospray mass spectrometry with maximum entropy data processing, *Analytical Chemistry*, **2008**, 80(18), 6915–6927.

57. Robinson, E.W.; Garcia, D.E.; Leib, R.D.; Williams, E.R., Enhanced mixture analysis of poly(ethylene glycol) using high-field asymmetric waveform ion mobility spectrometry combined with Fourier transform ion cyclotron resonance mass spectrometry, *Analytical Chemistry*, **2006**, 78(7), 2190–2198.

58. Bagal, D.; Zhang, H.; Schnier, P.D., Gas-phase proton-transfer chemistry coupled with TOF mass spectrometry and ion mobility-MS for the facile analysis of poly(ethylene glycols) and PEGylated polypeptide conjugates, *Analytical Chemistry*, **2008**, 80(7), 2408–2418.

59. Kanu, A.B.; Dwivedi, P.; Tam, M.; Matz, L.; Hill, H.H., Ion mobility-mass spectrometry, *Journal of Mass Spectrometry*, **2008**, 43(1), 1–22.

60. Trimpin, S.; Plasencia, M.; Isailovic, D.; Clemmer, D.E., Resolving oligomers from fully grown polymers with IMS-MS, *Analytical Chemistry*, **2007**, 79(21), 7965–7974.

61. Koster, S.; Duursma, M.C.; Boon, J.J.; Nielen, M.W.F.; de Koster, C.G.; Heeren, R.M.A., Structural analysis of synthetic homo- and copolyesters by electrospray ionization on a Fourier transform ion cyclotron resonance mass spectrometer, *Journal of Mass Spectrometry*, **2000**, 35(6), 739–748.

62. Chen, R.; Li, L., Lithium and transition metal ions enable low energy collision-induced dissociation of polyglycols in electrospray ionization mass spectrometry, *Journal of the American Society for Mass Spectrometry*, **2001**, 12(7), 832–839.

63. Giordanengo, R.; Viel, S.; Allard-Breton, B.; Thevand, A.; Charles, L., Tandem mass spectrometry of poly(methacrylic acid) oligomers produced by negative mode electrospray ionization, *Journal of the American Society for Mass Spectrometry*, **2009**, 20(1), 25–33.

64. Jackson, A.T.; Slade, S.E.; Thalassinos, K.; Scrivens, J.H., End-group characterisation of poly(propylene glycol)s by means of electrospray ionisation-tandem mass spectrometry (ESI-MS/MS), *Analytical and Bioanalytical Chemistry*, **2008**, 392(4), 643–650.

65. Chen, R.; Yu, X.L.; Li, L., Characterization of poly(ethylene glycol) esters using low-energy collision-induced dissociation in electrospray ionization mass spectrometry, *Journal of the American Society for Mass Spectrometry*, **2002**, 13(7), 888–897.

66. Stenson, A.C.; Landing, W.M.; Marshall, A.G.; Cooper, W.T., Ionization and fragmentation of humic substances in electrospray ionization Fourier transform-ion cyclotron resonance mass spectrometry, *Analytical Chemistry*, **2002**, 74(17), 4397–4409.

67. Guttman, C.M.; Wetzel, S.J.; Blair, W.R.; Fanconi, B.M.; Girard, J.E.; Goldschmidt, R.J.; Wallace, W.E.; VanderHart, D.L., NIST-sponsored interlaboratory comparison of polystyrene molecular mass distribution obtained by matrix-assisted laser desorption/ionization time-of-flight mass spectrometry: statistical analysis, *Analytical Chemistry*, **2001**, 73(6), 1252–1262.

68. Guttman, C.M.; Wetzel, S.J.; Flynn, K.M.; Fanconi, B.M.; VanderHart, D.L.; Wallace, W.E., Matrix-assisted laser desorption/ionization time-of-flight mass spectrometry interlaboratory comparison of mixtures of polystyrene with different end groups: statistical analysis of mass fractions and mass moments, *Analytical Chemistry*, **2005**, 77(14), 4539–4548.

69. Chen, R.; Tseng, A.M.; Uhing, M.; Li, L., Application of an integrated matrix-assisted laser desorption/ionization time-of-flight, electrospray ionization mass spectrometry and tandem mass spectrometry approach to characterizing complex polyol mixtures, *Journal of the American Society for Mass Spectrometry*, **2001**, 12(1), 55–60.

CHAPTER **2**

IONIZATION PROCESSES AND DETECTION IN MALDI-MS OF POLYMERS

Renato Zenobi

Department of Chemistry and Applied Biosciences, ETH Zurich, Zurich, Switzerland

2.1 THE MATRIX-ASSISTED LASER DESORPTION/ IONIZATION (MALDI) METHOD AND MALDI INSTRUMENTATION

One of the powerful soft ionization methods that is in widespread use for the analysis of high molecular weight, nonvolatile molecules is MALDI. MALDI was largely developed in the laboratory of Franz Hillenkamp at the University of Münster, Germany [1–3]. A similar method was developed almost simultaneously by Koichi Tanaka et al. at Shimadzu Research Laboratories in Kyoto, Japan [4]. Hillenkamp's method, which is very widely used today, works by embedding the sample in an organic matrix to facilitate desorption and ionization of the sample upon irradiation by a pulsed UV laser. Tanaka's paper reported the use of finely dispersed metal powder in a glycerol matrix for the same purpose. The main field of application of MALDI is the analysis of biopolymers such as peptides, proteins, or DNA. Another very important application is the chemical analysis of polymers, which is covered in this book and in a number of other publications [5–9]. In both cases, molecular ions well above 100,000 Da can be generated without fragmentation.

Figure 2.1 shows a typical linear MALDI time-of-flight (TOF) mass spectrometer. It consists of a sample holder, laser optics, a high-voltage acceleration stage, a drift tube, and detection electronics for the arriving ions. The sample is deposited at the tip of the sample holder that is introduced into the high vacuum of the ion source through an interlock. The pulse of a nitrogen laser ($\lambda = 337\,\text{nm}$) is focused onto the sample to initiate desorption and ionization; a small part of the laser pulse is split off to trigger the data acquisition. The ions formed are all accelerated to the same kinetic energy and are mass separated by virtue of their transit time ("TOF") through a field-free drift tube. Detection is usually done by ion-to-electron conver-

MALDI Mass Spectrometry for Synthetic Polymer Analysis, Edited by Liang Li
Copyright © 2010 John Wiley & Sons, Inc.

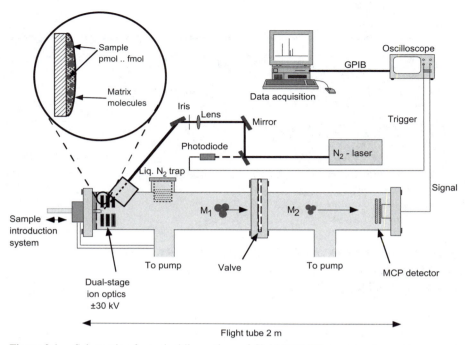

Figure 2.1.   Schematic of a typical linear time-of-flight MALDI mass spectrometer.
The valve separating the detector from the source region is optional but useful, for example,
for maintenance or cleaning of the ion source. See color insert.

sion using a multichannel plate (MCP) detector, whose signal is read by a digital
oscilloscope.

There are other variants of MALDI mass spectrometers. Other mass analyzers
such as Fourier transform ion cyclotron resonance (FT-ICR) mass spectrometers,
ion traps, quadrupole-based instruments, sector mass spectrometers, or hybrid instruments can be used. The TOF mass analyzer, however, is by far the most frequently
used, because it lends itself naturally to ion production by a short laser pulse. The
modest mass resolution of linear TOF instruments can be greatly improved using
appropriate ion optics, delayed extraction, and a reflectron TOF mass analyzer. All
of these options are incorporated in modern commercial instruments.

The precise way in which the ions are produced can also vary. Many different
laser lines have been used for MALDI, including infrared radiation at 2.94 μm
(Er:YAG laser) and 10.6 μm (CO_2 laser) with special matrices, but 337 nm (N_2 laser)
and 355 nm (third harmonic of the Nd:YAG laser) are the most popular, successful,
and affordable options, and are extensively used in the field of polymer MALDI
mass spectrometry (MS). Some mass spectrometers require a continuous or quasi-
continuous ion production or are designed to operate with atmospheric pressure ion
sources. In the former case, lasers with high repetition rates, up to 2 kHz, have been
used quite successfully. The latter requirement is easily met by atmospheric pressure

MALDI sources [10]; these have also become commercially available in the past few years.

There are several peculiarities about the MALDI process that warrant a deeper understanding of the mechanism for ion formation: (1) the yield of primary (matrix) ions in the absence of added cationization agents is relatively low, on the order of 10^{-4}; (2) MALDI is known to produce predominantly singly charged ions (incidentally, this is precisely the reason why MALDI is preferred over electrospray ionization (ESI) for the chemical analysis of complex polymer samples; ESI would result in overlapping oligomer and charge state distributions); (3) mutual signal suppression has been observed when analyzing mixtures by MALDI-MS. It can occur, for example, that a basic peptide in a mixture of other, less basic peptides, completely dominates the MALDI mass spectrum; (4) MALDI works at both UV and IR wavelengths, yielding similar spectra; and (5) only a limited selection of "good" matrices exists for a given analytical problem; apart from some basic requirements for the matrix (vacuum stable; good optical absorption; rapid sublimation when energized with a laser pulse), the choice is largely empirical. Any model for MALDI ion formation must take these observations into account and be capable of explaining them. In general, ionization in MALDI is complex, potentially involving a number of different processes. Furthermore, it would be naïve to expect that one mechanism is capable of explaining all different situations (UV MALDI and IR MALDI; the wide variety of MALDI matrices and sample preparations used; analysis of all different classes of compounds such as polynucleotides, peptides/proteins, sugars, and polymers).

In addition to this intellectual challenge, there are several practical reasons motivating researchers to obtain better understanding and control of MALDI ion formation. These include improvement of the ion yield ($10^{-4} \rightarrow$??), a better control of fragmentation, the need to obtain guidelines for matrix selection, to apply MALDI to new classes of compounds, and to obtain a more uniform MALDI response, for example, in the analysis of tryptic peptide mixtures or polydisperse oligomer distributions in the field of polymer analysis.

2.2   PRACTICE OF POLYMER MALDI

Before dealing with mechanistic aspects of MALDI ion formation in detail, we will briefly outline the experimental protocols used for MALDI analysis of polymers. The first observation concerns the laser wavelengths used for polymer MALDI: the vast majority of studies have been carried out with UV lasers (337 and 355 nm). IR MALDI plays only a minor role in the analysis of polymers. Another observation is that besides matrices that are popular for bioanalytical purposes (including 4-hydroxy-α-cynocinnamic acid [HCCA]; 2,5-dihydroxybenzoic acid [DHB]; and 2-(4-hydroxyphenylazo)benzoic acid [HABA]), there are a number of matrices that appear to be particularly well suited for the preparation of polymer MALDI samples. These matrices include trans-3-indoleacrylic acid (IAA), 9-nitroanthracene, 1,8,9-trihydroxyanthracene (dithranol), all-trans retinoic acid (RA), pentafluorobenzoic acid, pentafluorocinnamic acid, and 7,7,8,8-tetracyanoquinodimethane [8, 9]. The use of these specialized matrix compounds is normally driven by the need of compatibility of the polymer sample and the matrix in terms of solubility and co-crystallization (see Chapter 6). For the same reason, unusual solvents are employed for sample preparation, such as tetrahydrofuran (THF), chlorobenzene, or 1,1,1,3,3,3-hexafluoro-2-propanol (hexafluoroisopropanol). In some cases, it is impossible to find a common solvent for the sample and the matrix, or the sample itself is not soluble at all. Solvent-free sample preparation methods have been developed to deal with this difficulty [11, 12]; see also Chapter 7 of this book. It turns out that an intimate physical mixture of matrix and sample is sufficient for generating characteristic sample ions in a soft fashion. This observation also needs to be taken into account when considering possible scenarios for ion formation. Finally, polymer MALDI samples are very often spiked with cationization agents, such as alkali (Li, Na, K, Cs), Cu, or Ag salts.

2.3 IONIZATION MECHANISMS IN MALDI-MS

Despite the widespread use of MALDI for polymer MS and especially for bioanalytical purposes, the understanding and control of the ion formation is still incomplete. However, great progress has been made toward a deeper understanding of the underlying principles over the last decade [13–16]. The experimental phenomena listed above must be explained by any theory of MALDI ion formation. While some attempts have been made to explain desorption/ion formation in MALDI as one comprehensive process [14], by far the most successful model, in particular within the realm of polymer analysis, is to consider primary and secondary ionization processes in MALDI separately (Figure 2.2) [13, 15]. Primary events happen early, in the solid or near-solid matrix. Secondary ionization involves subsequent reactions in the expanding plume. Secondary processes are largely ion–molecule reactions involving primary ions and matrix neutrals. Both single molecules and aggregates are important.

The motivation for separating primary and secondary events in MALDI is one of time scale. The laser pulse typically lasts 3–5 ns (N_2 or Nd:YAG lasers), but the time required for expansion to collision-free densities is much longer, many microseconds. With the possible exception of preformed ions that are liberated later (e.g., by cluster evaporation, vide infra), the primary ions will be generated during the laser pulse or within the excited state lifetime of the matrix (also only a few nanoseconds). In the expanding plume, however, reactions between ions and neutrals will continue as long as there are collisions. If the number and energies of these collisions is high enough, any thermodynamically favorable processes can proceed to equilibrium. Note that the time scales of primary and secondary reactions may or may not be similar; the key consideration is the sequence of the reactions.

The ideas that have been put forth as possible primary and secondary ionization processes for UV MALDI are summarized in Table 2.1. Among the primary ionization processes, disproportionation, excited state proton transfer, thermal ionization, and multiphoton ionization have been found to be less likely, based mostly on thermodynamic measurements of MALDI matrices. The more likely initial proc-

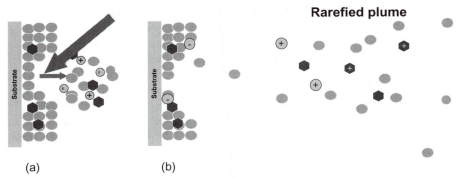

Figure 2.2. Schematic of the primary and secondary ionization processes in MALDI. The fat arrow represents the laser pulse, the gray circles represent the matrix, the dark hexagons represent the sample. (a) Within a few tens of nanoseconds of the laser pulse, initial charges are generated, mostly matrix and cations/anions from additives in the samples (e.g., cationization agents in the case of polymers). A phase of frequent collisions and charge transfer reactions in the dense plume follows. (b) Within a fraction of a microsecond after the laser pulse, the plume has spread, analyte ions have been charged by ion–molecule reactions in the plume, and, after several μsm the dilution finally brings chemical reactions in the plume to a halt. The ions formed are extracted from the source region and detected.

TABLE 2.1. Primary and Secondary Ionization Processes in UVMALDI

Primary ionization processes	disproportionation reactions
	excited state proton transfer
	desorption (liberation) of preformed ions
	exciton formation and ionization by energy pooling
	thermal ionization
	rupture of matrix and statistical charging of the resulting clusters/particles
	multiphoton ionization
Secondary ionization processes	**gas-phase proton transfer**
	gas-phase cation transfer
	electron transfer
	charge ejection

Notes: The processes given in boldface are believed to be the dominant mechanisms. For more detailed information, see References 13 and 15.

esses leading to the first charged molecules are in boldface in the table. "Rupture of matrix and statistical charging of the resulting clusters/particles" implies a kind of frictional process, similar to charging of small microdroplets in methods such as thermospray or inkjet printing that leads to a fraction of clusters and particles carrying a charge following the laser ablation step. Among the secondary processes listed, only charge ejection—for example, production of a singly charged analyte ion from a doubly charged precursor ion by loss of a proton—is not very likely. The other three processes are all known to occur frequently. A primary process occurring in MALDI that has been underestimated for a long time is the formation of clusters (matrix–matrix and matrix–analyte clusters) and even fairly large particles. Their existence has, however, been clearly shown [17–20]. Clusters, on the one hand, may be at the basis of the low overall ion yield in MALDI because many molecules that could potentially be ionized remain "hidden" in the cluster and go undetected; on the other hand, clusters provide a "reaction vessel" for processes such as ion production by exciton recombination (formation of charge when two excitons meet) [21] and other processes [16]. Finally, cluster evaporation from the sample well after the laser pulse may be responsible for delayed ion formation in MALDI. In the realm of polymer MALDI, liberation of preformed ions, often originating from the salt/cationization agent spiked into the sample, is an important primary charge formation process. These ions would then react on in secondary (plume) processes by cation transfer to the sample. Evidence for gas-phase cationization of the sample in polymer MALDI is given below.

Initially, many of the processes listed in Table 2.1 could not be discounted or confirmed easily due to the lack of thermodynamic data on MALDI matrices. Such data are, however, becoming increasingly available, and allow clear statements to be made in many cases. Take, for example, multiphoton ionization of matrix as one possible mechanism of initial charge generation. For DHB, it has been found that the ionization potential (IP) of the free molecule is 8.054 eV [22], which would require 3 photons of the laser. At the range of fluences used in MALDI, a 3-photon ionization process is highly unlikely. On the other hand, the IP of DHB clusters is reduced. For example, the IP of a cluster of 10 DHB molecules drops to 7.82 eV [23], and the complex of DHB with 4 prolines was found to have an IP of 7.0 eV, corresponding to a reduction of over 1 eV from the lowest IP constituent, DHB [24]. Two-photon ionization as a primary process becomes accessible with the usual 337 nm nitrogen or 355 nm Nd:YAG lasers when IPs are below 7.4 or 7.0 eV, respectively. The reduced IP of clusters may thus be important for primary mechanisms because it is known that clusters make up a significant mass fraction of a MALDI plume.

Another important recent finding is that photoelectrons can be emitted by laser irradiation of a metallic sample holder that is not covered completely by a compact layer of matrix, that is, if the metal is partially visible to the laser. These photoelectrons lead to a distortion of the charge balance in the plume, to reduction of multiply charged positive ions, and to neutralization of singly charged ions [25, 26]. The effect is most pronounced in field-free MALDI sources such as TOF instruments operated with delayed extraction. The predominant reactions of photoelectrons in the MALDI plume are with matrix to create negative matrix ions and with multiply charged cations, including matrix and metal ions. The presence of photoelectrons can partially explain the predominance of singly charged ions in MALDI, and the low overall (positive) ion yield: if photoelectron emission is avoided, more multiply charged signals [25], and an increase of the ion yield by up to 100× [26] have been observed. Nonmetallic sample holders can be used for achieving this, and in fact, some vendors indeed offer MALDI targets with special plastic coatings, albeit probably unaware of the fact that this suppresses photoelectron emission. Interestingly, the presence of photoelectrons does not always lead to a clear predominance of negative ions. The reasons for this may be that photoelectrons escape readily from the source area, and that commercial instruments are optimized for transmission and detection of positive ions.

Both matrix ionization potentials and the presence of photoelectrons are mentioned here because they have a direct effect on polymer ion formation in MALDI. For example, it has been shown that Cu(II) is easily reduced to Cu(I) either by free electrons or by neutral matrix if photoelectrons are absent [27]. The reaction

$$matrix + Cu^{++} \rightarrow matrix^+ + Cu^+$$

is exothermic by 10.91 eV for nicotinic acid, by 12.24 eV for DHB, and by 12.4 eV for dithranol, rendering the Cu(I) species the most probable cationization agent in polymer samples spiked with Cu(II) salts.

Gas-phase cation transfer as the most important secondary process in UV MALDI of polymers has not been widely studied, but the evidence supporting this mechanism is nevertheless quite solid [28–33]. The species that carry the initial charge can be free cations from an added cationization agent, an ion pair, or a cationized matrix species. One line of experiments used variation of the sample preparation procedure to either allow or prevent the formation of cationized polymer already in the solid-state sample. Consider, for example, the data on polystyrene (PS) in Figure 2.3. MALDI mass spectra of PS 5000 in dithranol matrix with Cu(II) salt as cationizing agent are shown. In Figure 2.3a, matrix and PS were deposited first from THF solution, and the copper salt was applied from aqueous solution afterward. Since water does not redissolve dithranol and PS, preformed copper adducts of PS are not possible using this sample preparation procedure. In Figure 2.3b, the components were crystallized together from THF, so preformed adduct ions are possible. The spectra show essentially no difference in intensity, suggesting that cationization occurred predominantly in the gas phase [29]. Similar conclusions were drawn in a study by Hoberg et al. using layered samples [28].

Another line of evidence for gas-phase cationization comes from cation selectivities of different polymer samples. For example, Wang et al. [32] compared the cationization of poly(ethylene glycol) (PEG; $HO–(CH_2-CH_2-O)_n–H$), poly(propylene glycol) (PPG; $HO–(CH_2-CH(CH_3)-O)_n–H$), and poly(tetrahydrofuran) (PTHF; $HO–(CH_2-CH_2-CH_2-CH_2-O)_n–H$) by equimolar amounts of Li^+ and Cs^+ present in the samples. These two cationization agents were chosen not only because their large mass difference allows easy distinction, but also because the polymers themselves were found to contain Na and K impurities. The data for PPG in IAA and HABA matrix are shown in Figure 2.4a,b. The selectivities for Li^+ increased from PEG to PPG to PTHF, showing a trend that differed from that in solution for the same cations. From this, it was concluded that cationization of the polyethers investigated takes place in the gas phase.

There is also a dependence of the cationization efficiency on the matrix (see the difference in Figure 2.4a,b [32]) and on the counter-ion of the cationization agents [34]. These effects are less well understood, but are generally explained in terms of the lattice energy that needs to be overcome to make the cation available in the gas phase. In conclusion of this section, the gas-phase cationization mechanism is universally accepted as the major secondary ion formation pathway in polymer MALDI.

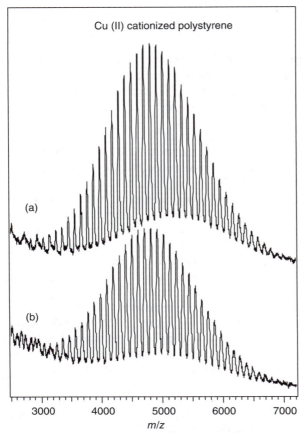

Figure 2.3.   MALDI mass spectra of polystyrene (PS) 5000 in dithranol matrix with Cu(II) salt as cationizing agent. Different sample preparation methods were applied to elucidate the cationization mechanism. In (a), matrix and PS were deposited first from THF solution, and the copper salts was applied from aqueous solution afterward. Since water does not redissolve dithranol and PS, preformed copper adducts of PS are not possible. In (b), the components were crystallized together from THF, so preformed adduct ions are possible. The lack of difference between the spectra suggests that cationization occurred predominantly in the gas phase. Reproduced with permission from Reference 29. Copyright IM Publications.

2.4   IMPLICATIONS OF THE MALDI IONIZATION MECHANISM FOR POLYMER ANALYSIS

2.4.1   Choice of MALDI Matrix

The choice of the MALDI matrix is probably less crucial for some polymers such as PEG, but very crucial for others such as PS and polydienes because the sample, the matrix, and the cationization agent must be soluble in a common solvent and

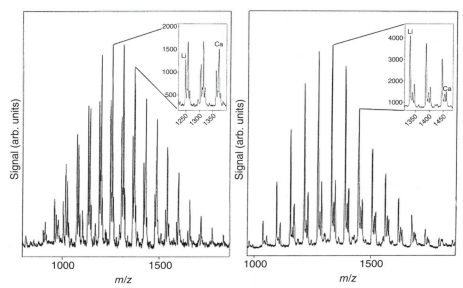

Figure 2.4. Left panel: MALDI-TOF mass spectrum obtained using a MALDI sample containing PPG 1200, equimolar CsCl/LiCl, and matrix IAA. Right panel: MALDI-TOF mass spectrum obtained using a MALDI sample containing PPG 1200, equimolar CsCl/LiCl, and matrix HABA. The peaks labeled with "Li" and "Cs" belong to the same PPG oligomer; the large difference in mass is due to the mass difference of these two cations. Reprinted with permission from Reference 32. Copyright 2000 by the American Society for Mass Spectrometry.

must co-crystallize properly. These conditions are sometimes not met at all, for example, due to solubility issues of high-molecular-weight polymers, which limits the choice of matrix and explains the somewhat exotic substances used for MALDI of polymer samples. The solid–solid sample preparation method presents a viable alternative, and it is expected that common MALDI matrices should be applicable in this case. Other factors, aside from solubility and crystallization, also play some roles in the selection of a proper matrix. For more information on the topic of matrix selection, the reader is referred to the earlier chapters on sample preparation and to Chapter 6.

2.4.2 Cationization Agents

The proper choice of the cationization agent is the key issue for MALDI of polymers. Polyethers are easily cationized by alkali metal ions; polymers containing aromatic moieties can be cationized with Ag^+ or Cu^+. In addition to these cations, which are by far the most widely used, a wide variety of salts have been employed for cationization of polymer samples [35]. Because of its poor affinity to metal cations, polyethylene (PE) has posed particular challenges in this respect, and it has been difficult to analyze PE by MALDI-MS for a long time. Yalcin et al. [36] used laser desorption/ionization without matrix, adding a metal particle suspension to the dissolved PE sample prior to drying it on the sample holder. With this sample preparation method, they were able to observe signals up to $m/z \approx 5000$. A gas-phase cationization mechanism was found even in this case [37]. A different strategy was followed by Wallace and coworkers. They derivatized the unsaturated end group of various PE samples with a triphenylphosphine bromide, creating a "positive charge tag" on the sample. As shown in Figure 2.5, this was quite successful; signals up to $m/z \approx 15,000$ could be observed. The ionization mechanism was not studied in detail, but it is highly probable that this presents an example of preformed ion desorption.

2.5   MASS DISCRIMINATION IN MALDI-MS OF POLYMERS

It was noticed by a number of research groups that MALDI-based molecular weight distributions determined for polymers often do not agree with results obtained from other methods such as light scattering or size exclusion chromatography (SEC) [38–41]. The agreement was found to be particularly poor for polymers with high polydispersity. Several explanations have been offered in the literature, some of which are related to the ion formation pathway. Unless one deals with a mixture of different polymers, differences in the MALDI ionization efficiency, for example, due to an increasing cation affinity with increasing molecular weight of an oligomer, is expected to be a minor contribution to mass discrimination because of the essentially identical chemical nature of oligomers in a polymer molecular weight distribution. Other effects are probably more important, the most obvious of which is related to the poor detection efficiency of higher-mass oligomers by conventional ion-to-electron conversion detectors such as MCPs or other electron multiplication detectors used in TOF MS. It is well known that the ion-to-electron conversion efficiency is strongly dependent on the impact velocity of the ions on the detector surface. This is illustrated in Figure 2.6 [39]: distinctly different mass spectra were obtained for a PMMA 1520 sample measured with 5 kV acceleration voltage (top) and 20 kV acceleration voltage (bottom). The lower the acceleration voltage and the higher the mass, the lower the impact velocity of the ions on the detector surface, which leads to a suppression of high-mass oligomers. This effect is particularly evident at the 5 kV acceleration voltage.

Novel detectors, for example, superconducting tunnel junction (STJ) detectors may overcome this problem. For example, Gervasio et al. [42] showed TOF mass spectra detected by an aluminum STJ detector of equimolar mixtures of PEG 6000 and PEG 35,000 and compared these measurements with a standard ion-to-electron conversion detector. While the standard detector strongly discriminated against the PEG 35,000, the spectrum measured with the STJ detector shows roughly equal response for the two samples. A drawback of the STJ detector, however, is its slow response time, and thus poor mass resolution. The mass resolution in the data shown in Reference 42 was far less than 100, and the mass accuracy was poor. Commercial instrumentation incorporating the STJ technology uses a special algorithm to precisely determine the flight time, and reaches a mass resolution of around 1000.

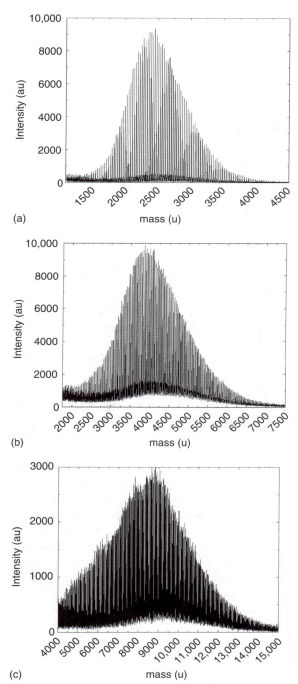

(a)

(b)

(c)

Figure 2.5.   MALDI-TOF mass spectra of polyethylene narrow mass fractions:
(a) linear high-density polyethylene LEA-51; (b) standard reference material (SRM) 2885; and
(c) SRM 1482. The data are shown without smoothing or background correction. Reproduced
with permission from Reference 46. Copyright 2002, American Chemical Society.

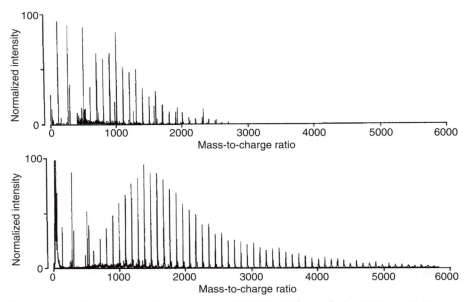

Figure 2.6.   The distinctly different spectra of PMMA 1520 acquired at 5 kV acceleration voltage (top) and 20 kV acceleration voltage (bottom). Reproduced with permission from Reference 39. Copyright 1996, American Chemical Society.

Another explanation for mass discrimination that has been given is due to the different laser power requirements to generate ions from the lower-mass and higher-mass components of a broad oligomer distribution. Figure 2.7 illustrates this point [38]: the mass spectra of an equimolar mixture of poly(styrene) standards with M_p = 5500 Da and M_p = 46,000 Da in dithranol were recorded at a laser power increasing from (a) to (d). Quite obviously, higher laser power is needed to efficiently desorb the 46,000 Da PS oligomers; the peak area of the high-molecular-weight component increases more than that of the low-molecular-weight component when the laser power is increased. This was found to be true up to an optimum point where the peak area of the 46,000 Da component did not increase further. If a laser power far above the optimum was used, the low-molecular-weight distribution became broader and shifted to lower mass (see Figure 2.8). Obviously, too high a laser power caused fragmentation. With increasing peak area of the high-molecular-weight component, the intensity of the doubly charged peak visible at m/z = 23,000 also increased in intensity; that is, part of the ion current was "diverted" to this channel.

Other instrumental factors, as well as the sample preparation [40, 41], have been shown to contribute to mass discrimination effects. It has also been hypothesized that the observed molecular weight distribution depends on the cationization efficiency [31]. For example, for a polyether (PS), it can easily be imagined that higher-mass oligomers with more oxygens (styrene units) available for cation attachment would be more easily ionized and thus overrepresented in the data. Once the

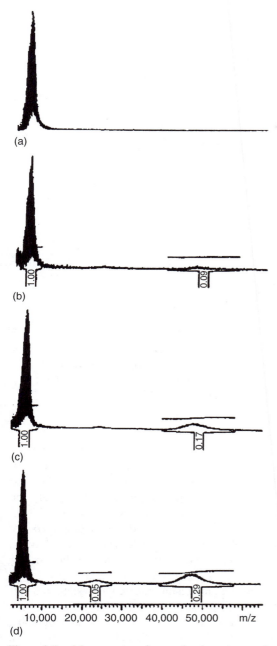

(a)

(b)

(c)

10,000 20,000 30,000 40,000 50,000   m/z

(d)

Figure 2.7.   Mass spectra of an equimolar mixture of poly(styrene) (PS) standards with M_p = 5500 Da and M_p = 46,000 Da in dithranol at increasing laser power from (a) to (d). At 23,000 Da, the doubly charged ion of the larger PS sample is visible. Reproduced with permission from Reference 38. Copyright 1996 John Wiley & Sons Ltd.

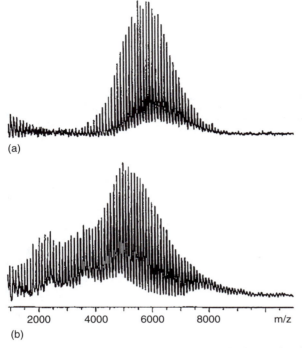

Figure 2.8.   Influence of the laser power on the low-molecular-weight distribution (polystyrene in dithranol): (a) low laser power, (b) high laser power. Reproduced with permission from Reference 38. Copyright 1996, John Wiley & Sons Ltd.

polyether chain is long enough to wrap around the cation, the addition of further units does not appreciably increase the cationization efficiency. There is also work suggesting that the Ag ion is sandwiched between the phenyl rings in PS (also explaining problems experienced in ionizing alpha-methyl styrene); statistically, the ionization efficiency should increase very slowly once the minimum number of units present is reached. Unfortunately, cation affinity data for different oligomers are largely unavailable, and on the other hand, these effects, if present, are outweighed by the discrimination against high-mass oligomers by the instrumental factors discussed above.

Shimada et al. [43, 44] have offered an interesting molecular system to quantitatively determine the mass dependent response of MALDI-TOF measurements done on polymer distributions. Using preparative supercritical fluid chromatography, they were able to isolate individual oligomers from PS and PEG samples. An example of such an isolation procedure is shown in Figure 2.9.

They used the fractions to prepare equimolar mixtures of oligomers with a fairly large spacing in mass, covering the entire mass range between 400 and 3000 Da. These artificial samples thus represent a polymer with a very high polydispersity. The data [43, 44] clearly reveal that mass discrimination is not solely a

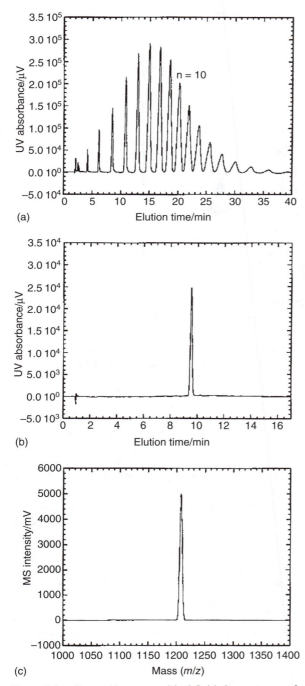

Figure 2.9. Preparative supercritical fluid chromatogram for (a) A-1000 polysyrene and (b) analytical supercritical fluid chromatogram for separated uniform oligomer (n = 10). (c) MALDI-TOF mass spectrum of uniform separated polystyrene oligomer (n = 10) at 2.6 × threshold laser power. Reproduced with permission from Reference 43. Copyright 1996, John Wiley & Sons Ltd.

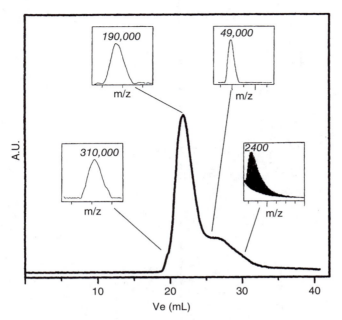

Figure 2.10.   Size-exclusion chromatogram of poly(dimethylsiloxane) sample in THF. The insets display the MALDI-TOF mass spectra of selected fractions. Reproduced with permission from Reference 45. Copyright 1996, John Wiley & Sons Ltd.

question of molecular weight (detector) effects, but is also laser power and cationization dependent.

For avoiding mass discrimination, Montaudo et al. recommend to work with prefractionation followed by MALDI-MS for polymer samples that exhibit broad molecular weight distributions [8, 9, 45]. These authors observe that for a polydispersity exceeding 1.1 … 1.2, straight MALDI-MS fails to yield reliable molecular weight distributions, in general giving values that are deemed too low. Offline or online fractionation by SEC has been shown to overcome the difficulty for a number of cases (see Chapter 11). MALDI mass spectra of SEC fractions can, for example, be used to calibrate the retention time axis of this separation method, which is a problem in itself. Figure 2.10 illustrates the procedure. It shows the SEC trace of a polydimethylsiloxane sample, together with MALDI mass spectra of four fractions. Based on the MALDI data, the SEC curve could be calibrated in terms of absolute molar masses, and the molecular mass average of the sample could be computed from the SEC curve [9, 45].

2.6   CONCLUSIONS

Great progress has been made in a detailed understanding of the MALDI ion formation processes. Dividing the process into primary (charge generation) and secondary (ion–molecule reaction) processes provides a useful framework for understanding and planning MALDI experiments. Recently, significant advances to understanding MALDI ionization have been made. These include new thermodynamic data on MALDI matrices [47, 48], new proposals for describing the MALDI ion formation process (for a recent review, see [49]) or extensions of older ideas [21, 50], and new experimental data [51], including studies on positive vs. negative ion formation [52, 53]. Not surprisingly, most of this work is targeting the description of MALDI of peptides, and we note that some of the concepts and data may be of little relevance for polymer MALDI. Important primary processes in the realm of polymer MALDI are liberation of charges in the form of tagged oligomers and metal cations from salts spiked into the samples. The most important secondary process in polymer MALDI is cationization in the gas phase. For positive ion mode, photoelectrons released from the metal sample support may be deleterious, especially if delayed extraction in a TOF instrument is used. Either a thick sample, covering the metal, or a nonmetallic support should be used. Negative ion MALDI-MS has very little practical significance in the field of polymer MS.

Quantitation in MALDI is known to be difficult, but a quantitative response over a wide mass range is precisely what is desired for polymer analysis. Among the largest obstacles are inconsistency and inhomogeneity of the sample preparation. The most important practical conditions are miscibility of solutions of analyte, matrix, and cationization agent. If this condition cannot be met, a solid–solid sample preparation can be employed. Mass discrimination can be caused by several factors, at the level of cationization, desorption, and detection, or result from fragmentation. Generally, a combination of these factors are present in a given experiment, usually leading to suppression of higher-mass components in an oligomer distribution. Pre-fractionation of high polydispersity samples and the use of MALDI data recorded on fractions appear to be the best strategies to circumvent mass discrimination.

REFERENCES

1. Karas, M.; Bachmann, D.; Hillenkamp, F., *Anal. Chem.*, **1985**, 57, 2935.
2. Karas, M.; Bachmann, D.; Bahr, U.; Hillenkamp, F., *Int. J. Mass Spectrom. Ion Proc.*, **1987**, 78, 53.
3. Karas, M.; Hillenkamp, F., *Anal. Chem.*, **1988**, 60, 2299.
4. Tanaka, K.; Waki, H.; Ido, Y.; Akita, S.; Yoshida, Y.; Yoshida, T., *Rapid Comm. Mass Spectrom.*, **1988**, 2, 151.
5. Chaudhary, A.K.; Critchley, G., Diaf, A.; Beckmann, E.J.; Russell, A.J., *Macromol.*, **1996**, 29, 2213.
6. Belu, A.M.; De Simone, J.M.; Linton, R.W.; Lange, G.W.; Friedman, R.M., *J. Am. Soc. Mass Spectrom.*, **1996**, 7, 11.
7. Räder, H.J.; Schrepp, W., *Acta Polym.*, **1998**, 49, 272.
8. Montaudo, G.; Montaudo, M.S.; Samperi, F., Matrix-Assisted Laser Desorption/Ionization Mass Spectrometry (MALDI-MS). In *Mass Spectrometry of Polymers* (Montaudo, G.; Lattimer, R.P., editors). Boca Raton, FL: CRC Press, **2002**, pp. 419.
9. Montaudo, G.; Samperi, F.; Montaudo, M.S.; Carroccio, S.; Puglisi, C., *Eur. J. Mass Spectrom.*, **2005**, 11, 1.
10. Laiko, V.V.; Moyer, S.C.; Cotter, R.J., *Anal. Chem.*, **2000**, 72, 5239.
11. Skelton, R.; Dubois, F.; Zenobi, R., *Anal. Chem.*, **2000**, 72, 1707.
12. Trimpin, S.; Rouhanipour, A.; Az, R.; Raeder, J.J.; Muellen, K., *Rapid Commun. Mass Spectrom.*, **2001**, 15, 1364.

13. Zenobi, R.; Knochenmuss, R., *Mass Spectrom. Rev.*, **1998**, 17, 337.

14. Karas, M.; Glückmann, M.; Schäfer, J., *J. Mass Spectrom.*, **2000**, 35, 1.

15. Knochenmuss, R.; Zenobi, R., *Chem. Rev.*, **2003**, 103, 441.

16. Karas, M.; Krüger, R., *Chem. Rev.*, **2003**, 103, 427.

17. Handschuh, M.; Nettesheim, S.; Zenobi, R., *J. Chem. Phys.*, **1998**, 108, 6548.

18. Livadaris, V.; Blais, J.-C.; Tabet, J.-C., *Eur. J. Mass Spectrom.*, **2000**, 6, 409.

19. Krutchinsky, A.N.; Chait, B.T., *J. Am. Soc. Mass Spectrom.*, **2002**, 13, 129.

20. Alves, S.; Kalberer, M.; Zenobi, R., *Rapid Commun. Mass Spectrom.*, **2003**, 17, 2034.

21. Knochenmuss, R.; *J. Mass Spectrom.*, **2002**, 37, 867.

22. Karbach, V.; Knochenmuss, R., *Rapid Commun. Mass Spectrom.*, **1998**, 12, 968.

23. Lin, Q.; Knochenmuss, R., *Rapid Commun. Mass Spectrom.*, **2001**, 15, 1422–1426.

24. Kinsel, G.R.; Knochenmuss, R.; Setz, P.; Land, C.M.; Goh, S.K.; Archibong, E.F.; Hardesty, J.H.; Marynick, D.A., *J. Mass Spectrom.*, **2002**, 37, 1131.

25. Frankevich, V.; Zhang, J.; Dashtiev, M.; Zenobi, R., *Rapid Commun. Mass Spectrom.*, **2003**, 17, 2343.

26. Frankevich, V.E.; Zhang, J.; Friess, S. D.; Dashtiev, M.; Zenobi, R., *Anal. Chem.*, **2003**, 75, 6063.

27. Zhang, J.; Frankevich, V.; Knochenmuss, R.; Friess, S.D.; Zenobi, R., *J. Am. Soc. Mass Spectrom.*, **2003**, 14, 42.

28. Hoberg, A.-M.; Haddleton, D.M.; Derrick, P.J., *Eur. Mass Spectrom.*, **1997**, 3, 471.

29. Knochenmuss, R.; Lehmann, E.; Zenobi, R., *Eur. Mass Spectrom.*, **1998**, 4, 421.

30. Rashidezadeh, H.; Guo, B., *J. Am. Soc. Mass Spectrom.*, **1998**, 9, 724.

31. Rashidezadeh, H.; Wang, Y.; Guo, B., *Rapid Commun. Mass Spectrom.*, **2000**, 14, 439.

32. Wang, Y.; Rashidazadeh, H.; Guo, B., *J. Am. Soc. Mass Spctrom.*, **2000**, 11, 639.

33. Hanton, S.D.; Owens, K.G.; Chavez-Eng, C.; Hoberg, A.-M.; Derrick, P.J., *Eur. J. Mass Spectrom.*, **2005**, 11, 23.

34. Hoberg, A.-M.; Haddleton, D.M.; Derrick, P.J.; Jackson, A.T.; Scrivens, J.H., *Eur. Mass Spectrom.*, **1998**, 4, 435.

35. Mwelase, S.R.; Bariyanga, J., *J. Mol. Struct.*, **2002**, 608, 235.

36. Yalcin, T.; Wallace, W.E.; Guttmann, C.M.; Li, L., *Anal. Chem.*, **2002**, 74, 4750.

37. Chen, R.; Yalcin, T.; Wallace, W.E.; Guttmann, C.M.; Li, L. *J. Am. Soc. Mass Spectrom.*, **2001**, 12, 1186.

38. Martin, K.; Spickermann, J.; Räder, H.J.; Müllen, K., *Rapid Commun. Mass Spectrom.*, **1996**, 10, 1471.

39. Axelsson, J.; Scrivener, E.; Haddleton, D.M.; Derrick, P.J., *Macromol.*, **1996**, 29, 8875.

40. Schriemer, D.C.; Li, L., *Anal. Chem.*, **1997**, 69, 4169.

41. Schriemer, D.C.; Li, L., *Anal. Chem.*, **1997**, 69, 4176.

42. Gervasio, G.; Gerber, D.; Gritti, D.; Gonin, Y.; Twerenbold, D.; Vuilleumier, J.-L., *Nucl. Instrum. Meth. Phys. Res. A*, **2000**, 444, 389.

43. Shimada, K.; Lusenkova, M.A.; Sato, K.; Saito, T.; Matsuyama, S.; Nakahara, H.; Kinugasa, S., *Rapid Commun. Mass Spectrom.*, **2001**, 15, 277.

44. Shimada, K.; Nagahata, R.; Kawabata, S.-I.; Matsuyama, S.; Saito, T.; Kinugasa, S., *J. Mass Spectrom.*, **2003**, 38, 948.

45. Montaudo, M.S.; Puglisi, C.; Samperi, F.M., *Rapid Commun. Mass Spectrom.*, **1998**, 12, 519.

46. Lin-Gibson, S.; Brunner, L.; Vanderhart, D.L.; Bauer, B.J.; Fanconi, B.M.; Guttmann, C.M.; Wallace, W.E., *Macromol.*, **2002**, 35, 7149.

47. Chinthaka, S.D.M.; Chu, Y.; Rannulu, N.S.; Rodgers, M.T., *J. Phys. Chem. A* **2006**, 110, 1426.

48. Brancia, F.L.; Stener, M.; Magistrato, A., *J. Am. Soc. Mass Spectrom.* **2009**, 20, 1327.

49. Batoy, S.M.A.B.; Akhmetova, E.; Miladinovic, S.; Smeal, J.; Wilkins, C.L., *Appl. Spectrosc. Rev.* **2008**, 43, 485.

50. Knochenmuss, R.; McCombie, G.; Faderl, M., *J. Phys. Chem. A*, **2006**, 110, 12728.

51. Shroff, R.; Rulisek, L.; Doubsky, J.; Svatoš, A., *Proc. Natl. Acad. Sci. U.S.A.* **2009**, 106, 10092.

52. Dashtiev, M.; Wäfler, E.; Röhling, U.; Gorshkov, M.; Hillenkamp, F.; Zenobi, R., *Int. J. Mass Spectrom.* **2007**, 268, 122.

53. Liu, B.-H.; Lee, Y.T.; Wang, Y.-S., *J. Am. Soc. Mass Spectrom.* **2009**, 20, 1078.

CHAPTER 3

TIME-OF-FLIGHT MASS SPECTROMETRY FOR POLYMER CHARACTERIZATION

Liang Li and Randy M. Whittal

Department of Chemistry, University of Alberta, Edmonton, Alberta, Canada

3.1 INTRODUCTION

There are a variety of different mass analyzers available for detecting polymeric ions. One of the most commonly used mass analyzers for polymer characterization is a time-of-flight (TOF) instrument. TOF mass spectrometers offer high sensitivity for ion detection, unlimited mass range, and good mass resolving power. They can be conveniently combined with matrix-assisted laser desorption ionization (MALDI) which is a powerful method for ionizing a great variety of polymeric species. In MALDI, desorption of the compound of interest usually takes place under a vacuum inside the ion source of the instrument using a pulsed laser (~1–10 ns duration). A high continuous voltage is applied to the ion source to accelerate the ions out (see Figure 3.1a). MALDI is generally considered a soft-ionization method; that is, generally, only pseudomolecular ions (e.g., protonated ion or MH^+) or metal ion adducts (e.g., MAg^+) are observed. However, MALDI ions do receive enough internal energy to undergo unimolecular decay or to fragment over long periods (2–20 μs). In a TOF instrument, if an ion fragments after leaving the ion source, then the mass of the fragment can be determined using a technique called post-source decay (PSD) analysis. This technique offers a way to analyze the structure of the compound of interest and can be applied to polymer structure analysis. In this chapter, MALDI-TOF instrumentation and its major attributes for polymer characterization will be discussed.

3.2 WHY USE THE TOF ANALYZER?

Analysis of ions formed by MALDI can be done by using mass spectrometers such as magnetic sectors, Fourier transform ion cyclotron resonance (FT-ICR), quadrupole ion trap, and Orbitrap instruments. However, it is the TOF to which MALDI

MALDI Mass Spectrometry for Synthetic Polymer Analysis, Edited by Liang Li
Copyright © 2010 John Wiley & Sons, Inc.

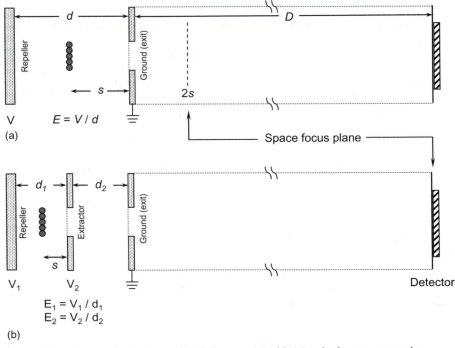

Figure 3.1. Schematic of a time-of-flight instrument with (a) a single-stage extraction source and (b) a dual-stage extraction source. Ions are accelerated in the source; all ions starting at a position s receive the same kinetic energy from the power supply. The ions enter a field-free region where they separate according to their velocity or mass-to-charge ratio; lighter ions (higher velocity) reach the detector first.

is most often coupled. In TOF mass spectrometers [1], all ions are accelerated by a voltage, V, to the same final kinetic energy as they exit the ion source; that is, for an ion of mass m

$$zV = \frac{mv^2}{2},$$ (Eq. 3.1)

where v is the ion's final velocity as it exits the ion source and z is the charge on the ion. The ions enter a field free region where they separate according to their mass-to-charge ratio (m/z) (see Figure 3.1a). The final velocity can be expressed as the length of the field free drift region, D, divided by the time, t, to traverse this region. Therefore, ion flight time, which is dependent on the square root of an ion's mass, can be expressed as

$$t = \left(\frac{m}{2zV}\right)^{1/2} D.$$ (Eq. 3.2)

A TOF mass spectrometer does not measure ion current continually. Ions are pulsed out of the ion source in discrete packets. Measurement of their flight time requires that the start point be known. For convenience, the start time is the time that the push-out pulse is turned on. MALDI is an inherently pulsed ionization method; that is, pulsed lasers are used for ion desorption, thus defining the start point. A TOF, equipped only with constant potential power supplies, provides a simple instrument for the analysis of MALDI formed ions.

Additionally, a TOF is the only mass analyzer with an unlimited mass range. MALDI generated ions greater than 1 MDa have been detected in TOF analyzers [2, 3]. Also, TOF instruments are highly sensitive mass spectrometers. With each laser pulse, all ions formed are detected; that is, scanning the mass analyzer, which reduces sensitivity, is not necessary.

TOF instruments are inexpensive to construct and operate. However, they often require relatively expensive electronic components and data collection systems. The electronic components include very stable high-voltage power supplies and fast-rising pulsed electronics. The data system includes a high-speed ion detector, high-speed amplifiers, fast digitizer, and rapid data processing/storage device. Sometimes, the resolution of a TOF can be limited by its detection system. However, recent advances in electronics and the introduction of digitizers with 1 ns/point or better temporal resolution have made this less likely. TOF instruments generate large data files (a spectrum collected at 1 ns/point out to a 50 μs flight time requires 100 kbytes of storage space). A good modern data handling and storage system can handle the quantity of data produced.

3.3   THE RESOLUTION PROBLEM

TOF analyzers were generally seen as low-resolution instruments in the past. However, rapid development in the past two decades or so in hardware designs, electronics, and data systems has resulted in significant improvement in mass resolving power of TOF mass analyzers. They can be considered as a mass analyzer capable of providing modest mass resolution for polymer analysis. Mass resolution, defined as $t/\delta t$ where δt is the full peak width at half maximum (fwhm) in time, of greater than 10,000 can be routinely achieved with state-of-the-art MALDI-TOF instruments.

Mass resolution of a TOF mass analyzer is governed by several factors. In TOF analysis, ideally, all ions formed in the source would start from the same point, simultaneously, with the same initial kinetic energy. In reality, this does not happen. Depending upon the ion source used, the resolution can be limited by the uncertainty in the time of ion formation, the uncertainty in ion position, the uncertainty in an ion's initial kinetic energy, or a combination of these factors. Ions formed in the gas phase, using electron impact ionization (EI), will have a spatial distribution equal to the width of the ionization beam and an initial kinetic energy distribution equal to the thermal energy distribution of the gas phase neutrals. If two ions, with the same energy, are formed at different points in the source, then the ion formed closer to the repeller (further from the detector) will catch up to the ion further from the repeller at the space-focus plane. In a single-stage ion source (a repeller and exit plate or ground plate, as shown in Figure 3.1a), if an ion is formed at a position s, then the space-focus plane is at a position $2s$. Placing a detector so close to the source is usually inconvenient because ions of different mass will not have time to separate sufficiently. A method used to correct for the initial spatial distribution was devised by Wiley and McLaren in 1955 [4]. In their method, a two-stage ion source was used (see Figure 3.1b). Adjusting the ratio of the electric field strengths in the second region, E_2, to the electric field strength in the first region, E_1, makes it possible to move the space-focus plane to the detector. The location of the space-focus plane is mass independent, but ions of different mass reach it at different times. Unfortunately, while a low value of E_1 favors moving the space-focus plane toward the detector, a high extraction potential is needed to minimize the effect of the initial kinetic energy distribution. Therefore, simultaneous space and energy focusing is not possible; a compromise is usually made. In addition, ions may be formed with their initial velocity directed toward or away from the detector. These ions will be accelerated to the same final kinetic energy but will reach the detector at different times, with the difference called the turnaround time [5].

Ions desorbed from a surface are directed forward (away from the surface) and have a minimal spatial distribution. Also, there is no effect of turnaround time. In MALDI, samples are usually deposited as thin films or small crystals and desorbed from a probe inserted through an opening in the repeller plate (see Chapter 2; Figure 2.1). The spatial contribution to the reduction in resolution is usually neglected. However, if inhomogeneous large crystals are formed on a MALDI plate, resolution may be reduced from the ion spatial contribution. In MALDI, pulsed lasers with 1–10 ns pulse widths are used for desorption. Thus, the temporal distribution of MALDI ions is low. The major contribution to low resolution in MALDI is the inherently broad initial kinetic energy distribution. Initial velocities are almost independent of mass; thus, initial kinetic energies increase approximately linearly with mass. For example, an insulin ion has an initial kinetic energy of ~11 eV ($1 \text{eV} = 1.60218 \times 10^{-19} \text{J}$) and a distribution of ~2 eV [6]. Fortunately, several methods have been devised to compensate for the initial kinetic energy distribution in TOF instruments.

3.4 SOLUTIONS TO THE RESOLUTION PROBLEM

Compensation for the initial kinetic energy distribution can be accomplished in several ways. The techniques include time-lag focusing (or delayed extraction) [4, 7–10], the ion mirror (or reflectron) [11–13], impulse-field focusing [14], velocity compaction [15], dynamic-field focusing [16], and post-source pulse focusing [17]. Of these, time-lag focusing and impulse-field focusing are in-source techniques, whereas the others are post-source methods. Among these techniques, reflectron and time-lag focusing have been most widely used in MALDI–TOF mass spectrometers.

3.4.1 The Ion Mirror (Reflectron)

As discussed above, the adjustments for the initial spatial distribution may be done in the source. Additionally, if the space-focus plane is the focal point of an ion

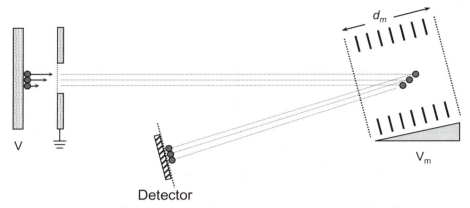

Figure 3.2.   Schematic of a single-stage ion mirror or reflectron time-of-flight instrument.

mirror, then the spatial distribution problem becomes an energy-focusing problem. Ion mirrors, introduced in 1973 by Mamyrin et al. [11], are commonly used to compensate for the initial spatial distribution or the initial kinetic energy distribution or both. The Mamyrin ion mirror is a two-stage design, where two-thirds of an ion's energy is reduced in a short retarding region and the ion's direction is reversed in a reflecting region. The reflecting region consists of a series of parallel plates with gradually increasing potentials applied; the potential of the last plate is usually set to match the source potential. A single-stage ion mirror was developed by Tang et al. [12] at the University of Manitoba. In their instrument, ion energy is reduced gradually across the entire length of the ion mirror. This requires that the single-stage ion mirror be much longer than the dual-stage alternative. However, the energy compensation of the single-stage reflectron is mass independent and a simple linear relationship exists between the flight time of parent ions and their unimolecular decay product ions; this is not true for the dual-stage instrument.

Figure 3.2 shows a single-stage ion mirror instrument. The mechanism of kinetic energy compensation is easily understood. Ions with greater kinetic energy penetrate more deeply into the reflecting region than ions of lower energy. Thus, the more energetic ions spend more time in the reflecting region; all ions of the same mass reach the detector simultaneously; the resolution is greatly improved.

A major reason to couple an ion mirror with a MALDI source is the application of the mirror to PSD analysis [18, 19]. Unimolecular decay in the field-free region forms fragment ions with the same velocity (different kinetic energy) as the parent ion. The fragment ion can be analyzed for its mass by reducing the potential applied to the ion mirror. The fragment ions penetrate the mirror and are brought to focus at the detector. The mass of the fragment is determined by the ratio of the source to ion mirror potential [19]. Cornish and Cotter developed an ion mirror that does not require scanning to analyze the mass of fragment ions [13]. All fragment ions are focused at the detector using a set of non-linear-gradient voltages applied to the ion mirror plates (i.e., curved reflectron). The detection sensitivity of fragment ions is higher than for other ion mirrors [20].

3.4.2 Time-Lag Focusing

Wiley and McLaren developed an energy compensation technique called time-lag focusing in 1955 [4]. As mentioned earlier, they used a two-stage extraction source for gas phase-generated ions that allow movement of the space-focus plane to the detector by selection of the proper ratio of E_2/E_1. Velocity focusing was accomplished by varying the delay between the time of ion formation and ion extraction. Ions were generated in the center of a field-free ion source and allowed to separate according to their initial energy (see Figure 3.3a). If one considers only ions moving toward the detector, then the initially more energetic ions move further from the repeller than the initially less energetic ions. Application of the extraction pulse imparts more energy to the ions closer to the repeller than to the extractor (Figure 3.3b). The initially less energetic ions closer to the repeller catch up to the initially more energetic ions at the detector plane (see Figure 3.3c). Of course, gas phase-generated ions are moving in all directions, which gives a spatial distribution and turnaround time. However, the turnaround time is minimized in the same way as the kinetic energy distribution. Ions closer to the repeller during extraction receive more energy and catch up to ions initially closer to the extractor. A compromise is made between the best space and velocity focusing. In addition, the optimal delay time is mass dependent. A thorough analysis of the mass dependence and range of focus was given by Erickson et al. [21] for a linear time-lag focusing TOF instrument for ions formed in the gas phase.

Equation 3.2 is a simplification of the true flight time of an ion. The equation only measures the time required to traverse the field-free region. In reality, the time spent to accelerate the ion needs to be included. Therefore, for a two-stage source as shown in Figure 3.1b, the total flight time, t, is the sum of the times spent in the first stage of the source, t_1, the second-stage, t_2, and the field-free region, t_D; that is,

$$t = t_1 + t_2 + t_D. \qquad \text{(Eq. 3.3)}$$

The flight time through each region can be calculated from basic equations from mechanical and electrical physics. The time in the first stage (or source) is given by

$$t_1 = \frac{v_f - v_0}{a}, \qquad \text{(Eq. 3.4)}$$

where v_f is the final velocity of the ion, v_0 is the initial velocity, and a is the acceleration of the ion. The acceleration can be expressed as

$$a = \frac{zE_1}{m}. \qquad \text{(Eq. 3.5)}$$

The final velocity and initial velocity can be expressed in energy, U_1, and U_0, respectively; that is,

$$v_f = \left(\frac{2U_1}{m}\right)^{1/2}, \quad v_0 = \left(\frac{2U_0}{m}\right)^{1/2}, \qquad \text{(Eq. 3.6)}$$

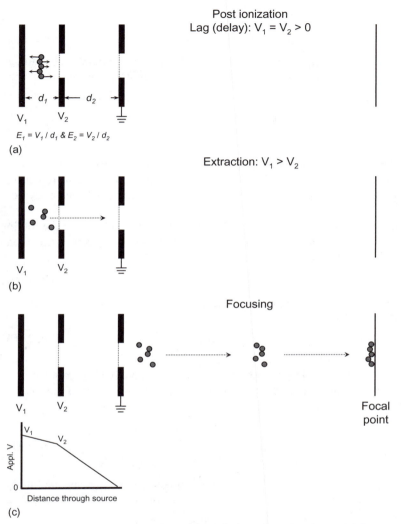

Figure 3.3.   Time-lag focusing in a time-of-flight mass spectrometer. (a) Ions generated in the gas phase in the center of a field-free region separate according to their initial velocity. Ions of the same mass with higher initial velocity (energy) move further from the center region. (b) As the extraction pulse is applied, the initially more energetic ions initially moving toward the detector receive less energy from the extraction pulse than the initially less energetic ions (moving toward the detector). Ions initially moving away from the detector (toward the repeller plate) receive more energy from the extraction pulse (C) The initially less energetic ions or ions initially moving away from the detector catch up to the initially more energetic ions or ions moving toward the detector at the detector or focus plane, greatly improving resolution (see text for details).

and U_1 can be written as

$$U_1 = U_0 + zsE_1, \tag{Eq. 3.7}$$

where s is the ion's position at the time the extraction pulse is applied. Substituting Equations 3.5, 3.6, and 3.7 into Equation 3.4 gives the flight time through the first stage as follows:

$$t_1 = \frac{(2\,m)^{1/2}}{zE_1} \left[(U_0 + zsE_1)^{1/2} \pm U_0^{1/2} \right]. \tag{Eq. 3.8}$$

In the first term containing U_0, the ion's energy as it exits the first region is independent of its initial direction. In the second term, the "+" and "−" signs represent ions that have an initial velocity directed away from and toward the detector, respectively [5].

The final velocity in the first region becomes the initial velocity of the ion in the second region. If d_2 is the distance the ion travels through the second stage, then the final kinetic energy that the ion receives as it exits the source is

$$U_2 = U_f = U_1 + zd_2E_2. \tag{Eq. 3.9}$$

Therefore, the flight time of the ion in the second stage is

$$t_2 = \frac{(2\,m)^{1/2}}{zE_2} \left[U_f^{1/2} - (U_0 + zsE_1)^{1/2} \right]. \tag{Eq. 3.10}$$

The flight time in the field-free region, given in Equation 3.2, can be restated as a function of the final kinetic energy, U_f, as the ion exits the second stage of the source, as follows:

$$t_D = \frac{(2m)^{1/2}}{2U_f^{1/2}} D. \tag{Eq. 3.11}$$

Equations 3.8, 3.10, and 3.11 are substituted into Equation 3.3 to give the total flight time for a linear instrument. For a single-stage ion mirror, one can consider the time spent in the ion mirror as t_m

$$t_m = \frac{(2m)^{1/2}}{2U_f^{1/2}} \cdot 4d_m = \frac{2(2m)^{1/2}\,d_m}{U_f^{1/2}}, \tag{Eq. 3.12}$$

where d_m is the mirror penetration depth [12]. Space focusing is achieved when $dt/ds = 0$ (see also Stein [22] for a more complete development). Wiley and McLaren showed that for given values of s, d, and D, a unique and mass independent value of E_2/E_1 exists if $U_0 = 0$. Of course, it is unlikely that $U_0 = 0$, and enhanced spatial focusing may be achieved using the method of Flory et al. [23]. Wiley and McLaren also showed that the optimum time lag, τ, is proportional to $m^{1/2}$. The optimum τ is found when dt/ds is negative ($U_0 \neq 0$), but this violates the space focus condition; thus, velocity focusing requires some sacrifice in space resolution.

3.4.3   Time-Lag Focusing and MALDI

In Equation 3.8, the initial energy was considered for ions traveling toward and away from the detector. Ions desorbed from a surface travel forward toward the detector; there is no turnaround time to consider. Thus, the second initial energy term, U_0, is negative. For ions desorbed into a field-free region, there will be an initial velocity distribution such that the initially faster ions will move further from the repeller than the initially slower ions (Figure 3.4a). Therefore, the position of any ion can be described as a function of the ion's initial velocity. If the extraction pulse is applied at a time, τ, after the desorption event and d_1 is the entire length of the first stage of the two-stage source, then the ions position, s, can described as

$$s = d_1 - v_0 \tau. \tag{Eq. 3.13}$$

Ions closer to the repeller have lower initial velocity than ions further from the repeller. Application of the extraction pulse imparts more acceleration to the initially slower ions, allowing them to catch up to the initially faster ions at the image plane (see Figure 3.4c). Expressing the ion's position as a function of initial velocity and time lag makes it more convenient to express initial energy as a function of velocity. Studies of the initial velocity of MALDI-formed ions have shown that all protein ions have approximately the same initial velocity, despite mass. Therefore, initial kinetic energy of the protein ions increases approximately linearly with mass. Consequently, a linear dependence of extraction pulse energy on mass should exist.

To find the optimum pulse voltage for a given mass, the total expression for flight time should be considered. Rewriting Equations 3.8, 3.10, and 3.11, and substituting them into Equation 3.3 gives the overall expression for flight time as follows:

$$t = \frac{(2m)^{1/2}}{zE_1}\left[(U_0 + zsE_1)^{1/2} - U_0^{1/2}\right] + \frac{(2m)^{1/2}}{zE_2}\left[(U_0 + zsE_1 + zd_2E_2)^{1/2} - (U_0 + zsE_1)^{1/2}\right]$$
$$+ \frac{(2m)^{1/2} D}{2(U_0 + zsE_1 + zd_2E_2)^{1/2}} + \frac{2(2m)^{1/2} d_m}{(U_0 + zsE_1 + zd_2E_2)^{1/2}} \tag{Eq. 3.14}$$

Recall that $U_0 = \frac{1}{2}mv_0^2$ and the dependence of initial position, s, upon initial velocity, v_0, and lag time, τ, from Equation 3.13. Thus, optimum velocity focusing is achieved as dt/dv approaches zero. This exact equation of ion flight time can be solved numerically to find the optimal τ for a given value of E_1 or the optimum E_1 for a given value of τ, providing that d_1, d_2, E_2, D, and d_m are known. Vestal and Juhasz have provided several general equations that can be applied to give first-, second-, or higher-order focusing conditions for single-stage and two-stage TOF instruments with and without an ion mirror [24]. We refer the reader to this excellent paper for a very thorough description of ion flight time in TOF instruments. The approximate equations can be solved to first or higher order to find the optimal lag time. Generally speaking, for each given delay, there is an optimum extraction pulse potential that minimizes the difference in flight time for ions of the same mass with different initial velocities. A minimum difference in the flight time of ions of the

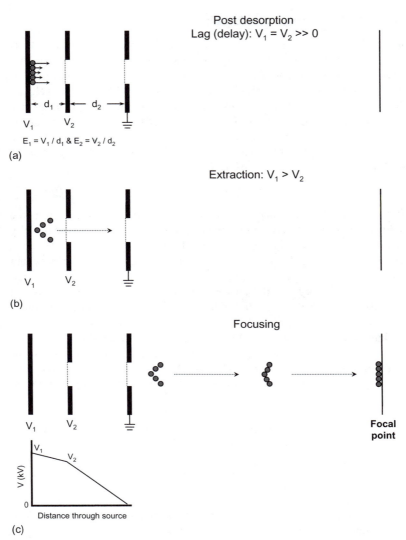

Post desorption
Lag (delay): $V_1 = V_2 \gg 0$

$E_1 = V_1 / d_1$ & $E_2 = V_2 / d_2$

(a)

Extraction: $V_1 > V_2$

(b)

Focusing

Focal point

(c)

Figure 3.4. Time-lag focusing in a MALDI time-of-flight mass spectrometer. (a) Ions are desorbed into a field-free region where they separate according to their initial velocity. Ions of the same mass with higher initial velocity (energy) move further from the repeller. (b) As the extraction pulse is applied, the initially more energetic ions receive less energy from the extraction pulse than the initially less energetic ions. (c) The initially less energetic ions catch up to the initially more energetic ions at the detector plane, greatly improving resolution.

same mass means the resolution is maximized. A shorter time lag requires a higher extraction potential to bring the ions into focus.

Besides the improved resolution, several other reasons make time-lag focusing combined with MALDI attractive. Decoupling ion desorption from ion extraction reduces the number of collisions that would otherwise occur in a continuous extraction source. Excessive collisions can cause peak broadening. Collisions can ionize neutral molecules above the surface of the MALDI probe leading to an unanticipated spatial distribution [25]. Additionally, the number of collisions increases with laser power and signal intensity. This means that ion flight time is laser power-dependent. With continuous extraction, longer flight times are observed or ions receive less imparted kinetic energy than anticipated as laser power increases. Zhou et al. showed that the kinetic energy deficit increases with mass [25]. Time-lag focusing greatly reduces this problem.

Time-lag focusing has been referred to in the literature since 1955 with various names, including pulsed extraction or pulsed ion extraction or delayed extraction or space/velocity correlation focusing. However, all of them are, in reality, time-lag focusing. Many research groups successfully apply the technique to several different ionization sources. For example, Cotter et al. used time-lag focusing for spatial and energy focusing to study such compounds as tetraalkylammonium halides, cyclosporin A, peptides, glycosides, and chlorophyll-a by laser desorption [26–29]; peptides by liquid secondary ion mass spectrometry (MS) [30]; and proteins by MALDI [4, 7–10, 31]. O'Malley and coworkers reported that time-lag focusing can be used to study metastable ion decomposition rates of ions formed by multiphoton ionization [32, 33]. Demirev et al. also showed the use of time-lag focusing to study metastable ion decomposition, including applications to peptide sequencing [34].

3.4.4 MALDI-TOF with an Ion Mirror and Time-Lag Focusing

Ion mirrors are normally used with TOF instruments to compensate for the initial kinetic energy distribution. However, improved resolution is limited to ~10 kDa. The metastable decomposition of MALDI-formed ions reduces the sensitivity of the pseudomolecular ion. Loss of small neutral molecules, such as water, carbon dioxide, and ammonia, from the pseudomolecular parent ion leads to overlap of the parent ion with the fragment ion. In a linear instrument, metastable ions formed in the field-free region have the same velocity as the parent ion. If postacceleration is low, then the fragment ions are indistinguishable from the parent ions; sensitivity and resolution are maintained. Time-lag focusing maintains good resolution at least up to ~30 kDa for proteins and up to ~55 kDa for polymers.

The analysis of metastable ions with short decay times or prompt fragmentation (collision induced) is possible with time-lag focusing but not in a continuous extraction instrument. Brown and Lennon have shown that fast metastable fragmentation analysis can be applied to peptide sequencing [35–37]. Unlike ion-mirror instruments, mass calibration of fragment ions is the same as for parent ions. A disadvantage of this technique is the requirement for very pure samples. In ion-mirror instruments, a mass gate can be used to select the parent ion of interest before metastable decay in the flight tube. Metastable decay takes place over a longer time;

the number of fragment ions is higher in ion-mirror instruments, possibly providing more comprehensive information [38].

3.5 MALDI-TOF FOR POLYMER ANALYSIS

The practice of polymer analysis by MS depends on the type of analytical problems to be addressed. MALDI-TOF instruments are powerful tools for polymer molecular weight determination. They are also useful for generating compositional information of a polymeric system, if the oligomers are resolved. MALDI-TOF with an ion mirror can be used to produce fragment ions for some polymers and thus some structural information about the backbone, side-chain, or end group can be obtained. To utilize a TOF mass spectrometer to its full capacity for polymer analysis, understanding the unique features and limitations of the instrument is important in deciding what instrumental conditions would be selected for a given analysis. Some of the instrumental parameters and their attributes for mass analysis are discussed below.

3.5.1 Mass Resolution and Accuracy

TOF mass spectrometers equipped with reflectron and time-lag focusing provide good mass resolving power and mass measurement accuracy. Typically, a low-mass oligomer (<5000 Da) can be isotopically resolved with a mass measurement accuracy of better than 100 ppm using external mass calibration. Better accuracy (i.e., <30 ppm) can be achieved by using internal mass calibration with a suitable calibrant added to the polymer sample. As polymer mass increases, the isotope peaks of an oligomer become difficult to resolve by TOF. However, if oligomer resolution can be achieved, it can be very useful for structural and compositional analysis of a polymeric material [39, 40].

Figure 3.5a–c shows mass spectra of narrow polydispersity poly(styrene) 18,700, 32,660, and 45,000 obtained with time-lag focusing MALDI-TOF [41]. The pulse voltage was set to optimize the resolution at the center of the distribution of each compound. In all cases, oligomer ions are well resolved. The mass resolution for poly(styrene) 18,700 is 725–905 fwhm, for poly(styrene) 32,660, the resolution is 870–1065 fwhm, and for poly(styrene) 45,000, the resolution is 550–670 fwhm. Figure 3.5d is an expansion of Figure 3.5c at the high-mass end of the distribution, illustrating that oligomer peaks, spaced by a mass of 104 Da, are well resolved at masses beyond 50,000 Da. It should be noted that in analyzing these relatively high-mass polymers the use of reflectron does not further improve the mass resolution obtained. Because of losses of high-mass ions in a reflectron TOF, likely due to metastable fragmentation and/or reduced efficiency of ion focusing, this type of sample is better handled by using a time-lag focusing MALDI-TOF instrument operated in linear mode.

As Figure 3.5 illustrates, the apparent mass resolution gradually decreases as the ion mass increases. The actual value of the upper mass limit where oligomers are resolved is dependent on several factors, including the number of components

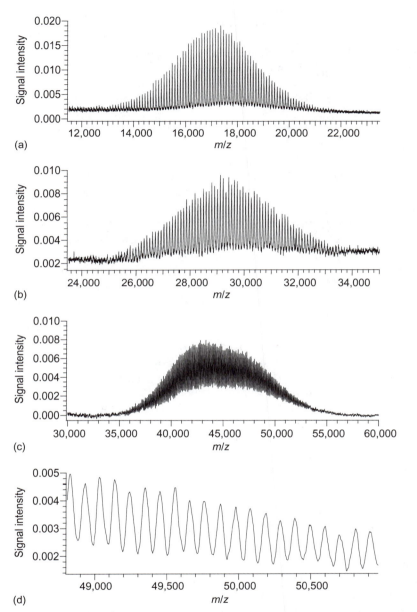

Figure 3.5.   MALDI mass spectra of (a) poly(styrene) 18,700, (b) poly(styrene) 32,660, (c) poly(styrene) 45,000, and (d) the expanded spectrum of poly(styrene) 45,000 using retinoic acid as the matrix with silver cationization. Reprinted with permission from the American Chemical Society from Whittal, R.M.; Schriemer, D.C.; Li, L., "Time-Lag Focusing MALDI Time-of-Flight Mass Spectrometry for Polymer Characterization: Oligomer Resolution, Mass Accuracy, and Average Weight Information." *Analytical Chemistry*, **1997**, 69(14), 2734–2741. Copyright (1997), American Chemical Society.

present in a polymer sample, the mass(es) of the repeat units(s), and the matrix/ sample preparation method. For the latter, whenever possible, the formation of two or more cation adducts should be avoided. The addition of a preferred cation suppresses other cationization and can improve the quality of the spectrum by reducing peak overlap.

The advantage of high-resolution and high-accuracy analysis of polymers is illustrated in the MALDI analysis of poly(ethylene glycol) 20,000 (Figure 3.6a) [41]. Despite the high mass of the polymer and low mass of the repeat unit (44 Da), oligomer peaks are still well resolved across the entire polymer distribution (Figure 3.6b). Figure 3.6a shows that this sample has a bimodal mass distribution, with one distribution centered at m/z ~23,000 and the other centered at m/z 18,500. Inspection of the low-mass region of Figure 3.6a shows that there is a second type of polymer present in the sample. Figure 3.6c is an expansion of a spectrum of poly(ethylene glycol) 20,000, obtained by adjusting the extraction pulse potential to achieve optimal resolution at m/z 12,000. The peaks at m/z 11,142 (calculated 11,142) and 11,935 (calculated 11,935) are from sodium-cationized $HO-(CH_2CH_2O)_n-H$, where n = 252 and 270, respectively. The peaks at m/z 11,126 and 11,919 likely correspond to $H-(CH_2CH_2O)_n-H$, with calculated masses of 11,126 and 11,919 Da, respectively. The sodiated form of this distribution was verified, as the addition of potassium chloride also generated oligomer mass increases of 16 Da. This second polymer distribution is observed in the low-mass region of the spectrum. Extensive studies of poly(ethylene glycol) fragmentation patterns suggest that it is not possible to produce predominant fragment ions that arise from the poly(ethylene glycol) parent ion with a loss of 16 or 16 + 44n Da (where n is an integer number of oligomers) [42]. The MALDI results shown in Figure 3.6 demonstrate that this sample contains two major poly(ethylene glycol) components with different end groups. Since limited information on the synthesis of this sample is provided by the supplier, the exact nature of the end group for the second distribution cannot be confirmed. Nevertheless, this example illustrates that some structural and compositional information can be obtained for polymers if sufficient mass resolution and mass accuracy can be achieved.

3.5.2 Sensitivity and Dynamic Range

Because the amount of material available for polymer analysis is usually not limited, one may underestimate the sensitivity issue in MALDI polymer characterization. In reality, the use of an MS instrument that provides high sensitivity and a wide dynamic range of ion detection is pivotal for the success of polymer analysis. This is true for the measurement of polymer average mass as well as for the determination of polymer composition [43–53]. With limited detection sensitivity, some oligomers within a polymer distribution or minor components in a polymer mixture may not be adequately detected, resulting in measurement errors or misrepresentation of the polymer sample.

Unlike biopolymers such as proteins and peptides, industrial polymers do not have a single exact molecular mass. They display a distribution of molecular masses. The molecular mass distribution and average mass of a particular polymeric

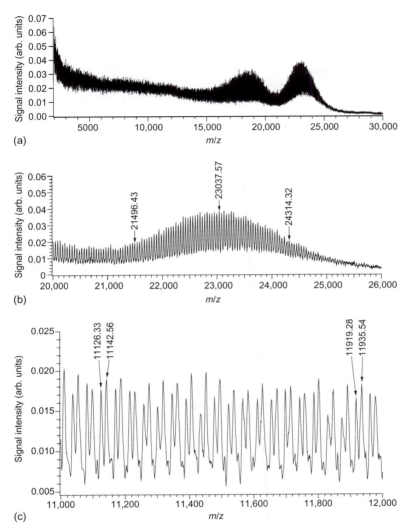

Figure 3.6.   (a) MALDI mass spectrum of poly(ethylene glycol) 20,000 using HABA as matrix with sodium cationization and time-lag focusing set to optimize the resolution of ions at m/z 23,000. (b) Expansion of (a) in the high-mass region of the spectrum. (c) Expanded MALDI mass spectrum of poly(ethylene glycol) 20,000 using HABA as matrix with sodium cationization and time-lag focusing extraction set to optimize the resolution of ions at m/z 12,000. Reprinted with permission from the American Chemical Society from Whittal, R.M.; Schriemer, D.C.; Li, L., "Time-Lag Focusing MALDI Time-of-Flight Mass Spectrometry for Polymer Characterization: Oligomer Resolution, Mass Accuracy, and Average Weight Information." *Analytical Chemistry*, **1997**, 69(14), 2734–2741. Copyright (1997), American Chemical Society.

system depends on the polymerization kinetics and mechanism. To measure the average molecular mass accurately, the MALDI method must be able to generate a mass spectrum that reflects the actual oligomer distribution as well as the relative amounts of all oligomers within the distribution. Sensitivity limitations, background interference, and/or mass discrimination can cause a change in the polymer distribution function, broadening or narrowing of the overall distribution, and/or truncation of detected oligomer peaks within a distribution (i.e., missing low- or high-mass tails). Any one of these variations can result in errors in average mass measurement.

Most commercial MALDI-TOF instruments provide good sensitivity for polymer analysis. However, if the instrument is not fine-tuned to its optimal performance, erratic results may be generated. In addition, a less optimized sample preparation method can lead to low detection sensitivity, as discussed in other chapters. In the absence of any information from a more sensitive detection method, Zhu et al. proposed to use polymer blends to evaluate the dynamic range of detection in MALDI-MS, thus gauging the accuracy of molecular mass measurement [51]. For example, in the case of polystyrene 7000, two bicomponent blends containing polystyrene and a second polymer component of either polystyrene 5050 or polystyrene 11,600 can be prepared. Polystyrene 5050 contains oligomers with the same masses as those of oligomers at the low-mass tail of the polystyrene 7000 distribution, whereas polystyrene 11,600 has peak overlap in the high-mass tail of polystyrene 7000. A sensitive method should be able to detect the mass spectral change after a small amount of a second component is added. With the addition of 5% polystyrene 5050 to polystyrene 7000, the optimized method can readily detect the additional oligomer peaks at the low-mass tail of the polystyrene 7000 distribution. The non-optimal method was not sensitive enough to extend the detection of oligomers to lower masses beyond those from polystyrene 7000. Similarly, very little signal from polystyrene 11,600 is detected in the blend containing polystyrene 7000 and polystyrene 11,600 in a mole ratio of 100 to 2. In contrast, the sensitive method can detect minor components in the mixture of polystyrene 7000 and polystyrene 11,600 even at a mole ratio of 100 to 1, which suggests that any oligomers with relative contents greater than 1% of the most abundant oligomer in the high-mass tail of the polystyrene 7000 distribution should be detectable in MALDI. For polystyrene 7000, the addition of any peaks with intensities less than 1% of the most abundant peak at both tails of the distribution does not affect the calculated M_n, M_w, and polydispersity values, within experimental precision. Thus, the molecular mass measurement results from the optimal method should be accurate.

It is worth noting that analyzing a polymer standard such as polystyrene 7000 by MALDI is an excellent way to gauge the sensitivity and detection dynamic range of an instrument. For a given sample preparation protocol, a poorly designed mass spectrometer or a mass spectrometer that is not optimized for its sensitivity will generate a low-quality spectrum characterized by poor baseline level, asymmetric oligomer distribution, and truncation of oligomer peaks at the tails of the distribution. In the case of analyzing polymer mixtures containing both high- and low-abundance polymers, an instrument with poor sensitivity will result in the failure of detecting the low-abundance polymers.

3.5.3 Time-Lag Focusing

Time-lag focusing is a mass-dependent initial energy compensation method. To optimize mass resolution for ions at different mass, the extraction pulse potential or time lag between ion desorption and ion extraction needs to be adjusted. Ions whose mass-to-charge ratio is less than the optimally focused mass-to-charge ratio receive more than the optimum energy from the extraction pulse; whereas ions with a mass-to-charge ratio that is higher than the optimally focused mass-to-charge ratio receive less than the optimum energy from the extraction pulse. In polymer analysis by MALDI, one of the major objectives is to measure the weight-average and number-average molecular mass, which is dependent upon the relative peak area of each oligomer and the oligomer mass. Thus, it is important to determine if the extraction pulse voltage affects the measurement of relative peak area.

Figure 3.7 shows the MALDI spectra of a blend of narrowly disperse poly(styrene) compounds to give an overall polydispersity of ~1.15. The sample used is a mixture of narrowly disperse poly(styrene) 3250, 5050, 7000, 9240, and 11,600. In Figure 3.7a, the extraction pulse potential was adjusted to optimize resolution for ions at m/z 4000. In Figure 3.7b, the extraction pulse potential was adjusted to optimize resolution for ions at m/z 11,000. The extraction pulse potential can significantly influence the shape of the distribution. However, the relative peak area of each oligomer remains approximately constant. In Figure 3.7a, the resolution decreases with increasing mass, requiring a drop in peak height to maintain the peak area. Table 3.1 summarizes the M_n and M_w values determined for the poly(styrene) blend at four extraction potentials [41]. The standard deviation in determining M_n and M_w at a given extraction potential is 1.6%. The average standard deviation between extraction potentials is 1.2% and 1.1% for M_n and M_w, respectively. Thus, the extraction potential does not have a statistically significant effect upon the relative peak area for the polydispersity studied here. The same conclusion holds for all narrowly disperse polymers studied with the time-lag focusing instrument used.

3.5.4 Mass Range

The upper mass range of a polymer that can be analyzed by MALDI-MS is dependent on the polymer type. For example, Danis et al. reported the detection of water-soluble poly(styrenesulfonic acid) with a molecular mass just under 400,000 [54] and the detection of a poly(methyl methacrylate) sample with a molecular mass of about 256,000 [55]. Multiply charged ions from a starburst polyamidoamine dendrimer with a molecular mass as high as 1.2 million has been reported by Savickas [56]. Yalcin et al. demonstrated that polybutadienes of narrow polydispersity with masses up to 300,000 Da and polyisoprenes of narrow polydispersity with masses up to 150,000 Da can be analyzed [57].

Schriemer and Li still hold the current record of detecting the highest masses of polymers by MALDI-MS using a conventional multichannel plate detector [2]. They have shown that accurate molecular mass determination of samples up to 1 million can be achieved from the singly charged polymeric species (see Figure 3.8). For a poly(styrene) with a molecular mass of approximately 1.5 million, signals

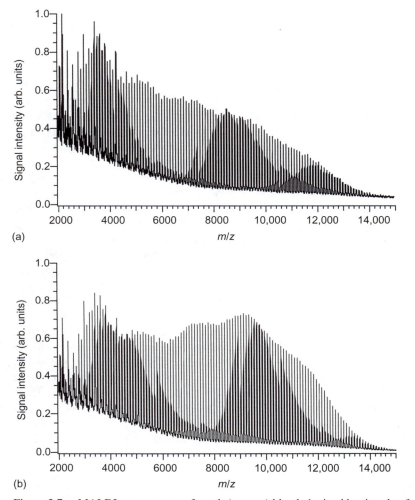

Figure 3.7.   MALDI mass spectra of a poly(styrene) blend obtained by time-lag focusing with the extraction pulse potential set to optimize resolution at (a) m/z 4000 and (b) m/z 11,000. The matrix used was retinoic acid with silver cationization.

TABLE 3.1. M_n and M_w Determined for a Blend of Poly(styrene) 3250, 5050, 7000, 9240, and 11,600 at Different Time-Lag Focusing Extraction Pulse Potentials

Extraction Pulse Optimized at: (m/z)	Observed M_n	Observed M_w
4,000	7,420	8,540
11,000	7,390	8,420
20,000	7,470	8,510
25,000	7,260	8,330

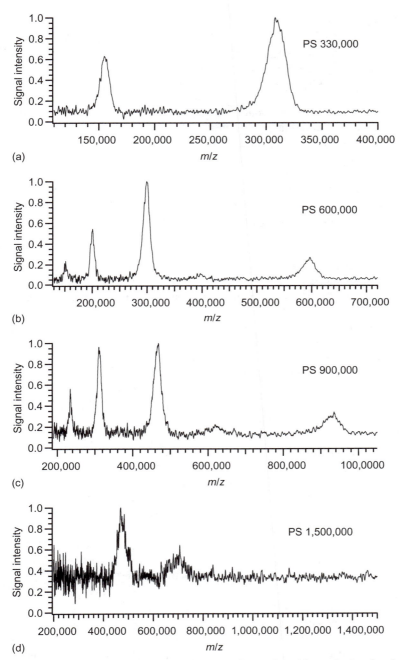

Figure 3.8.   Mass spectra of three poly(styrene) samples with nominal molecular weights of (a) 330,000, (b) 600,000, (c) 900,000, and (d) 1.5 million. Reprinted with permission from the American Chemical Society from Schriemer, D.C.; Li, L., "Detection of High Molecular Weight Narrow Polydisperse Polymers up to 1.5 million Daltons by MALDI Mass Spectrometry." *Analytical Chemistry*, **1996**, 68(17), 2721–2725. Copyright (1996), American Chemical Society.

corresponding to the multiply charged ions of the principal distribution are observed (Figure 3.8d). More recently, Aksenov and Bier have demonstrated the possibility of detecting polystyrene aggregates of greater than several mega Dalton by using MALDI-TOF equipped with a cryodetector [58].

The upper mass limit is set by the requirement of using higher matrix-to-polymer ratios with increasing polymer molecular mass, to the point where the instrument can no longer detect the small quantity of polymer present in the matrix host [2]. Thus, a combination of a highly sensitive MALDI instrument and a highly sensitive sample/matrix preparation method is required to detect high-mass polymers. From the instrument point of view, TOF mass analyzers have no theoretical mass limit and are commonly used for high-mass detection. However, microchannel plate (MCP) detectors are mostly used for detecting the ions in a TOF mass analyzer, which exhibit a strongly decreased detection sensitivity in the high-mass range. This is best illustrated in Figure 3.8a–c where mass spectra of three polystyrene samples with nominal molecular masses of 330,000, 600,000 and 900,000 are shown.

At the high-mass region shown in Figure 3.8, the adjacent oligomer peaks with a mass difference of 104 Da are unresolved, and consequently, the entire oligomer distribution appears as a broad peak. The salient features of these spectra are the appearance of the singly charged molecular ion peaks as well as peaks from multiply charged ions. For Figure 3.8b, the low-intensity peak at m/z ~400,000 is from the triply charged dimer of polystyrene 600,000. Likewise, the small peak at m/z ~620,000 in Figure 3.8c is from the triply charged dimer of polystyrene 900,000 (the mass of the dimer should be ~1.8 million Da). As the mass increases, it becomes clear that the multiply charged distributions begin to dominate the spectra. The relative decrease in the intensity of the principal distribution is probably mostly due to a steady drop in detector efficiency of the MCP, although a shift toward higher charge states with increasing mass has also been observed in MALDI [59].

Careful sample preparation is also crucial in analyzing high-mass polymers. In the preparation of polymer samples for MALDI, a general rule of thumb is to use a matrix-to-analyte molar ratio ranging between 500:1 and 10,000:1. In the MALDI analysis of high-molecular-mass polymers, this does not hold. When the experimenter experiences poor detection sensitivity, it is intuitively sensible that an increase in the amount of analyte would increase the signal strength. However, just the opposite is the case for the analysis of high-molecular-mass polymers [2]. For example, the analysis of polystyrene 900,000 was achieved from 10 fmol of total polymer loaded on the probe tip, with a matrix-to-analyte ratio (M/A) of ~8 × 10^6:1. With the high-mass polystyrenes, use of the molar matrix-to-analyte ratios in the range between 500:1 and 10,000:1 yielded no ion signals at all. It has been observed that with increasing molecular mass of the analyte, progressively lower molar amounts must be added to the probe tip for a given amount of matrix in order to achieve highest sensitivity [2]. The exact reason underlying this observation is unknown; however, it seems likely that for high-mass analytes, the volume ratio of analyte to matrix is the important parameter, rather than the molar ratio so as to obtain a suitable "solid solution" and thereby preclude polymer entanglement and/or the formation of regions of microcrystallinity.

Because it is necessary to decrease the molar amount of polymer loading as the molecular mass increases, this would suggest a practical limit to the mass range for polymer analysis by MALDI. As molecular mass increases, the sensitivity of the instrument is challenged on two fronts: decreased sensitivity due to loss in detector efficiency and decreased sensitivity from the requirement of lower (molar) sample loading. In the analysis of polystyrene, this limit appears to occur at ~1.5 million Da. To obtain the MALDI mass spectrum of polystyrene with a nominal mass of 1.5 million Da [2], 5 fmol of total polymer was loaded on the probe. Even a slight increase of the amount loaded to 15 fmol (for an identical amount of matrix) resulted in a total, reproducible signal suppression, and loading less than 5 fmol proved to be beyond the sensitivity of the instrument. This example illustrates that selection of a proper matrix-to-analyte ratio becomes increasingly critical as the molecular mass of the polymer increases. The window of matrix/analyte leading to observable polymer signal is extremely small as the upper mass limit is approached. In the case of polyisoprene with a nominal mass of 137,000 Da, this window corresponds to 50–500 fmol loaded (M/A = 1.6×10^6:1 to 1.6×10^5:1, with an optimum at 120 fmol loaded (M/A = 6.7×10^5:1). An attempted analysis of polyisoprene 144,800 proved unsuccessful. For polybutadiene 315,200, this window corresponds to 15–200 fmol (M/A = 5.3×10^6:1 to 4.0×10^5:1), with an optimum at 70 fmol loaded (M/A = 1.1×10^6:1). An attempt to analyze polybutadiene 400,000 failed to generate any signals.

The above discussion indicates that the upper mass limit achieved in MALDI analysis of polymers is determined by the requirement of increasing the matrix-to-analyte ratio as the polymer molecular mass increases. A MALDI instrument providing high detection sensitivity in conjunction with the use of a sensitive sample preparation method can be used to detect very high-mass polymers. The work of high-mass polystyrene analysis by MALDI-TOF-MS can be used as a benchmark for future instrumental and method development. The current upper mass limit of a TOF instrument appears to be about ~1.5 million Da for polymers. To push beyond this limit, more sensitive mass spectrometers, particularly with a sensitive detection system tailored for high-mass ions [60], needs to be developed. On the other hand, if a more sensitive sample preparation method than the current method used for polyisoprene and polybutadiene can be developed in the future, the upper mass limit may be extended up to ~1.5 million Da. Another critical parameter is the ion yield of the analyte species. Here, the choice of the best matrix and/or cationization reagent could play a decisive role.

3.5.5 MS/MS Capability

Tandem mass spectrometry (MS/MS) is important for structural and compositional analysis of polymers (see Chapter 5 for detailed discussion on the subject). With a reflectron TOF instrument, product ions of a mass-selected oligomer can be generated and analyzed for some polymeric systems via a PSD process [61–72]. While PSD spectra can be generated using conventional reflectron TOF instruments, it is not widely used for polymer structural analysis for reasons such as the difficulty of selecting a narrow mass range of a precursor ion for fragmentation, lack of control on the internal energy of the precursor ions, and relatively poor quality of product ion spectra for unambiguous structural assignments. However, in the past few years, major advances in MALDI tandem mass spectrometric instrumentation have taken place. MALDI-MS/MS can now be carried out in several commercially available mass spectrometric platforms, including TOF-TOF, quadrupole time-of-flight (QTOF), ion trap (IT), IT-TOF, and FT-ICR-MS. The availability of these commercial instruments is expected to make MALDI-MS/MS a truly powerful tool for polymer structural characterization.

3.6 CONCLUSIONS

MALDI-TOF mass spectrometers have become one of the most important tools for polymer characterization. TOF instruments have no theoretical mass limit and are capable of analyzing molecular ions of as high as 1 MDa. They are uniquely suited for analyzing high-mass singly charged polymeric ions commonly generated by the MALDI process. TOF mass analyzers are sensitive and have a good dynamic range for ion detection, allowing accurate determination of average molecular masses and mass distributions of many low- as well as high-mass polymers. MALDI-TOF equipped with an ion mirror or reflectron and time-lag focusing can provide mass resolving power of greater than 10,000 fwhm. Oligomer ions of a variety of polymers can be readily resolved for structural or compositional analysis. Mass measurement accuracy of 100 ppm with external calibration and 30 ppm with internal calibration can be routinely obtained using state-of-the-art equipment and thus, structural information on repeat units and end groups can sometimes be generated. An intact oligomer ion may be selected in a reflectron TOF mass spectrometer, and the fragment ions generated from the PSD of the oligomer ion can be analyzed. Other TOF configurations such as QTOF and TOF/TOF are capable of carrying out MS/MS analysis of polymer ions for structural analysis. MALDI-TOF mass spectrometers are simple to operate and maintain. However, for polymer characterization, mass spectrometers must be fully optimized to generate reliable results. Several parameters such as instrumental sensitivity can have a major effect on mass spectral results. Loading a large amount of sample to the instrument to offset a detection sensitivity problem of an instrument can produce erratic data, as average molecular mass analysis for polymers requires both accurate mass determination and relative abundance measurements of oligomers within the polymeric distribution. We envisage that MALDI-TOF instruments will continue to be improved in the near future, to enable routine characterization of many different types of polymers.

REFERENCES

1. Stephens, W.E., A pulsed mass spectrometer with time dispersion, *Physical Review*, **1946**, 69(11–12), 691.
2. Schriemer, D.C.; Li, L., Detection of high molecular weight narrow polydisperse polymers up to 1.5 million Daltons by MALDI mass spectrometry, *Analytical Chemistry*, **1996**, 68(17), 2721–2725.
3. Nelson, R.W.; Dogruel, D.; Williams, P., Detection of human-IGM at m/z-similar-to-1 MDA, *Rapid Communications in Mass Spectrometry*, **1995**, 9(7), 625.

4. Wiley, W.C.; McLaren, I.H., Time-of-flight mass spectrometer with improved resolution, *Review of Scientific Instruments*, **1955**, 26(12), 1150–1157.

5. Cotter, R.J., *Time-of-Flight Mass Spectrometry: Instrumentation and Applications in Biological Research*. Washington, DC: American Chemical Society, **1997**.

6. Beavis, R.C.; Chait, B.T., Velocity distributions of intact high mass polypeptide molecule ions produced by matrix assisted laser desorption, *Chemical Physics Letters*, **1991**, 181(5), 479–484.

7. Brown, R.S.; Lennon, J.J., Mass resolution improvement by incorporation of pulsed ion extraction in a matrix-assisted laser-desorption ionization linear time-of-flight mass-spectrometer, *Analytical Chemistry*, **1995**, 67(13), 1998–2003.

8. Colby, S.M.; King, T.B.; Reilly, J.P., Improving the resolution of matrix-assisted laser desorption/ ionization time-of-flight mass-spectrometry by exploiting the correlation between ion position and velocity, *Rapid Communications in Mass Spectrometry*, **1994**, 8(11), 865–868.

9. Whittal, R.M.; Li, L., High-resolution matrix-assisted laser desorption ionization in a linear time-of-flight mass-spectrometer, *Analytical Chemistry*, **1995**, 67(13), 1950–1954.

10. Vestal, M.L.; Juhasz, P.; Martin, S.A., Delayed extraction matrix-assisted laser-desorption time-of-flight mass-spectrometry, *Rapid Communications in Mass Spectrometry*, **1995**, 9(11), 1044–1050.

11. Mamyrin, B.A.; Karataev, V.I.; Shmikk, D.V.; Zagulin, V.A., Mass-reflectron: a new nonmagnetic time-of-flight high-resolution mass-spectrometer, *Zhurnal Eksperimentalnoi I Teoreticheskoi Fiziki*, **1973**, 64(1), 82–89.

12. Tang, X.; Beavis, R.; Ens, W.; Lafortune, F.; Schueler, B.; Standing, K.G., A secondary ion time-of-flight mass-spectrometer with an ion mirror, *International Journal of Mass Spectrometry and Ion Processes*, **1988**, 85(1), 43–67.

13. Cornish, T.J.; Cotter, R.J., A curved-field reflectron for improved energy focusing of product ions in time-of-flight mass-spectrometry, *Rapid Communications in Mass Spectrometry*, **1993**, 7(11), 1037–1040.

14. Marable, N.L.; Sanzone, G., High-resolution time-of-flight mass spectrometry: theory of the impulsed-focused time-of-flight mass spectrometer, *International Journal of Mass Spectrometry and Ion Physics*, **1974**, 13(3), 185–194.

15. Muga, M.L., Velocity compaction—theory and performance, *Analytical Instrumentation*, **1987**, 16(1), 31–50.

16. Yefchak, G.E.; Enke, C.G.; Holland, J.F., Models for mass-independent space and energy focusing in time-of-flight mass-spectrometry, *International Journal of Mass Spectrometry and Ion Processes*, **1989**, 87(3), 313–330.

17. Kinsel, G.R.; Johnston, M.V., Post source pulse focusing—a simple method to achieve improved resolution in a time-of-flight mass-spectrometer, *International Journal of Mass Spectrometry and Ion Processes*, **1989**, 91(2), 157–176.

18. Spengler, B.; Kirsch, D.; Kaufmann, R., Fundamental aspects of postsource decay in matrix-assisted laser desorption mass-spectrometry: 1. residual-gas effects, *Journal of Physical Chemistry*, **1992**, 96(24), 9678–9684.

19. Spengler, B.; Kirsch, D.; Kaufmann, R.; Jaeger, E., Peptide sequencing by matrix-assisted laser-desorption mass-spectrometry, *Rapid Communications in Mass Spectrometry*, **1992**, 6(2), 105–108.

20. Cotter, R.J.; Gardner, B.D.; Iltchenko, S.; English, R.D., Tandem time-of-flight mass spectrometry with a curved field reflectron, *Analytical Chemistry*, **2004**, 76(7), 1976–1981.

21. Erickson, E.D.; Yefchak, G.E.; Enke, C.G.; Holland, J.F., Mass dependence of time-lag focusing in time-of-flight mass-spectrometry—an analysis, *International Journal of Mass Spectrometry and Ion Processes*, **1990**, 97(1), 87–106.

22. Stein, R., Space and velocity focusing in time-of-flight mass spectrometers, *International Journal of Mass Spectrometry and Ion Physics*, **1974**, 14(2), 205–218.

23. Flory, C.A.; Taber, R.C.; Yefchak, G.E., Analytic expression for non-linear ion extraction fields which yield ideal spatial focusing in time-of-flight mass spectrometry, *International Journal of Mass Spectrometry and Ion Processes*, **1996**, 152(2–3), 169–176.

24. Vestal, M.; Juhasz, P., Resolution and mass accuracy in matrix-assisted laser desorption ionization time-of-flight, *Journal of the American Society for Mass Spectrometry*, **1998**, 9(9), 892–911.

25. Zhou, J.; Ens, W.; Standing, K.G.; Verentchikov, A., Kinetic-energy measurements of molecular-ions ejected into an electric-field by matrix-assisted laser desorption, *Rapid Communications in Mass Spectrometry*, **1992**, 6(11), 671–678.

26. Vanbreemen, R.B.; Snow, M.; Cotter, R.J., Time-resolved laser desorption mass-spectrometry: 1. desorption of preformed ions, *International Journal of Mass Spectrometry and Ion Processes*, **1983**, 49(1), 35–50.

27. Tabet, J.C.; Cotter, R.J., Time-resolved laser desorption mass-spectrometry: 2. measurement of the energy spread of laser desorbed ions, *International Journal of Mass Spectrometry and Ion Processes*, **1983**, 54(1–2), 151–158.

28. Tabet, J.C.; Jablonski, M.; Cotter, R.J.; Hunt, J.E., Time-resolved laser desorption: 3. the metastable decomposition of chlorophyll-a and some derivatives, *International Journal of Mass Spectrometry and Ion Processes*, **1985**, 65(1–2), 105–117.

29. Tabet, J.C.; Cotter, R.J., Laser desorption time-of-flight mass-spectrometry of high mass molecules, *Analytical Chemistry*, **1984**, 56(9), 1662–1667.

30. Olthoff, J.K.; Cotter, R.J., Liquid secondary ion mass-spectrometry: 1. molecular ion intensities as a function of primary ion pulse frequency, *Nuclear Instruments & Methods in Physics Research Section B-Beam Interactions with Materials and Atoms*, **1987**, 26(4), 566–570.

31. Spengler, B.; Cotter, R.J., Ultraviolet-laser desorption ionization mass-spectrometry of proteins above 100000 Daltons by pulsed ion extraction time-of-flight analysis, *Analytical Chemistry*, **1990**, 62(8), 793–796.

32. Martin, W.B.; O'Malley, R.M., The non-resonant multiphoton ionization and fragmentation of benzene, phenylacetylene and benzonitrile, *International Journal of Mass Spectrometry and Ion Processes*, **1984**, 59(3), 277–294.

33. Zimmerman, J.A.; O'Malley, R.M., Multiphoton ionization of aniline, aniline-15N and aniline-2,3,4,5,6-d5: ionization and fragmentation mechanisms, *International Journal of Mass Spectrometry and Ion Processes*, **1990**, 99(3), 169–190.

34. Demirev, P.; Olthoff, J.K.; Fenselau, C.; Cotter, R.J., High-mass ion fragmentation as a function of time and mass, *Analytical Chemistry*, **1987**, 59(15), 1951–1954.

35. Brown, R.S.; Lennon, J.J., Sequence-specific fragmentation of matrix-assisted laser-desorbed protein peptide ions, *Analytical Chemistry*, **1995**, 67(21), 3990–3999.

36. Brown, R.S.; Carr, B.L.; Lennon, J.J., Factors that influence the observed fast fragmentation of peptides in matrix-assisted laser desorption, *Journal of the American Society for Mass Spectrometry*, **1996**, 7(3), 225–232.

37. Brown, R.S.; Feng, J.H.; Reiber, D.C., Further studies of in-source fragmentation of peptides in matrix-assisted laser desorption-ionization, *International Journal of Mass Spectrometry and Ion Processes*, **1997**, 169/170, 1–18.

38. Kaufmann, R.; Chaurand, P.; Kirsch, D.; Spengler, B., Post-source decay and delayed extraction in matrix-assisted laser desorption/ionization/reflectron time-of-flight mass spectrometry: are there trade-offs? *Rapid Communications in Mass Spectrometry*, **1996**, 10(10), 1199–1208.

39. Schriemer, D.C.; Whittal, R.M.; Li, L., Analysis of structurally complex polymers by time-lag-focusing matrix-assisted laser desorption ionization time-of-flight mass spectrometry, *Macromolecules*, **1997**, 30(7), 1955–1963.

40. Whittal, R.M.; Li, L., Characterization of pyrene end-labeled polyethylene glycol by high resolution MALDI time-of-flight mass spectrometry, *Macromolecular Rapid Communications*, **1996**, 17(1), 59–64.

41. Whittal, R.M.; Schriemer, D.C.; Li, L., Time-lag focusing MALDI time-of-flight mass spectrometry for polymer characterization: oligomer resolution, mass accuracy, and average weight information, *Analytical Chemistry*, **1997**, 69(14), 2734–2741.

42. Lattimer, R.P., Tandem mass spectrometry of lithium-attachment ions from polyglycols, *Journal of the American Society for Mass Spectrometry*, **1992**, 3(3), 225–234.

43. McEwen, C.; Jackson, C.; Larsen, B., The fundamentals of characterizing polymers using MALDI mass spectrometry, *Polymer Preprints (American Chemical Society, Division of Polymer Chemistry)*, **1996**, 37(1), 314–315.

44. Guo, B.; Rashidzadeh, H., Major origins of mass discrimination encountered in the MALDI-TOF analysis of polydisperse polymers, *ANTEC 1998 Plastics: Plastics on My Mind*, Volume 2: *Materials* (pp. 2096–2100). Society of Plastics Engineers. http://knovel.com/web/portal/browse/display?_EXT_KNOVEL_DISPLAY_bookid=129&VerticalID=0.

45. Rashidzadeh, H.; Guo, B., Use of MALDI-TOF to measure molecular weight distributions of polydisperse poly(methyl methacrylate), *Analytical Chemistry*, **1998**, 70(1), 131–135.

46. Mize, T.H.; Simonsick, W.J., Jr.; Amster, I.J., Characterization of polyesters by matrix-assisted laser desorption/ionization and Fourier transform mass spectrometry, *European Journal of Mass Spectrometry*, **2003**, 9(5), 473–486.

47. Williams, J.B.; Chapman, T.M.; Hercules, D.M., Matrix-assisted laser desorption/ionization mass spectrometry of discrete mass poly(butylene glutarate) oligomers, *Analytical Chemistry*, **2003**, 75(13), 3092–3100.

48. Montaudo, G.; Montaudo, M.S.; Puglisi, C.; Samperi, F., Characterization of polymers by matrix-assisted laser desorption ionization-time of flight mass spectrometry: end group determination and molecular weight estimates in poly(ethylene glycols), *Macromolecules*, **1995**, 28(13), 4562–4569.

49. Jackson, A.T.; Yates, H.T.; Lindsay, C.I.; Didier, Y.; Segal, J.A.; Scrivens, J.H.; Critchley, G.; Brown, J., Utilizing time-lag focusing matrix-assisted laser desorption/ionization mass spectrometry for the end group analysis of synthetic polymers, *Rapid Communications in Mass Spectrometry*, **1997**, 11(5), 520–526.

50. Jackson, A.T.; Yates, H.T.; Lindsay, C.I.; Didier, Y.; Segal, J.A.; Scrivens, J.H.; Critchley, G.; Brown, J., Utilizing time-lag focusing ultraviolet-matrix-assisted laser desorption/ionization-mass spectrometry for the end group analysis of synthetic polymers, *Analusis*, **1998**, 26(10), M31–M35.

51. Zhu, H.; Yalcin, T.; Li, L., Analysis of the accuracy of determining average molecular weights of narrow polydispersity polymers by matrix-assisted laser desorption ionization time-of-flight mass spectrometry, *Journal of the American Society for Mass Spectrometry*, **1998**, 9(4), 275–281.

52. Jaber Arwah, J.; Kaufman, J.; Liyanage, R.; Akhmetova, E.; Marney, S.; Wilkins Charles, L., Trapping of wide range mass-to-charge ions and dependence on matrix amount in internal source MALDI-FTMS, *Journal of the American Society for Mass Spectrometry*, **2005**, 16(11), 1772–1780.

53. Wetzel, S.J.; Guttman, C.M.; Flynn, K.M.; Filliben, J.J., Significant parameters in the optimization of MALDI-TOF-MS for synthetic polymers, *Journal of the American Society for Mass Spectrometry*, **2006**, 17(2), 246–252.

54. Danis, P.O.; Karr, D.E., Analysis of poly(styrenesulfonic acid) by matrix-assisted laser desorption/ionization time-of-flight mass spectrometry, *Macromolecules*, **1995**, 28(25), 8548–8551.

55. Danis, P.O.; Karr, D.E., A facile sample preparation for the analysis of synthetic organic polymers by matrix-assisted laser desorption/ionization, *Organic Mass Spectrometry*, **1993**, 28(8), 923–925.

56. Savickas, P.J., Characterization of Starburst Dendrimers by Matrix-Assisted UV Laser Desorption Mass Spectrometry, 6th Sanibel Conference on Mass Spectrometry, Sanibel, FL, **1994**.

57. Yalcin, T.; Schriemer, D.C.; Li, L., Matrix-assisted laser desorption ionization time-of-flight mass spectrometry for the analysis of polydienes, *Journal of the American Society for Mass Spectrometry*, **1997**, 8(12), 1220–1229.

58. Aksenov, A.A.; Bier, M.E., The analysis of polystyrene and polystyrene aggregates into the mega Dalton mass range by cryodetection MALDI TOF MS, *Journal of the American Society for Mass Spectrometry*, **2008**, 19(2), 219–230.

59. Knochenmuss, R.; Zenobi, R., MALDI ionization: the role of in-plume processes, *Chemical Reviews*, **2003**, 103(2), 441–452.

60. Wenzel, R.J.; Matter, U.; Schultheis, L.; Zenobi, R., Analysis of megadalton ions using cryodetection MALDI time-of-flight mass spectrometry, *Analytical Chemistry*, **2005**, 77(14), 4329–4337.

61. Havlicek, V.; Kieburg, C.; Novak, P.; Bezouska, K.; Lindhorst, T.K., Structure analysis of trivalent glyco-clusters by post-source decay matrix-assisted laser desorption/ionization mass spectrometry, *Journal of Mass Spectrometry*, **1998**, 33(7), 591–598.

62. Przybilla, L.; Rader, H.J.; Mullen, K., Post-source decay fragment ion analysis of polycarbonates by matrix-assisted laser desorption/ionization time-of-flight mass spectrometry, *European Journal of Mass Spectrometry*, **1999**, 5(2), 133–143.

63. Nonami, H.; Wu, F.; Thummel, R.P.; Fukuyama, Y.; Yamaoka, H.; Erra-Balsells, R., Evaluation of pyridoindoles, pyridylindoles and pyridylpyridoindoles as matrices for ultraviolet matrix-assisted laser desorption/ionization time-of-flight mass spectrometry, *Rapid Communications in Mass Spectrometry*, **2001**, 15(23), 2354–2373.

64. Przybilla, L.; Francke, V.; Raeder, H.J.; Muellen, K., Block length determination of a poly(ethylene oxide)-b-poly(p-phenylene ethynylene) diblock copolymer by means of MALDI-TOF mass spectrometry combined with fragment-ion analysis, *Macromolecules*, **2001**, 34(13), 4401–4405.

65. Adhiya, A.; Wesdemiotis, C., Poly(propylene imine) dendrimer conformations in the gas phase: a tandem mass spectrometry study, *International Journal of Mass Spectrometry*, **2002**, 214(1), 75–88.

66. Fournier, I.; Marie, A.; Lesage, D.; Bolbach, G.; Fournier, F.; Tabet, J.C., Post-source decay time-of-flight study of fragmentation mechanisms of protonated synthetic polymers under matrix-assisted laser desorption/ionization conditions, *Rapid Communications in Mass Spectrometry*, **2002**, 16(7), 696–704.

67. Keki, S.; Nagy, M.; Deak, G.; Miklos, Z.; Herczegh, P., Matrix-assisted laser desorption/ionization mass spectrometric study of bis(imidazole-1-carboxylate) end-functionalized polymers, *Journal of the American Society for Mass Spectrometry*, **2003**, 14(2), 117–123.

68. Laine, O.; Trimpin, S.; Raeder, H.J.; Muellen, K., Changes in post-source decay fragmentation behavior of poly(methyl methacrylate) polymers with increasing molecular weight studied by matrix-assisted laser desorption/ionization time-of-flight mass spectrometry, *European Journal of Mass Spectrometry*, **2003**, 9(3), 195–201.

69. Hanton, S.D.; Parees, D.M.; Owens, K.G., MALDI PSD of low molecular weight ethoxylated polymers, *International Journal of Mass Spectrometry*, **2004**, 238(3), 257–264.

70. Hoteling, A.J.; Owens, K.G., Improved PSD and CID on a MALDI TOFMS, *Journal of the American Society for Mass Spectrometry*, **2004**, 15(4), 523–535.

71. Berndt, U.E.C.; Zhou, T.; Hider, R.C.; Liu, Z.D.; Neubert, H., Structural characterization of chelator-terminated dendrimers and their synthetic intermediates by mass spectrometry, *Journal of Mass Spectrometry*, **2005**, 40(9), 1203–1214.

72. Rizzarelli, P.; Puglisi, C.; Montaudo, G., Sequence determination in aliphatic poly(ester amide)s by matrix-assisted laser desorption/ionization time-of-flight and time-of-flight/time-of-flight tandem mass spectrometry, *Rapid Communications in Mass Spectrometry*, **2005**, 19(17), 2407–2418.

CHAPTER *4*

POLYMER ANALYSIS WITH FOURIER TRANSFORM MASS SPECTROMETRY

Sabine Borgmann and Charles L. Wilkins

Department of Chemistry and Biochemistry, University of Arkansas,
Fayetteville, Arkansas

4.1 INTRODUCTION

This chapter discusses the use of Fourier transform mass spectrometry (FTMS) for analysis of synthetic polymers. It is intended to provide a general background about the technique itself, as well as a representative profile of the current literature beginning in 2001. Because of the volume of literature regarding FTMS and polymer analysis as well as space limitations, necessarily not all articles will be discussed. For a comprehensive review of polymer analysis with FTMS before 2001, consult the review by Pastor et al. [1]. Section 4.1 begins with an introduction of the topics covered in this chapter. Section 4.2 is a tutorial on FTMS in general, aimed at providing readers fundamental FTMS theory and a basis for understanding the limitations and benefits of FTMS for analysis of polymeric samples. Section 4.3 highlights recent work dedicated to polymers performed on FTMS instruments between 2001 and the present.

What makes FTMS one of the methods of choice for polymer analysis? For an analytical chemist selecting a technical platform on which to perform polymer analysis, several practical aspects need to be considered such as cost, throughput and time requirements of the analysis, instrument availability, and compatibility with either online or offline sample purification. However, the most essential parameter is how much information can be obtained with the selected technique. Being a mass spectrometry (MS) method, FTMS obviously provides information about the molecular weight distribution (MWD) of polymers. There has been extensive work done on polymer characterization with MS [2–9], so why use FTMS? To address the overall question properly, the needs and requirements from a polymer scientist's perspective have to be considered.

Synthetic polymers comprise a wide variety of chemical compounds (Figure 4.1). In addition to the diversity of polymers themselves, they are often

MALDI Mass Spectrometry for Synthetic Polymer Analysis, Edited by Liang Li
Copyright © 2010 John Wiley & Sons, Inc.

heterogeneous and contain a number of additives and low-mass oligomers. Homopolymers can be either polar or nonpolar. Copolymers can be divided into random copolymers (randomly distributed repeat units), alternating copolymers (alternating repeat units), block copolymers (alternating building units), and graft copolymers (branched).

Addressing the complex analysis of intact synthetic polymers implies a need to investigate MWD, functionality type distribution (FTD), chemical composition distribution (CCD), sequence and block-length distribution, and architecture distribution (Figure 4.1). In order to measure most of these distributions with a reasonable effort, the analytical method(s) need(s) to be general, accurate, easy, fast, and cheap. MS has shown to become more and more important for the analysis of synthetic polymers [7, 8, 10–12] because it is capable of analyzing, with respect to composition, molecular weight and structure. "Soft" ionization techniques such as electrospray ionization (ESI) and matrix-assisted laser desorption/ionization (MALDI) have successfully been used to investigate MWD and FTD. These methods are, in principle, applicable to investigate CCD, the distribution of sequence and block length, and overall structure of polymers. Another important issue is the investigation of polymer additives that comprise molecules, such as antioxidants, UV absorbers, stabilizers, and dyes. There was a basic paper about polymer additives published by Asamoto et al. in 1990 using laser desorption (LD)-FTMS [13]. In this respect, the high resolving power of FTMS analysis enables reliable identification of polymer impurities.

When it comes to the point where a polymer scientist wants to have delineated all the different distributions of a complex polymeric sample in a single experiment, that is where FTMS can play a role. The main advantage of FTMS over other MS

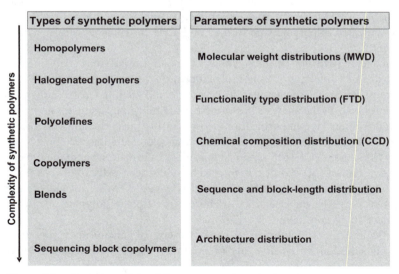

Figure 4.1. Overview of the different types of synthetic polymers (left) and parameters of interest related to analyzing synthetic polymers (right).

techniques for polymer analysis is its high-mass resolving power, the ability to do accurate mass measurements, and the capability for multistage analysis (MS^n), which helps to identify (unambiguously) the composition of the investigated polymer. For example, FTMS generally can provide isotopic resolution: in previous reports, 50 oligomers of a >20kDa poly(ethylene glycol) (PEG) [14] and approximately 200 oligomers of the copolymer dipropoxylated bisphenol A/adipic acid/isophthalic acid (DAI 12) [15] were observed. Therefore, using FTMS as an analytical method in polymer science is fruitful because it allows understanding the structure of the particular synthetic polymers in detail, and later can be used to correlate the information obtained with the properties of the polymers investigated. The information provided by the FTMS analysis may serve as feedback to develop and/or optimize synthesis strategies to produce polymers with desired properties. Mass spectrometric techniques provide rapid and accurate information that is compatible with the needs of modern (combinatorial and high-throughput-oriented) approaches. Table 4.1 summarizes the advantages and disadvantages of applying FTMS for characterizing synthetic polymers. Detailed explanations of the technique and its features are included in Section 4.2.

Summarizing, FTMS is especially convenient for analytical problems that require very high-mass measurement accuracy, comprise very complex mixtures, or

TABLE 4.1. Benefits and Limitations of FTMS for Polymer Analysis (Refer to Section 4.2 for a Detailed Explanation of the Phenomena)

Benefits	Limitations
Highest mass resolving power of all mass spectrometers [152]	Limited dynamic mass range [153, 154]
Internal and external sources in different FTMS-instrument designs available	Strict ultrahigh vacuum (UHV) required
Standard combination with "soft" ionization methods such as MALDI and ESI for high-mass polymers; with LD for materials and surface characterization	Influenced by space-charge effects [38, 113, 155–158] and molecular reactions of ions
Stable mass calibration available by using superconducting magnets and suitable calibration approaches	Harmonics and sidebands (artifacts) observable in mass spectra
Nondestructive ion detection which allows ion remeasurement, thus high-performance detection of polymer ions	Overall performance of the mass spectra depending on many parameters (excitation, trapping, detection) that need to be carefully considered especially for polymer samples
Extensive options for performing ion chemistry and MS^n experiments to rule out structural features of polymers	Not all standard ion manipulation techniques are suitable for MS^n experiments due to the inert nature of polymers
Low limit of detection (is at a signal-to-noise ratio of 3:1 roughly 100 ions, provided the ions are excited to travel at half of the maximum cyclotron radius [158])	Somewhat limited throughput due to the time requirements of high-resolution FTMS spectra

when very low amounts of polymer samples are available. FTMS is a vital tool for the study of polymers. For the present discussion, the different mass range categories as the following are arbitrarily defined as: low mass up to 5000, middle mass 5000–10,000, and high mass greater than 10,000. (Note: These are not equally populated categories.) A lot of mass spectrometric work has successfully been done for low-mass and middle-mass polymers, but high-mass polymers are still challenging. To date, the main experimental challenges are posed by copolymers and hydrocarbon polymers [16–20].

4.2 FTMS TUTORIAL

This section is devoted to a discussion of FTMS. Numerous excellent reviews have been published on FTMS [21–38]. Contained in these are the basic theory of the method [31, 39–42], discusssion of excitation and detection techniques [43, 44], ion trap cell design [45], a theoretical treatment of cell designs [45], and applications of FTMS for polymer analysis [34, 38, 46–48].

The development of FTMS has been far-reaching since its introduction in 1974 by Comisarow and Marshall [49], and its applications range from chemical to biological questions. The historical development of FTMS is described in detail by Comisarow and Marshall [50]. Here, a few milestones during the development of FTMS are mentioned. Lawrence and Livingstone were the first to explore the general principle of ion cyclotron resonance (ICR) in 1932 and in their experiments accelerated protons to MeV kinetic energies [51]. Penning implemented an electrostatical confinement perpendicular to the magnetic field in 1936, which later came to be called the "Penning trap." The first mass spectrometer employing ICR was called the omegatron and was pioneered by Hipple et al. in 1949 and Sommer et al. in 1951 [52, 53]. So-called ICR spectrometers [54–56] produce mass spectra by the absorption of radio-frequency (RF) energy by analytes placed in a strong magnetic field while scanning either the frequency or the magnetic field. Therefore, these instruments have much in common with nuclear magnetic resonance (NMR) spectrometers. In 1966, Fourier transformation was introduced for spectral analysis of NMR spectra by Ernst and Anderson [57]. Application of Fourier transform analysis for MS was performed for the first time in 1974 by Comisarow and Marshall's Fourier transformation of the collected time-domain transient after exciting the cyclotron motion of methane ions trapped in a trapped-ion cell [49]. With their follow-up work [49, 58, 59], they established Fourier transform ion cyclotron resonance-mass spectrometry (FT-ICR-MS, as they entitled the technique) as a high-resolution MS technique being able to perform studies of gas-phase chemistry of ions. The second FTMS instrument was built by Bill Horsley at Nicolet Instrument Corporation, Mountain View, California and the first published spectra obtained with this apparatus were published by Wilkins in 1978 together with a new name, FTMS, for the technique in order to punctuate the difference between FTMS and conventional scanning ICR MS which has limited resolution and mass range [60]. The first FTMS instrument explicitly built for analytical purposes was produced in the laboratories of Wilkins and Gross [61]. Since then, numerous other FTMS instruments have been built [62–64]. The two key patents in FTMS instrumentation are those of McIver [65] who patented the trapped-ion cell and the one from Comisarow and Marshall [59] who patented the FTMS technique itself. Since the early developments of FTMS, the technique focus has shifted from studies of pure ion physics to complex chemical and biological investigations over a period of about 30 years [66]. Great instrument improvements have been made with respect to ion optics, ICR cell designs, superconducting magnets, data acquisition software and hardware [42]. Excellent reviews discussing these have been published [22–38, 54].

In FTMS, ions are confined within a magnetic (or electromagnetic) field. Perpendicular to the magnetic field, ions experience a cyclotron motion that keeps them orbiting in small circular paths. Figure 4.2 shows a schematic representation of a modern FTMS instrument equipped for external ionization with MALDI. For details about the MALDI process, refer to Section 4.3.2. An FT mass spectrometer typically consists of an ion source (internal or external), an ion focusing optic for the latter, an ICR analyzer cell within the homogenous magnetic field, and a computer system to analyze and process the data.

A simple FTMS experiment consists of the following events: quenching ions to empty the analyzer cell, ionizing the analyte, guiding the ions toward the analyzer cell, introducing ions into the analyzer cell, development of stable cyclotron motion of the ions, selecting the desired range of mass-to-charge (m/z) ratio, exciting the ions in their cyclotron motion, detecting the ions, storing the data obtained, and performing the FT. An important feature of such an instrument is an ultrahigh vacuum (UHV) system, which employs either turbomolecular, diffusion, or highly efficient cryopumping systems in order to provide pressures as low as 10^{-11} Torr. Maintaining UHV efficiently is especially important in order to allow the ions to maintain their cyclotron orbits and to permit extended observation times. In addition, cooling and collisional gases are often employed to assist in controlling the internal energy of the ion.

In order to avoid confusion, two definitions should be kept in mind. Separation of two mass spectral peaks with a full width of a spectral peak at half-maximum peak height (full width half maximum [FWHM]) is defined as *resolution* ($\Delta m_{50\%}$).

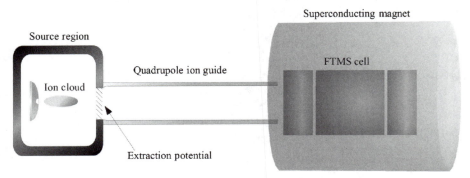

Figure 4.2. Schematic representation of a MALDI-FTMS instrument with external ionization. See color insert.

Consequently, *resolving power* is described as $m/\Delta m_{50\%}$ and will increase with decreasing mass-to-charge ratio [35]. It is important to note that mass resolution in FTMS is directly related to analysis time and is therefore inversely related to pressure. The difference between the measured and theoretical (calculated) mass for a certain ion is referred to as *mass accuracy*. Modern commercial FTMS instruments using an automatic gain control to control the number of ions admitted to the analyzer cells easily can provide mass measurement accuracies better than two parts per million (ppm).

4.3   IONIZATION PROCESSES

Ionization of the polymer is the first event required for obtaining FTMS spectra. There are a variety of ionization methods available. Here, only those having the highest relevance to polymer analysis such as ESI, MALDI, and field desorption/field ionization (FD/FI) will be discussed. Historically, a problem encountered while analyzing high-mass polymers was the production of intact ions in the gas phase. To circumvent this limitation, direct LD ionization was successfully applied for a variety of molecules [67–70].

4.3.1   ESI

The introduction of ESI [71] has altered the analysis of high-mass molecules which benefit from production of multiply charged ions with accordingly lower *m/z* values. The processes involved in ESI are well understood [72, 73]. Basically the analyte is dissolved in an appropriate solvent system. The mixture is conducted through the end of a charged capillary with an applied high voltage. As a consequence, a spray consisting of the now charged analyte and solvent molecules is obtained. Between the end of the capillary and the first skimmer electrode, the solvent molecules evaporate, enabling the introduction of analyte molecules into the FTMS instrument. Therefore, ESI spectra often consist of a series of Gaussian-distributed mass-to-charge ratio peaks of multiply charged ions related to a single analyte molecule. Figure 4.3 depicts this process. However, with various small molecules, singly charged ions can predominate.

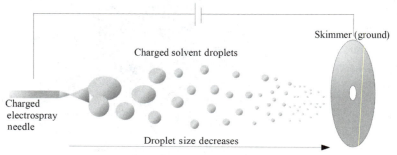

Figure 4.3.   Schematic representation of the ESI ionization process. See color insert.

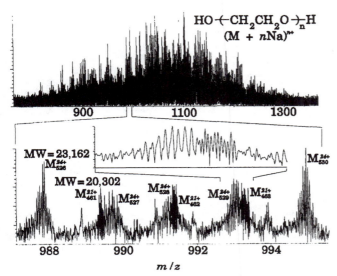

Figure 4.4.   Broadband ESI-FTMS spectrum (30 spectra co-added) of PEG 20,000, with regions expanded once and twice. (Reproduced from Reference 74 with permission of the copyright holder.)

ESI has proven to be a powerful technique for exploration of bioanalytical questions. However, polymer research has had a relatively limited amount of investigation. A drawback of the ESI process is that the production of charge states is not trivial to control. Therefore, a decrease in relative ion abundances is observed because the number of trapped ions is propagated over a relative large mass-to-charge ratio range and results in greatly reduced effective sensitivities obtained for each particular charge state. An example published by McLafferty [74] makes the problem of the resulting complex spectra obvious. Analysis of a PEG 20,000 yielded 5,000 peaks from isotopic clusters and indicated the presence of 65 oligomers in 12 charge states (Figure 4.4). The fact that data interpretation is very complex and expensive can be seen in Reference 14 but, according to the authors, adequate results can be obtained.

4.3.2   MALDI

Since its introduction [75–78], MALDI-MS has been employed for a wide variety of analytes [79] including synthetic polymers [5, 79, 80]. In fact, MALDI ionization has become common for investigation of polymers by MS [3–5, 7, 8, 80–83]. For readers who are interested in the very early developments, the excellent tutorial about LD-FTMS applied to polymer characterization published by Campana et al. [47] should be consulted.

MALDI is comprised of several steps (Figure 4.5). First of all, the analyte molecules need to be co-crystallized with a matrix on a sample probe tip in molar

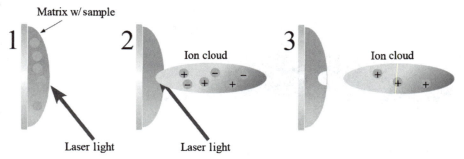

Figure 4.5. Schematic representation of the MALDI ionization process. See color insert.

matrix-to-analyte ratios typically ranging from 1 : 1 to greater than 10,000 : 1. The type of matrix is usually selected so that its absorption characteristics match the wavelength of the laser used in the FTMS. In addition, a successful matrix requires a suitable solvent capable of dissolving both the analyte and the matrix. Here, a number of aspects of sample preparation are briefly addressed. In general, aerospray techniques [20] are faster and result in more homogeneous surfaces. This method also appears to result in smaller crystal formation compared to those derived from the dried-droplet method. Thus, the aerospray method seems to provide more reproducible results.

Following sample preparation, the laser beam desorbs ions from the co-crystallized sample into the gas phase. The MALDI process is not yet fully understood, but it is believed that the matrix plays the role of adsorbing the laser energy and gives the analyte a lift into the gas phase while being ejected. Even though only a small percentage of the analyte molecules are ionized, it is easily possible to further manipulate and subsequently detect them by FTMS. Generation of high-mass ions by internal MALDI in conjunction with detection by means of FTMS was shown successfully for the first time by Castoro et al. [32]. An excellent review of MALDI was published by Knochenmuss et al. [84]. In particular, synthetic polymers almost exclusively cationize by the formation of metal ion adducts [4, 84–87] which explains why addition of metal salts tends to improve the quality of MALDI spectra. A recent article by Zhang and Zenobi that addresses matrix-dependent cationization is highly recommended [86]. A desirable prerequisite for performing MALDI-FTMS is that no mass discrimination exists in the MALDI or cationization process, the conduction of ions to the ICR cell, the excitation, or detection. The topic of mass discrimination during MALDI is still discussed. A short time after ionization, an ion packet of mostly singly charged analyte ions has been formed. The ion packet needs to be manipulated and trapped effectively for detection before being pumped away with neutral molecules. Summarizing, the major advantage of MALDI ionization is the ease of sample preparation without the need of complicated separation methods. In addition, mass spectra obtained from MALDI ionization profit from a relatively simple pattern generated by the presence of primarily singly charged ions.

4.3.3 FD/FI (Field Desorption/Field Ionization)

FI [88] was combined with MS in 1954 by Inghram and Gomer [89]. FD MS employs an FI emitter on which analytes are deposited and desorbed into the gas phase by use of a high electric field (10^7–10^8 V/cm) with thermal support. FI and FD have been applied to a wide range of samples, among them hydrocarbons [90] and polymers [91, 92]. FI has the following advantages for analyzing complex mixtures: It produces true molecular ions ($M^{+\bullet}$) by electron tunneling instead of cationized molecular ions obtained by forming adducts with a proton or another ion. Moreover, the ionization efficiencies for nonpolar molecules are higher compared to other ionization methods. Because no matrix is required, cleaner mass spectra can sometimes be obtained.

4.3.4 Internal versus External Ionization

The instrumental design differs significantly for internal and external ionization FTMS. A major difference of the two ionization types is time scale of the ionization-to-detection events: 100–200 ms for internal ionization versus 1000–5000 ms for external ionization. For internal ionization, the gated trapping technique is the standard. Internal ionization combined with MALDI has been intensively studied for polymer research [20, 93, 94], but an ESI interface was also designed for internal ionization [95]. To date, a lot of research has been performed on external ion sources because they facilitate interfacing with a variety of techniques such as fast atom bombardment (FAB), secondary ion mass spectrometry (SIMS), ESI, and MALDI. They allow tuning of instrumental parameters independently from optimization of ionization and detection. Ion optics such as electrostatic lenses [96], electrostatic ion guides [97] and RF multipole ion guides [98–100], or an ion conveyor [101] are needed to transfer ions from an external ion source to the ICR analyzer cell. For both ionization types, it is important to note that the most obvious form of mass discrimination occurs at the stage of ion transmission to the ICR cell and trapping. Therefore, the main issue contributing to possible polymer mass discrimination is the trapping protocol because analytes might not be properly "sampled" if the trap is closed prematurely (biasing for lower-mass species) or left open too long (biasing for higher-mass species).

4.4 DETECTION PRINCIPLE

In order to detect ions via FTMS, an analyzer cell, also called an ICR cell, is needed. This cell is placed in the (ideal) homogeneous region of a strong magnetic field with typical field strengths of 3–12 Tesla [102, 103] (T; $1 T = 1 N/Am = 10^4 G$ with N = Newton, A = Ampere, m = meter, and G = Gauss). The magnet of a typical FTMS instrument has a wide bore through its middle to accommodate the ICR cell. The field lines of the magnetic field B should be parallel oriented to the ions injected from the ion source.

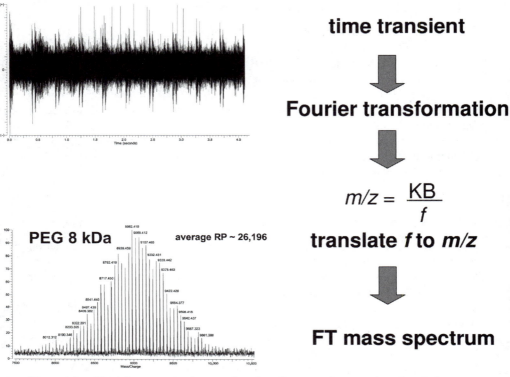

Figure 4.6. Obtaining mass-to-charge ratio values from raw time-transient data during an FTMS experiment.

4.4.1 Ion Motion within the Analyzer Cell

Detection itself takes place by measuring the "image" currents on the detector plates induced from the coherent ion packets repeatedly passing them at their specific orbits and cyclotron frequencies, respectively, defined by their mass-to-charge ratio values. Detection in FTMS stands out because ion "image" currents are recorded, which means the detection technique itself is nondestructive. The AC currents are generated by the movement of coherently excited ion packets within a magnetic field under UHV at the detector electrodes. The ion motion is commonly referred to as ion cyclotron motion [35] and may last for several seconds to hours dependent on the vacuum conditions. The detected waveform motion is subjected to an FT to measure ion frequencies to the precision of one thousandth of a Hertz. During the detection event, the transient free induction decay (FID) is detected, representing the sum of the harmonic "image" currents over time. By applying FT on the FID, the frequencies of the different ion packets are obtained. Equation 4.1 allows the calculation of the corresponding mass-to-charge ratio values from the FT obtained frequency values. The process is visualized in Figure 4.6. In Figure 4.6, K, is simply a constant, incorporating the conversion factors necessary to account for units employed for f and B. The related amplitudes within the resulting mass spectrum represent the abundances of the constituent ions:

$$f_c = zB/2\pi m \qquad\qquad \text{(Eq. 4.1)}$$

Equation 4.1. Fundamental equation describing the relationship between the mass-to-charge ratio (m/z) of the ions and their cyclotron frequency f_c (with $f_c = v_{xy}^2/r2\pi$) within the magnetic field B.

Note that the cyclotron frequency f_c is inversely proportional to mass-to-charge ratio (m/z). The higher the mass-to-charge ratio is, the smaller is the f_c. In addition, Equation 4.1 explains the uniqueness of FTMS in comparison to all other mass spectrometric techniques. There is no focusing of translational energies necessary because the velocity of the ions can be neglected due to the fact that their kinetic energy is small compared to the energy uptake from the RF excitation field (Note that the complex ion motion (Figure 4.7) is important with respect to accurate frequency-to-mass calibration).

For an understanding of FTMS experiments, it is necessary to consider the motion of ions within the analyzer cell in more detail. Once ions enter the (ideal) homogeneous magnetic field, they begin to follow their corresponding cyclotron orbits. Using a simplified model, the initial orbital radius is defined by the balance between the Lorentz force F_L and the centrifugal force F_Z (Equation 4.2 and Figure 4.7). Ions with the same mass-to-charge ratio but opposite charges would move with the same cyclotron frequency but in opposite directions if they were both present in the ICR cell:

$$F_L = zv \qquad\qquad \text{(Eq. 4.2)}$$

Equation 4.2. Lorentz force F_L: force of a magnetic field B on a moving charge z with a velocity.

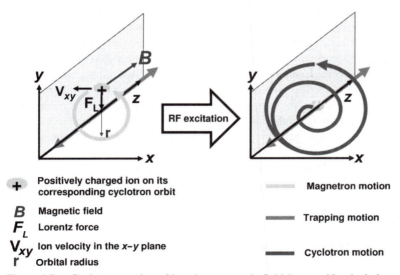

RF excitation

+   Positively charged ion on its corresponding cyclotron orbit

B Magnetic field
F_L Lorentz force
V_{xy} Ion velocity in the x–y plane
r Orbital radius

Magnetron motion

Trapping motion

Cyclotron motion

Figure 4.7.   Cyclotron motion of ions in a magnetic field B caused by the balance between Lorentz force F_L and the centrifugal force F_Z. Schematic simplified representation of the ion motions in the ICR cell. See color insert.

$$vF_z = m\, v_{xy}^2 / r \qquad \text{(Eq. 4.3)}$$

Equation 4.3. Centrifugal force F_Z with ion mass m, ion velocity v_{xy} in the x–y plane and the orbital radius r.

However, the overall ion motions within the analyzer cell are more complex. In addition to the stable magnetron motion of the ions in a plane perpendicular to the magnetic field, there is also the z-directional trapping and the cyclotron motions to be considered. Ions are contained within the analyzer cell with the help of trapping plates, which induce an oscillatory motion in the z-plane independent of the strength of the magnetic field. The magnetron motion itself does not allow an accurate monitoring of the "image" currents on the detector plates because (1) the ions are statistically distributed within the analyzer cell, (2) their the magnetron radii are too small to avoid space-charge effects between the ions upon Coulomb interaction, and (3) their orbital radii are too small to be detected.

To circumvent the above-mentioned drawbacks of the magnetron motions, ions are excited to greater orbits with the help of an additional applied oscillating electric field on the excitation plates (Equation 4.4). The resulting motion is called cyclotron motion and allows the formation of coherent ion packets. The excitation event only takes place if the frequency of the applied RF field is identical with the cyclotron frequency of the ions in order to enable energy absorption by them to result in increased velocities and orbital radii without changing their cyclotron frequency:

$$E_t = E_0 \cos f_c t \qquad \text{(Eq. 4.4)}$$

Equation 4.4. Applied RF field E_t for exciting ions from their magnetron to their cyclotron motion.

Excitation and detection techniques were reviewed by Schweikhard [43]. The amplitude of the RF image currents is proportional to the number of ions and its frequency is characteristic of the mass [104]. Several excitation modes have been developed such as stored waveform inverse Fourier transform (SWIFT) excitation [28], arbitrary waveform, frequency-sweep, or "CHIRP" (sweeping a range of frequencies) excitation [105, 106]. Technically, it is not trivial to cover wide mass-to-charge ratio ranges during an excitation. For narrow band excitation, the rectangular pulse mode and heterodyne mode has been employed, resulting in higher performance FT mass spectra of at least one order of magnitude compared to broadband excitations.

Note that the ion-detection principle is nondestructive. The ion packets can be "stored" in their cyclotron orbits for up to several hours as long as UHV ($\leq 10^{-8}$ Torr) is maintained. During a typical detection period, the ions make about 10^4–10^8 cycles between the detector plates. With the help of such an elongated FID recording within the analyzer cell, the mass spectra obtained are characterized by a significantly improved sensitivity and extraordinary high values of resolution and mass accuracy.

Typically, the design of an analyzer cell is adapted to its versatile tasks during the excitation and detection process. It generally consists of a front and back trapping

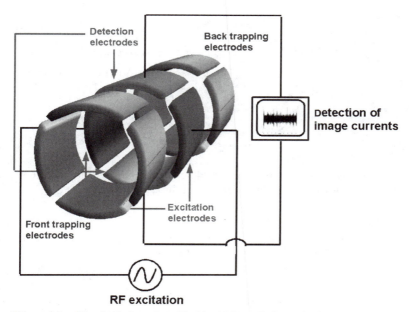

Figure 4.8. Simple design of a cylindrical ICR cell. See color insert.

electrode, two opposed excitation electrodes, and two opposed detection electrodes (Figure 4.8).

As the reader might anticipate, there is no "ideal" analyzer cell existing, because for all actual designs there is an obvious discrepancy between optimal needs and physical limitations. Therefore, most ICR cells are a compromise between mass accuracy, ion capacity, mass resolving power, and complexity of the design. Choosing a suitable analyzer cell depends upon the type of analyte, the analytical question, and the availability.

A major driving force for the development of different geometries of analyzer cells is to reduce the space-charge effects within the FT-ICR cell in order to avoid resulting frequency shifts of the involved ions. Generally, the geometry of an analyzer cell defines the actual feasible cyclotron radii of the ions to be detected. In Figure 4.9, the differences of basic features of a cubic and a cylindrical analyzer cell are summarized. Obviously, from a geometric point of view, the cylindrical ICR cell exploits the space within the bore of the magnet more optimally in order to generate greater cyclotron orbits of the excited ions (Figure 4.9a,b). Very recently, Hawkridge et al. have determined that the optimal postexcitation radius is located between ca. 60% and 70% of the maximum radius of an ICR analyzer cell [107], which is in accordance with the theoretical optimal radius for cylindrical FTMS cells [108] calculated to be 60%. For excitation of ions within an analyzer cell, one must take into account that the orbits of ions can only be increased so much in order that wall collisions of the lightest ions of the mass range of interest are avoided. Furthermore, the ion packets can pass the detector plates in a cylindrical cell design for a longer

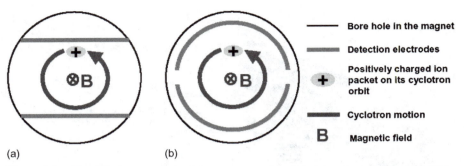

(a)   (b)

Figure 4.9.   Simplified comparison of the geometry of a cubic (a) and a cylindrical (b) ICR cell within a bore hole of a superconducting magnet. See color insert.

period of time at a constant distance allowing detection of better transients from the induced "image currents" compared to a cubic cell design.

A more detailed comparison of the advantages and disadvantages of different analyzer-cell geometries is summarized in Table 4.2.

As already mentioned, FTMS stands out because ions are created, stored, and analyzed nondestructively, which enables one to perform a variety of ion manipulations (e.g., ion activation, quadrupolar axialization, remeasurement) during the course of an FTMS experiment. Williams et al. introduced multiple remeasurement of ions for FTMS [109]. There are a variety of dissociation techniques established such as collision-induced dissociation (CID), multiple excitation collisional activation (MECA), sustained off-resonance irradiation for collision-induced dissociation (SORI-CID), very low-energy collision-induced dissociation (VLE-CID), nozzle-skimmer (NS) dissociation, surface-induced dissociation (SID), electron-induced dissociation (EID), photodissociation (PD), and blackbody infrared radiative dissociation (BIRD). Applications to polymer dissociation are reviewed in two comprehensive articles [110, 111].

FTMS provides researchers with very accurate mass-to-charge ratio values at high resolution given that the calibration has been done correctly. There are two ways to do so, either internal or external calibration. An internal calibration is an approach where a known compound with known mass, the calibrant, is mixed with the analyte and both are measured simultaneously. However, in external calibration, calibrant and analyte are analyzed individually but under identical conditions. For external calibration, it is important to ensure that the calibrant is of the same mass range as the analyte and the number of trapped ions of analytes and calibrants within the ICR cell are equivalent. The mass spectrum of the calibrant is calibrated first using a suitable calibration equation to fit the expected mass-to-charge ratio values of the known. The calibration constants obtained are then applied to calibrate the FTMS spectra of the analyte. Although internal calibration provides approximately three times better accuracy than external calibration, external calibration seems to be more feasible taking the complexity of FTMS spectra obtained from polymers into account. Extensive work has been accomplished in the area of mass calibration in FTMS [38, 42, 93, 101, 112–114]. Nikolaev and coworkers have addressed the

TABLE 4.2. Comparison of Different Types of ICR Cells for FTMS

Analyzer-Cell Designs	Advantages	Disadvantages
Cubic Designs		
Cubic FTMS cells [159–161]		Only a theoretical optimal radius of 0.75 a/2 with a = distance between the opposed detector plates is provided [108]
Dual cubic FTMS cell	Consists of a source cell separated with a conductance limit from the analyzer cell; employed in an internal ionization detection is performed within 100 to 200 ms	
Cylindrical Designs		
Cylindrical FTMS cells [35, 162, 163]	A theoretical optimal radius of 0.6 r with r = radius of the analyzer cell was determined [108].	
Close-ended cylindrical FTMS cells		Only a restricted access is permitted resulting in a limited external ion injection and, thus, a reduced sensitivity [153]. Mass-dependent ion loss and mass discrimination during excitation events may be observed [154].
Open-ended cylindrical FTMS cells [48, 164–167]	This design overcomes most of the above-mentioned restrictions such as the linear flight space is tripled over that of the most cubic cells and less mass discrimination is observed.	
Open-ended cylindrical FTMS cells with capacitive RF coupling between the three sections [48, 164–168]		
Open-ended cylindrical FTMS cell with capacitive RF coupling operating at 77 to 438 K [169]	This design allows definition and control of the temperature of ions.	
"Infinity" FTMS cell [170]	The end caps are segmented to linearize the excitation potential.	This design has never been realized.

fundamental theory in detail [115, 116]. Recently, an excellent review by Zhang et al. covered mass calibration comprehensively [38].

4.5 RECENT DEVELOPMENTS IN POLYMER RESEARCH USING FTMS

One of the first parameters a polymer scientist is interested in is the MWD of a polymer. MS allows rapid and accurate analysis and will reveal information about the repeat unit(s), end groups, additives, impurities, and synthetic side products. Due to its outstanding resolving power, FTMS is ideal for analysis of polymer mixtures. This section is a survey of recent polymer analysis applications of FTMS.

There are numerous examples of such applications. A good starting point for considering those is to read a number of recent review articles.

- Ionization: Montaudo et al. summarized the latest developments in MALDI techniques for polymers in a review in 2005 [83].
- FTMS for polymer characterization: Zhang et al. addressed several important aspects in their recent review about accurate mass measurements [38].
- Ion manipulation: Laskin and Futrell reviewed extensively the activation of large ions [111]. Tandem MS was reviewed by Sleno and Volmer [110].
- Book: The development of MS of polymers is extensively addressed in the book *Mass Spectrometry of Polymers* edited by Montaudo and Lattimer [117].

In the following, the application of FTMS analysis to characterize polar and nonpolar polymers, copolymers, end groups, and structural patterns of polymers for the period from 2001 to the beginning of 2006 is emphasized.

Analysis of polymers by FTMS requires a careful selection of all parameters during the whole process. As highlighted by Jaber et al. [94], the matrix-to-analyte ratio plays an important role when an internal ionization source is employed for MALDI-FTMS of an equimolar mixture of four PEG polymers (PEG 2000, PEG 4000, PEG 6000, PEG 8000) with 2,5-dihydroxy benzoic acid (DHB) as matrix. As the matrix-to-analyte ratio is increased, the signal of the ions with higher mass increases and the signal for lower-mass ions decreases. In Figure 4.10, it is shown that laser-power dependence of the signal follows the same trend. Flight time effects are another important issue, especially with so-called "internal source" instruments. As Figure 4.11 demonstrates, with a similar mixture of PEG polymers, appropriate choice of gated trapping delay times can result in complete discrimination in favor of one or more components [118].

Hydrocarbon (nonpolar) polymers are still a major challenge in FTMS analysis because they do not have functional groups and are chemically inert, resulting in difficulties with respect to sample preparation, ionization, and mass spectral analysis. This problem was first addressed in MALDI by using chemical ionization agents such as silver [6, 16, 17, 19, 91, 94, 119, 120] for hydrocarbons analyzed by FTMS [94, 121]. Several approaches were attempted for improving ionization during the MALDI ionization of hydrocarbons, including chemical modification of the

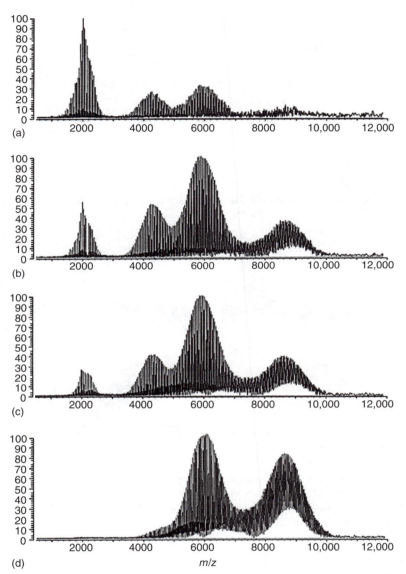

Figure 4.10.   Dependence of ion abundance on the influence of laser energy. Spectrum (a) = 0.79 mJ, Spectrum (b) = 0.93 mJ, Spectrum (c) = 1.08 mJ, Spectrum (d) = 1.35 mJ. The matrix-to-analyte ratio is 20,000 : 1 : 1 : 1 : 1. The trapping time was 125 ms; the retarding potential was 9.5 V; the trapping potential was 2.0 V on both the front and back plate. (Reproduced from Reference 94 with permission of the copyright holder.)

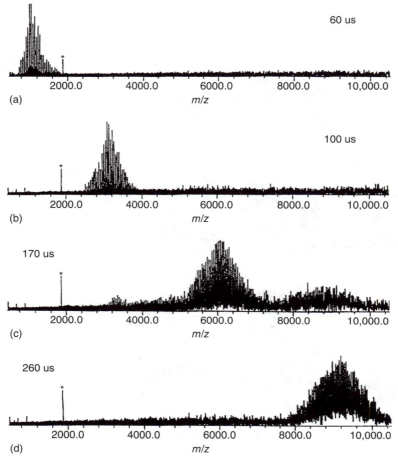

Figure 4.11.   MALDI-FT mass spectra of a mixture of PEG 1000, PEG 3000, PEG 6000, and PEG 8000 recorded for various decelerating times: (a) 60, (b) 100, (c) 170, and (d) 260 μs. The asterisk indicates a noise peak. (Reproduced from Reference 118 with permission of the copyright holder.)

polymers prior to analysis employing bromine [122], phosphonium [112] or sulfonate [123] salts. Promising results were obtained, but the general drawback is that these methods may comprise multiple steps, and thus are time-consuming and may affect the molecular structure of the analyte. Other sample preparation techniques were investigated [124–128], which consist of mixing the hydrocarbon polymer, the matrix, and the cationization agent before grinding them with a mortar and pestle or a ball mill. The mixture is then applied to the MALDI target plate, for example, by means of a double-sided sticky tape [127]. The resulting spectra are comparable to spectra resulting from the solvent-based approach. However, impressive results were obtained with the conventional solvent-based techniques. Jaber et al. very

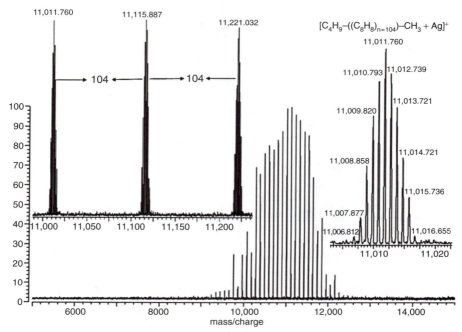

Figure 4.12.   A 9.4 T MALDI-FTMS spectrum of polystyrene with an average molecular weight of 10,000 Da. The left inset displays a partial spectrum of the repeat unit of the styrene monomer. The right inset represents the partial spectrum of isotopically resolved oligomers. (Reproduced from Reference 129 with permission of the copyright holder.)

recently reported observation of silver-cationized hydrocarbons with a mass range up to 12 kDa in a comparative study using an external MALDI 9.4 T FTMS and a reflectron time-of-flight (TOF)-MS [94]. Figure 4.12 shows an ultraviolet MALDI-FTMS spectrum of polystyrene which highlights the mass accuracy and resolving power of this method [129]. Furthermore the issue of fragmentation observed by Chen et al. [130] and Yalcin et al. [131] was addressed by this work. FTMS spectra show no low-mass fragments as observed in TOF, presumably due to the different time frames of the measurements [94].

A different approach has been employed by Schaub et al. who decided to use FD FTMS for analyzing nonpolar molecules in complex mixtures [132, 133]. Although these mixtures are not polymer mixtures, they clearly reveal the potential for polymer applications. The instrumentation and method is explained in detail [133]. Figure 4.13 shows a broadband spectrum of a U.S. Gulf of Mexico crude oil they reported. Remarkably, 588 assignments to unique chemical compositions with an average resolving power of 442,000 were made [134]. Continuous-flow sample introduction for field desorption/ionization MS was also introduced [135].

Quite a bit of work has been done in the so-called petroleomics area using, for example, ESI as ionization method for petroleum crude oils [36, 133, 134, 136]. An illustration from this complex field of research is given in Figure 4.14 [42]. In

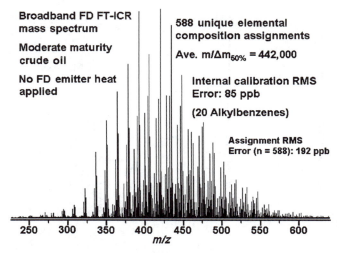

Broadband FD FT-ICR
mass spectrum

Moderate maturity
crude oil

No FD emitter heat
applied

588 unique elemental
composition assignments

Ave. $m/\Delta m_{50\%}$ = 442,000

Internal calibration RMS
Error: 85 ppb

(20 Alkylbenzenes)

Assignment RMS
Error (n = 588): 192 ppb

250   300   350   400   450   500   550   600
m/z

Figure 4.13.   Broadband FD-FTMS spectrum of a U.S. Gulf of Mexico crude oil. The spectrum is the sum of nine time-domain transients from 10 s external ion accumulations. (Reproduced from Reference 134 with permission of the copyright holder.)

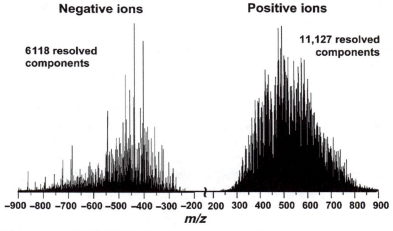

Negative ions   **Positive ions**

6118 resolved
components

11,127 resolved
components

−900 −800 −700 −600 −500 −400 −300 −200 200 300 400 500 600 700 800 900
m/z

Figure 4.14.   ESI 9.4 T FTMS spectrum of a crude oil. Positive-ion and negative-ion spectra are combined in the display. Average mass resolving power is approximately 350,000, allowing for resolution of thousands of different elemental compositions. (Reproduced from Reference 42 with permission of the copyright holder).

their 2004 review [121], Marshall and Rodgers point out many advantages as well as one major disadvantage of ESI. Only polar compounds, which represent a minor part of the overall chemical composition, are ionized from the crude oils. Thus, there is further need to extend the current methodologies to more nonpolar compounds.

A unique feature of FTMS is that it is a trapped-ion technique. This allows one to perform MS^n studies revealing insights into structural features of the polymers of interest within a reasonable time scale [34]. A comprehensive tutorial regarding tandem MS was published by Sleno and Volmer in 2004 [110]. A general problem encountered with respect to dissociation is that polymers are highly stable in the gas phase, leading to limited success for most common dissociation methods. In addition, the amount of energy that can be accommodated by an ion is limited by the strength of the magnetic field and the size of the analyzer cell. SORI-CID and quadrupolar axialization seem to be most promising because they improve the collection efficiency of ions. Laskin and Futrell extensively describe the activation and dissociation of large ions in FTMS in their 2005 review [111]. It is important to note that with increasing size of large ions, the dissociation rates decrease significantly. Consequently, more sophisticated methods for dissociating large molecular ions are required in order to obtain structural information from high-molecular-weight synthetic polymers. For distinguishing isobaric fragment ions, for example, structure determination using SORI-CID can be applied [15, 20].

One particular research interest is copolymer characterization, which requires the investigation of the molar mass of each type of repeat unit, the average chain length of each polymer type, and the variation of composition as a function of molecular weight. Cerda et al. reported a study using electron-capture dissociation (ECD)/FTMS in a modified 6 T Finnegan FTMS instrument for sequencing specific copolymer oligomers proving the outstanding capabilities of FTMS for structural analysis [137]. Their results for sequencing the commercial copolymer "PEG-block-poly (propylene glycol) (PPG)-block-PEG" are displayed in Figure 4.15. Undoubtedly, high-mass resolution is required to detect and assign individual oligomers from copolymers.

Part of structural analysis is the end-group analysis. End groups are of particular interest for polymer scientists due to their impact on the chemical, physical, and mechanical properties of the polymer. A traditional approach for performing end-group analysis is the comparison of the derivatized and underivatized form of the polymer of interest. de Koster et al. employed high-resolution FTMS together with graphical methods to determine the end groups of various PEGs in an external MALDI 7 T FTMS [138]. Mathematical methods were also used to calculate the end-group functionalities of polymers by means of a regression and averaging method and result in reduction of the FTMS data complexity [139]. van Rooij et al. stated that a complete mass spectrum of the polymer is needed for this approach and the whole method is only valid for masses up to 10,000 Da. More recently, end-group analysis has been addressed by Mize et al. analyzing homo- and co-polyesters by means of MALDI-FTMS [93]. Here, it should be noted that errors in assigning repeat units from isobarically resolved FTMS spectra are significantly reduced compared with those from isotopically resolved spectra. To visualize this, Figure 4.16 displays high resolution and ultrahigh resolution of poly[dipropoxylated bisphenol A-alt-(adipic acid-co-isophtalic acid)] (U415) FTMS spectra [93].

FTMS is also a useful tool to monitor reactions in the gas phase as performed by Guo et al. [140]. ESI coupled with FTMS analysis has made great progress in recent years [87, 103, 121, 141–146]. Nevertheless, overall, more work has been done with MALDI of polymers. Each method discussed has specific merits and limitations which are summarized in Table 4.3.

Generally, there are reasonable limitations observed for FTMS with respect to efficient ionization of the polymers, quantitation (only relative ion abundance is

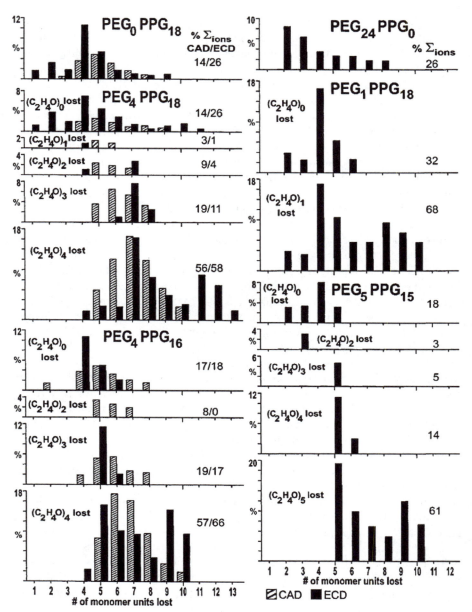

Figure 4.15. CAD (collision activated dissociation) (A, C ions, striped bars) and ECD (electron capture dissociation) (A ions, solid bars) spectra of MS-separated $(PEG_xPPG_y + 2H)^{2+}$ ions showing relative ion abundances as a function of monomer unit lost; separate bar graphs are shown for each x value for $(C_2H_4O)_x$ lost. (Reproduced from Reference 136 with permission of the copyright holder.)

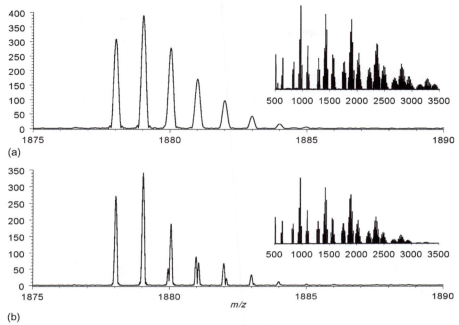

Figure 4.16.   Examples of (a) high-resolution and (b) ultrahigh-resolution mass spectra of U415. At the higher level of resolution, assignments of the cyclic and linear oligomers is straightforward and can be automated. Insets show the entire mass range for each mass spectrum. The bottom mass spectrum demonstrates the resolution of isobaric isotope peaks from species that differ in molecular weight by 2 amu. (Reproduced from Reference 93 with permission of the copyright holder.)

provided in the spectra), space-charge effects due to high number of ions in a broad-band mass-to-charge range, and partly complex data interpretation. Nevertheless, MALDI-FTMS has proven obtaining high-resolution mass spectra makes accurate mass assignments in complex mixtures possible [42, 146–148].

4.6 CONCLUSION

It is obvious that FTMS is a vital polymer analysis tool and a significant alternative to the numerous less specific analytical methods in polymer science. The aim of the present chapter was to highlight the accomplishments, versatility, and accuracy of FTMS for analyzing polymers in order to further promote the use of this technique. Some progress has been made on high-throughput approaches for polymer analysis, in general [149, 150]. Only rare reports can be found for MS coupled with high-throughput screening. But since the introduction of FTMS applications in combinatorial chemistry [151], there might be more interest in accomplishing at least a moderately improved throughput for characterizing polymers by means of FTMS.

TABLE 4.3. Comparison of ESI and MALDI-FTMS for Polymer Analysis

	ESI-FTMS	MALDI-FTMS
Ionization process	"Soft" ionization method with little or no fragmentation even for high-mass ions	
	From complex polymeric mixtures principally, only the most acidic or basic molecules ionize to negative and positive ions, respectively.	More easily tunable for analytes difficult to ionize
Sample preparation	Easy sample preparation; polymer of interest needs to be soluble in a solvent that can be electrosprayed.	Minimal sample pretreatment is necessary; easy straightforward sample preparation; sample preparation methods for insoluble polymers available [127, 128, 171].
Experimental complexity	Moderate to challenging	
Sensitivity	High sensitivity	
Accuracy and resolution	High-mass accuracy and high resolution; high resolving power for ions below m/z 1000; mass assignment accuracy <10 ppm [148]	
Data interpretation	Multiple-charged ions complicate spectral interpretation for polymers with masses above a few kDa.	Straightforward; computational approaches are needed for complex mixtures.
Reproducibility and quantitation	Reproducible characterization and quantitation of polymers possible	
Copolymer analysis	Realized with high performance	
Interpretation of mass spectra	Multiple-charged ions complicate the analysis of polymers above a few kDa.	Singly charged ions simplify rapid analysis.
Capability for MS^n experiments	Broad and easy to perform	
Compatibility with other analytical techniques	Can be easily interfaced to liquid separation methods	
Throughput	Moderate	
Costs	Expensive equipment	

ACKNOWLEDGMENTS

The authors gratefully acknowledge financial support from National Science Foundation grants CHE-00-91868, CHE-99-82045, and CHE-04-55134. The author also wishes to thank Dr. J.J. Jones for helpful discussions and for providing some of the figures.

REFERENCES

1. Pastor, S.; Wilkins, C.L, *Mass Spectrometry of Polymers Vol. 9* (Lattimer, R.P., editor). Boca Raton, FL: CRC Press LLC, **2001**, pp. 389–417.
2. Wu, K.J.; Odom, R.W., Characterizing synthetic polymers by MALDI MS, *Anal. Chem.*, **1998**, 70, 456A–461A.
3. Räder, H.J.; Schrepp, W., MALDI-TOF mass spectrometry in the analysis of synthetic polymers, *Acta Polym.*, **1998**, 49, 272–293.
4. Nielen, M.W.F., MALDI time-of-flight mass spectrometry of synthetic polymers, *Mass Spectrom. Rev.*, **1999**, 18, 309–344.
5. Montaudo, G.; Carroccio, S.; Montaudo, A.S.; Puglisi, C.; Samperi, F., Recent advances in MALDI mass spectrometry of polymers, *Macromol. Symp.*, **2004**, 218, 101–112.
6. Deery, M.J.; Jennings, K.R.; Jasieczek, C.B.; Haddleton, D.M.; Jackson, A.T.; Yates, H.T.; Scrivens, J.H., A study of cation attachment to polystyrene by means of matrix-assisted laser desorption/ionization and electrospray ionization-mass spectrometry, *Rapid Commun. Mass Spectrom.*, **1997**, 11, 57–62.
7. McEwen, C.N.; Peacock, P.M., Mass spectrometry of chemical polymers, *Anal. Chem.*, **2002**, 74, 2743–2748.
8. Peacock, P.M.; McEwen, C.N., Mass spectrometry of synthetic polymers, *Anal. Chem.*, **2004**, 76, 3417–3427.
9. Przybylski, M., *Polymer and Biopolymer Mass Spectrometry*. Weinheim, Germany: Wiley-VCH, **2004**.
10. Smith, P.B.; Pasztor, A.J.; McKelvy, M.L.; Meunier, D.M.; Froelicher, S.W.; Wang, F.C.Y., Analysis of synthetic polymers and rubbers, *Anal. Chem.*, **1997**, 69, R95–R121.
11. Smith, P.B.; Pasztor, A.J.; McKelvy, M.L.; Meunier, D.M.; Froelicher, S.W.; Wang, F.C.Y., Analysis of synthetic polymers and rubbers, *Anal. Chem.*, **1999**, 71, 61R–80R.
12. Montaudo, G., Mass spectrometry of synthetic polymers—mere advances or revolution? *Trends Polym. Sci.*, **1996**, 4, 81–86.
13. Asamoto, B.; Young, J.R.; Citerin, R.J., Laser desorption/Fourier transform ion cyclotron resonance mass spectrometry of polymer additives, *Anal. Chem.*, **1990**, 62, 61–70.
14. O'Connor, P.B.; McLafferty, F.W., Oligomer characterization of 4–23 kDa polymers by electrospray Fourier transform mass spectrometry, *J. Am. Chem. Soc.*, **1995**, 117, 12826–12831.
15. Koster, S.; Duursma, M.C.; Boon, J.J.; Nielen, M.W.F.; de Koster, C.G.; Heeren, R.M.A., Structural analysis of synthetic homo- and copolyesters by electrospray ionization on a Fourier transform ion cyclotron resonance mass spectrometer, *J. Mass Spectrom.*, **2000**, 35, 739–748.
16. Kahr, M.S.; Wilkins, C.L., Silver-nitrate chemical ionization for analysis of hydrocarbon polymers by laser-desorption Fourier-transform mass-spectrometry, *J. Am. Soc. Mass Spectrom.*, **1993**, 4, 453–460.
17. Walker, K.L.; Kahr, M.S.; Wilkins, C.L.; Xu, Z.F.; Moore, J.S., Analysis of hydrocarbon dendrimers by laser-desorption time-of-flight and Fourier-transform mass-spectrometry, *J. Am. Soc. Mass Spectrom.*, **1994**, 5, 731–739.
18. Guan, S.H.; Marshall, A.G.; Scheppele, S.E., Resolution and chemical formula identification of aromatic hydrocarbons and aromatic compounds containing sulfur, nitrogen, or oxygen in petroleum distillates and refinery streams, *Anal. Chem.*, **1996**, 68, 46–71.
19. Pastor, S.J.; Wilkins, C.L., Analysis of hydrocarbon polymers by matrix-assisted laser desorption/ionization Fourier transform mass spectrometry, *J. Am. Soc. Mass Spectrom.*, **1997**, 8, 225–233.

20. Pastor, S.J.; Wilkins, C.L., Sustained off-resonance irradiation and collision-induced dissociation for structural analysis of polymers by MALDI-FTMS, *Int. J. Mass Spectrom.*, **1998**, 175, 81–92.
21. Lehmann, W.D.; Beckey, H.D.; Schulten, H.R., Qualitative and quantitative-analyses of bansyl derivatives of dopamine and some of its metabolites in urine samples by electron-impact and field desorption mass-spectrometry, *Anal. Chem.*, **1976**, 48, 1572–1575.
22. Wilkins, C.L.; Gross, M.L., Fourier-transform mass-spectrometry for analysis, *Anal. Chem.*, **1981**, 53, 1661A–1676A.
23. Johlman, C.L.; White, R.L.; Wilkins, C.L., Applications of Fourier-transform mass-spectrometry, *Mass Spectrom. Rev.*, **1983**, 2, 389–415.
24. Wanczek, K.P., Ion-cyclotron resonance spectrometry—a review, *Int. J. Mass Spectrom. Ion Proc.*, **1984**, 60, 11–60.
25. Nibbering, N.M.M., Principles and applications of Fourier-transform ion-cyclotron resonance mass-spectrometry—plenary lecture, *Analyst*, **1992**, 117, 289–293.
26. Cody, R.B.; Kinsinger, J.A.; Ghaderi, S.; Amster, I.J.; McLafferty, F.W.; Brown, C.E., Developments in analytical Fourier-transform mass-spectrometry, *Anal. Chim. Acta*, **1985**, 178, 43–66.
27. Comisarow, M.B., Fundamental aspects and applications of Fourier-transform ion-cyclotron resonance spectrometry, *Anal. Chim. Acta*, **1985**, 178, 1–15.
28. Marshall, A.G., Fourier-transform ion-cyclotron resonance mass-spectrometry, *Acc. Chem. Res.*, **1985**, 18, 316–322.
29. Laude, D.A.; Johlman, C.L.; Brown, R.S.; Weil, D.A.; Wilkins, C.L., Fourier-transform mass-spectrometry—recent instrumental developments and applications, *Mass Spectrom. Rev.*, **1986**, 5, 107–166.
30. Russell, D.H., An evaluation of Fourier-transform mass-spectrometry for high mass applications, *Mass Spectrom. Rev.*, **1986**, 5, 167–189.
31. Marshall, A.G.; Schweikhard, L., Fourier-transform ion-cyclotron resonance mass-spectrometry—technique developments, *Int. J. Mass Spectrom. Ion Proc.*, **1992**, 118, 37–70.
32. Castoro, J.A.; Köster, C.; Wilkins, C., Matrix-assisted laser desorption/ionization of high-mass molecules by Fourier-transform mass spectrometry, *Rapid Commun. Mass Spectrom.*, **1992**, 6, 239–241.
33. Holliman, C.L.; Rempel, D.L.; Gross, M.L., Detection of high mass-to-charge ions by Fourier-transform mass-spectrometry, *Mass Spectrom. Rev.*, **1994**, 13, 105–132.
34. Dienes, T.; Pastor, S.J.; Schürch, S.; Scott, J.R.; Yao, J.; Cui, S.L.; Wilkins, C.L., Fourier transform mass spectrometry—advancing years (1992 mid 1996), *Mass Spectrom. Rev.*, 1996, 15, 163–211.
35. Marshall, A.G.; Hendrickson, C.L.; Jackson, G.S., Fourier transform ion cyclotron resonance mass spectrometry: a primer, *Mass Spectrom. Rev.*, **1998**, 17, 1–35.
36. Klein, G.C.; Rodgers, R.P.; Teixeira, M.A.G.; Teixeira, A.; Marshall, A.G., Petroleomics: electrospray ionization FT-ICR mass analysis of NSO compounds for correlation between total acid number, corrosivity, and elemental composition, *Abstr. Pap. Am. Chem. Soc.*, **2003**, 225, U844.
37. Su, Y.; Wang, H.Y.; Guo, Y.L.; Xiang, B.R.; An, D.K., Effect of damping mechanisms on mass calibration equation for MALDI/FTMS, *Chin. J. Chem.*, **2005**, 23, 1053–1059.
38. Zhang, L.K.; Rempel, D.; Pramanik, B.N.; Gross, M.L., Accurate mass measurements by Fourier transform mass spectrometry, *Mass Spectrom. Rev.*, **2005**, 24, 286–309.
39. Marshall, A.G.; Verdun, F.R., *Fourier Transforms in NMR, Optical and Mass Spectrometry*. 1st edition. Amsterdam, The Netherlands: Elsevier, **1990**.
40. Asamoto, B., *FT-ICR/MS: Analytical Applications of Fourier Transform Ion Cyclotron Resonance Mass Spectrometry*. New York: VCH, **1991**.
41. Comisarow, M.B., Fundamental aspects of FT-ICR and applications to chemistry, *Hyperfine Interact.*, **1993**, 81, 171–178.
42. Hughey, C.A.; Rodgers, R.P.; Marshall, A.G., Resolution of 11,000 compositionally distinct components in a single electrospray ionization Fourier transform ion cyclotron resonance mass spectrum of crude oil, *Anal. Chem.*, **2002**, 74, 4145–4149.
43. Schweikhard, L., Excitation and detection geometries for Fourier-transform mass-spectrometry, *Rapid Commun. Mass Spectrom.*, **1994**, 8, 10–13.

44. Guan, S.; Wahl, M.C.; Marshall, A.G., Broadband axialization in an ion cyclotron resonance ion trap, *J. Chem. Phys.*, **1994**, 100, 6137–6140.

45. Campbell, V.L.; Guan, Z.Q.; Vartanian, V.H.; Laude, D.A., Cell geometry considerations for the Fourier-transform ion-cyclotron resonance mass-spectrometry remeasurement experiment, *Anal. Chem.*, **1995**, 67, 420–425.

46. Brenna, J.T.; Creasy, W.R.; Zimmerman, J., Laser Fourier-transform mass-spectrometry for polymer characterization, *Adv. Chem. Ser.*, **1993**, 129–154.

47. Campana, J.E.; Sheng, L.S.; Shew, S.L.; Winger, B.E., Polymer analysis by photons, sprays, and mass spectrometry, *Trac—Trends Anal. Chem.*, **1994**, 13, 239–247.

48. Easterling, M.L.; Mize, T.H.; Amster, I.J., MALDI FTMS analysis of polymers: improved performance using an open-ended cylindrical analyzer cell, *Int. J. Mass Spectrom.*, **1997**, 169, 387–400.

49. Comisarow, M.B.; Marshall, A.G., Fourier-transform ion-cyclotron resonance spectroscopy, *Chem. Phys. Lett.*, **1974**, 25, 282–283.

50. Comisarow, M.B.; Marshall, A.G., The early development of Fourier transform ion cyclotron resonance (FT-ICR) spectroscopy, *J. Mass Spectrom.*, **1996**, 31, 581–585.

51. Lawrence, E.O.; Livingstone, M.S., The production of high-speed light ions without the use of high voltages, *Phys. Rev.*, **1932**, 40, 19–30.

52. Hipple, J.A.; Sommer, H.; Thomas, H.A., A precise method of determining the Faraday by magnetic resonance, *Phys. Rev.*, **1949**, 76, 1877–1878.

53. Sommer, H.; Thomas, H.A.; Hipple, J.A., The measurement of e/m by cyclotron resonance, *Phys. Rev.*, **1951**, 82, 697–702.

54. Lehman, T.A.; Bursey, M.M., *Ion Cyclotron Resonance Spectrometry*. New York: Wiley-Interscience, **1976**.

55. McIver, R.T.A., Trapped ion analyzer cell for ion cyclotron resonance spectroscopy, *Rev. Sci. Instrum.*, **1970**, 41, 555–558.

56. Gross, M.L.; Wilkins, C.L., Ion cyclotron resonance spectrometry—recent advances of analytical interest, *Anal. Chem.*, **1971**, 43, 65A–68A.

57. Ernst, R.R.; Anderson, W.A., Application of Fourier transform spectroscopy to magnetic resonance, *Rev. Sci. Instrum.*, **1966**, 37, 93–102.

58. Comisarow, M.B.; Marshall, A.G., Resolution-enhanced Fourier-transform ion-cyclotron resonance spectroscopy, *J. Chem. Phys.*, **1975**, 62, 293–295.

59. Comisarow, M.B.; Marshall, A.G., Theory of Fourier-transform ion-cyclotron resonance mass-spectroscopy: 1. fundamental equations and low-pressure line-shape, *J. Chem. Phys.*, **1976**, 64, 110–119.

60. Wilkins, C.L., Fourier-transform mass-spectrometry, *Anal. Chem.*, **1978**, 50, 493A–500A.

61. Ledford, E.B.; Ghaderi, S.; White, R.L.; Spencer, R.B.; Kulkarni, P.S.; Wilkins, C.L.; Gross, M.L., Exact mass measurement by Fourier-transform mass-spectrometry, *Anal. Chem.*, **1980**, 52, 463–468.

62. Hunter, R.L.; McIver, R.T., Conceptual and experimental basis for rapid scan ion-cyclotron resonance spectroscopy, *Chem. Phys. Lett.*, **1977**, 49, 577–582.

63. Allemann, M.; Kellerhals, H.; Wanczek, K.P., A new Fourier-transform mass-spectrometer with a superconducting magnet, *Chem. Phys. Lett.*, **1980**, 75, 328–331.

64. Allemann, M.; Kellerhals, H.; Wanczek, K.P., High magnetic-field Fourier-transform ion-cyclotron resonance spectroscopy, *Int. J. Mass Spectrom. Ion Proc.*, **1983**, 46, 139–142.

65. McIver, R.T., United State Patent, 3,742,212, **1973**.

66. Roempp, A.; Taban, I.M.; Mihalca, R.; Duursma, M.C.; Mize, T.H.; McDonnell, L.A.; Heeren, R.M.A., Examples of Fourier transform ion cyclotron resonance mass spectrometry developments: from ion physics to remote access biochemical mass spectrometry, *Eur. J. Mass Spectrom.*, **2005**, 11, 443–456.

67. Conzemius, R.J.; Capellen, J.M., A review of the applications to solids of the laser ion-source in mass-spectrometry, *Int. J. Mass Spectrom. Ion Proc.*, **1980**, 34, 197–271.

68. Posthumus, M.A.; Kistemaker, P.G.; Meuzelaar, H.L.C.; Tennoeverdebrauw, M.C., Laser desorption-mass spectrometry of polar non-volatile bio-organic molecules, *Anal. Chem.*, **1978**, 50, 985–991.

69. Wilkins, C.L.; Weil, D.A.; Yang, C.L.C.; Ijames, C.F., High mass analysis by laser desorption Fourier-transform mass-spectrometry, *Anal. Chem.*, **1985**, 57, 520–524.
70. Ijames, C.F.; Wilkins, C.L., First demonstration of high resolution laser desorption mass spectrometry of high mass organic ions, *J. Am. Chem. Soc.*, **1988**, 110, 2687–2688.
71. Fenn, J.B., Electrospray wings for molecular elephants (Nobel lecture), *Angew. Chem. Int. Ed.*, **2003**, 42, 3871–3894.
72. Gaskell, S.J., Electrospray: principles and practice, *J. Mass Spectrom.*, **1997**, 32, 677–688.
73. Lorenz, S.A.; Maziarz, E.P.; Wood, T.D., Electrospray ionization Fourier transform mass spectrometry of macromolecules: the first decade, *Appl. Spectrosc.*, **1999**, 53, 18A–36A.
74. McLafferty, F.W., High-resolution tandem FT mass spectrometry above 10 KDa, *Acc. Chem. Res.*, **1994**, 27, 379–386.
75. Karas, M.; Bachman, D.; Bahr, U.; Hillenkamp, F., Matrix-assisted ultraviolet laser desorption of non-volatile compounds, *Int. J. Mass Spectrom. Ion Proc.*, **1987**, 78, 53–68.
76. Karas, M.; Hillenkamp, F., Laser desorption ionization of proteins with molecular masses exceeding 10,000 Daltons, *Anal. Chem.*, **1988**, 60, 2299–2301.
77. Tanaka, K.; Waki, H.; Ido, Y.; Akita, S.; Yoshida, Y., Protein and polymer analysis up to m/z 100,000 by laser ionization time-of-flight mass spectrometry, *Rapid Commun. Mass Spectrom.*, **1988**, 2, 151–153.
78. Tanaka, K., The origin of macromolecule ionization by laser irradiation (Nobel lecture), *Angew. Chem. Int. Ed.*, **2003**, 42, 3860–3870.
79. Stump, M.J.; Fleming, R.C.; Gong, W.H.; Jaber, A.J.; Jones, J.J.; Surber, C.W.; Wilkins, C.L., Matrix-assisted laser desorption mass spectrometry, *Appl. Spectrosc. Rev.*, **2002**, 37, 275–303.
80. Murgasova, R.; Hercules, D.M., MALDI of synthetic polymers—an update, *Int. J. Mass Spectrom.*, **2003**, 226, 151–162.
81. Macha, S.F.; Limbach, P.A., Matrix-assisted laser desorption/ionization (MALDI) mass spectrometry of polymers, *Curr. Opin. Solid State Mat. Sci.*, **2002**, 6, 213–220.
82. Pasch, H.; Schrepp, W., *MALDI-ToF Mass Spectrometry of Synthetic Polymers*. Heidelberg, Germany: Springer, **2003**.
83. Montaudo, G.; Samperi, F.; Montaudo, M.S.; Carroccio, S.; Puglisi, C., Current trends in matrix-assisted laser desorption/ionization of polymeric materials, *Eur. J. Mass Spectrom.*, **2005**, 11, 1–14.
84. Knochenmuss, R.; Lehmann, E.; Zenobi, R., Polymer cationization in matrix-assisted laser desorption/ionization, *Eur. J. Mass Spectrom.*, **1998**, 4, 421–426.
85. Fujii, T., Alkali-metal ion/molecule association reactions and their applications to mass spectrometry, *Mass Spectrom. Rev.*, **2000**, 19, 111–138.
86. Zhang, J.; Zenobi, R., Matrix-dependent cationization in MALDI mass spectrometry, *J. Mass Spectrom.*, **2004**, 39, 808–816.
87. Hanton, S.D.; Owens, K.G.; Chavez-Eng, C.; Hoberg, A.M.; Derrick, P.J., Updating evidence for cationization of polymers in the gas phase during matrix-assisted laser desorption/ionization, *Eur. J. Mass Spectrom.*, **2005**, 11, 23–29.
88. Mueller, E.W.; Tsong, T.T., *Field Ion Microscope*. New York: Elsevier, **1969**.
89. Inghram, M.G.; Gomer, R., Mass spectrometric analysis of ions from the field microscope, *J. Chem. Phys.*, **1954**, 22, 1279–1280.
90. Gross, J.H.; Vekey, K.; Dallos, A., Field desorption mass spectrometry of large multiply branched saturated hydrocarbons, *J. Mass Spectrom.*, **2001**, 36, 522–528.
91. Carr, R.H.; Jackson, A.T., Preliminary matrix-assisted laser desorption ionization time-of-flight and field desorption mass spectrometric analyses of polymeric methylene diphenylene diisocyanate, its amine precursor and a model polyether prepolymer, *Rapid Commun. Mass Spectrom.*, **1998**, 12, 2047–2050.
92. Evans, W.J.; DeCoster, D.M.; Greaves, J., Evaluation of field desorption mass spectrometry for the analysis of polyethylene, *J. Am. Soc. Mass Spectrom.*, **1996**, 7, 1070–1074.
93. Mize, T.H.; Simonsick, W.J.; Amster, I.J., Characterization of polyesters by matrix-assisted laser desorption/ionization and Fourier transform mass spectrometry, *Eur. J. Mass Spectrom.*, **2003**, 9, 473–486.

94. Jaber, A.J.; Kaufman, J.; Liyanage, R.; Akhmetova, E.; Marney, S.; Wilkins, C.L., Trapping of wide range mass-to-charge ions and dependence on matrix amount in internal source MALDI-FTMS, *J. Am. Soc. Mass Spectrom.*, **2005**, 16, 1772–1780.

95. Belov, M.E.; Nikolaev, E.N.; Anderson, G.A.; Udseth, H.R.; Conrads, T.P.; Veenstra, T.D.; Masselon, C.D.; Gorshkov, M.V.; Smith, R.D., Design and performance of an ESI interface for selective external ion accumulation coupled to a Fourier transform ion cyclotron mass spectrometer, *Anal. Chem.*, **2001**, 73, 253–261.

96. Kofel, P.; Allemann, M.; Kellerhals, H.; Wanczek, K.P., Time-of-flight ICR spectrometry, *Int. J. Mass Spectrom. Ion Proc.*, **1986**, 72, 53–61.

97. Limbach, P.A.; Marshall, A.G.; Wang, M., An electrostatic ion guide for efficient transmission of low-energy externally formed ions into a Fourier-transform ion-cyclotron resonance mass-spectrometer, *Int. J. Mass Spectrom. Ion Proc.*, **1993**, 125, 135–143.

98. McIver, R.T.; Hunter, R.L.; Bowers, W.D., Coupling a quadrupole mass-spectrometer and a Fourier-transform mass-spectrometer, *Int. J. Mass Spectrom. Ion Proc.*, **1985**, 64, 67–77.

99. Huang, Y.L.; Guan, S.H.; Kim, H.S.; Marshall, A.G., Ion transport through a strong magnetic field gradient by RF-only octupole ion guides, *Int. J. Mass Spectrom. Ion Proc.*, **1996**, 152, 121–133.

100. Belov, M.E.; Gorshkov, M.V.; Udseth, H.R.; Anderson, G.A.; Smith, R.D., Zeptomole-sensitivity electrospray ionization—Fourier transform ion cyclotron resonance mass spectrometry of proteins, *Anal. Chem.*, **2000**, 72, 2271–2279.

101. Colburn, A.W.; Giannakopulos, A.E.; Derrick, P.J., The ion conveyor: an ion focusing and conveying device, *Eur. J. Mass Spectrom.*, **2004**, 10, 149–154.

102. Bruker Daltonics Publication, 40 Manning Road, Manning Park, Billerica, MA 01821, USA, **2004**.

103. Barrow, M.P.; Headley, J.V.; Peru, K.M.; Derrick, P.J., Fourier transform ion cyclotron resonance mass spectrometry of principal components in oil sands naphthenic acids, *J. Chromatogr. A*, **2004**, 1058, 51–59.

104. McMahon, T.B., Characterization of intensity oscillations in trapped ion cyclotron-resonance spectra, *Int. J. Mass Spectrom. Ion Proc.*, **1978**, 26, 359–367.

105. Hanson, C.D.; Castro, M.E.; Russell, D.H., Phase synchronization of an ion ensemble by frequency sweep excitation in Fourier-transform ion-cyclotron resonance, *Anal. Chem.*, **1989**, 61, 2130–2136.

106. Marshall, A.G.; Roe, D.C., Theory of Fourier- transform ion-cyclotron resonance mass-spectroscopy: 5. response to frequency-sweep excitation, *J. Chem. Phys.*, **1980**, 73, 1581–1590.

107. Hawkridge, A.M.; Nepomuceno, A.I.; Lovik, S.L.; Mason, C.J.; Muddiman, D.C., Effect of post-excitation radius on ion abundance, mass measurement accuracy, and isotopic distributions in Fourier transform ion cyclotron resonance mass spectrometry, *Rapid Commun. Mass Spectrom.*, **2005**, 19, 915–918.

108. Gorshkov, M.V.; Marshall, A.G.; Nikolaev, E.N., Analysis and elimination of systematic-errors originating from coulomb mutual interaction and image charge in Fourier-transform ion-cyclotron resonance precise mass difference measurements, *J. Am. Soc. Mass Spectrom.*, **1993**, 4, 855–868.

109. Williams, E.R.; Henry, K.D.; McLafferty, F.W., Multiple remeasurement of ions in Fourier-transform mass-spectrometry, *J. Am. Chem. Soc.*, **1990**, 112, 6157–6162.

110. Sleno, L.; Volmer, D.A., Ion activation methods for tandem mass spectrometry, *J. Mass Spectrom.*, **2004**, 39, 1091–1112.

111. Laskin, J.; Futrell, J.H., Activation of large ions in FT-ICR mass spectrometry, *Mass Spectrom. Rev.*, **2005**, 24, 135–167.

112. Byrd, H.C.M.; Bencherif, S.A.; Bauer, B.J.; Beers, K.L.; Brun, Y.; Lin-Gibson, S.; Sari, N., Examination of the covalent cationization method using narrow polydisperse polystyrene, *Macromolecules*, **2005**, 38, 1564–1572.

113. Taylor, P.K.; Amster, I.J., Space charge effects on mass accuracy for multiply charged ions in ESI-FTICR, *Int. J. Mass Spectrom.*, **2003**, 222, 351–361.

114. O'Connor, P.B.; Costello, C.E., Internal calibration on adjacent samples (InCAS) with Fourier transform mass spectrometry, *Anal. Chem.*, **2000**, 72, 5881–5885.

115. Boldin, I.A.; Nikolaev, E.N., Misphasing of ion motion in quadratic potential induced by space-periodic disturbance, *Eur. J. Mass Spectrom.*, **2008**, 14, 1–5.

116. Nikolaev, E.N.; Heeren, R.M.A.; Popov, A.M.; Pozdneev, A.V.; Chingin, K.S., Realistic modeling of ion cloud motion in a Fourier transform ion cyclotron resonance cell by use of a particle-in-cell approach, *Rapid Commun. Mass Spectrom.*, **2007**, 21, 3527–3546.

117. Pastor, S.J.; Wilkins, C.L., Laser Fourier Transform Mass Spectrometry (FT-MS), in *Mass Spectrometry of Polymers* (Montaudo, G.; Lattimer, R.P., Eds.), Boca Raton, FL: CRC Press, **2002**, pp. 389–417.

118. Dey, M.; Castoro, J.A.; Wilkins, C.L., Determination of molecular weight distributions of polymers by MALDI-FTMS, *Anal. Chem.*, **1995**, 67, 1575–1579.

119. Dean, P.A.; Omalley, R.M., Au+, Ag+ and Cu+ attachment to polybutadiene in laser desorption at 1064-Nm, *Rapid Commun. Mass Spectrom.*, **1993**, 7, 53–57.

120. Mowat, I.A.; Donovan, R.J., Metal-ion attachment to nonpolar polymers during laser-desorption ionization at 337-Nm, *Rapid Commun. Mass Spectrom.*, **1995**, 9, 82–90.

121. Hughey, C.A.; Rodgers, R.P.; Marshall, A.G.; Walters, C.C.; Qian, K.N.; Mankiewicz, P., Acidic and neutral polar NSO compounds in Smackover oils of different thermal maturity revealed by electrospray high field Fourier transform ion cyclotron resonance mass spectrometry, *Org. Geochem.*, **2004**, 35, 863–880.

122. Bauer, B.J.; Wallace, W.E.; Fanconi, B.M.; Guttman, C.M., "Covalent cationization method" for the analysis of polyethylene by mass spectrometry, *Polymer*, **2001**, 42, 9949–9953.

123. Ji, H.N.; Sato, N.; Nakamura, Y.; Wan, Y.N.; Howell, A.; Thomas, Q.A.; Storey, R.F.; Nonidez, W.K.; Mays, J.W., Characterization of polyisobutylene by matrix-assisted laser desorption ionization time-of-flight mass spectrometry, *Macromolecules*, **2002**, 35, 1196–1199.

124. Marie, A.; Fournier, F.; Tabet, J.C., Characterization of synthetic polymers by MALDI-TOF/MS: investigation into new methods of sample target preparation and consequence on mass spectrum finger print, *Anal. Chem.*, **2000**, 72, 5106–5114.

125. Pruns, J.K.; Vietzke, J.P.; Strassner, M.; Rapp, C.; Hintze, U.; Konig, W.A., Characterization of low molecular weight hydrocarbon oligomers by laser desorption/ionization time-of-flight mass spectrometry using a solvent-free sample preparation method, *Rapid Commun. Mass Spectrom.*, **2002**, 16, 208–211.

126. Przybilla, L.; Brand, J.D.; Yoshimura, K.; Rader, H.J.; Mullen, K., MALDI-TOF mass spectrometry of insoluble giant polycyclic aromatic hydrocarbons by a new method of sample preparation, *Anal. Chem.*, **2000**, 72, 4591–4597.

127. Trimpin, S.; Rouhanipour, A.; Az, R.; Rader, H.J.; Mullen, K., New aspects in matrix-assisted laser desorption/ionization time-of-flight mass spectrometry: a universal solvent-free sample preparation, *Rapid Commun. Mass Spectrom.*, **2001**, 15, 1364–1373.

128. Trimpin, S.; Grimsdale, A.C.; Rader, H.J.; Mullen, K., Characterization of an insoluble poly(9,9-diphenyl-2,7-fluorene) by solvent-free sample preparation for MALDI-TOF mass spectrometry, *Anal. Chem.*, **2002**, 74, 3777–3782.

129. Jaber, A.J.; Wilkins, C.L., Hydrocarbon polymer analysis by external MALDI Fourier transform and reflectron time of flight mass spectrometry, *J. Am. Soc. Mass Spectrom.*, **2005**, 16, 2009–2016.

130. Chen, R.; Yalcin, T.; Wallace, W.E.; Guttman, C.M.; Li, L., Laser desorption ionization and MALDI time-of-flight mass spectrometry for low molecular mass polyethylene analysis, *J. Am. Soc. Mass Spectrom.*, **2001**, 12, 1186–1192.

131. Yalcin, T.; Wallace, W.E.; Guttman, C.M.; Li, L., Metal powder substrate-assisted laser desorption/ionization mass spectrometry for polyethylene analysis, *Anal. Chem.*, **2002**, 74, 4750–4756.

132. Schaub, T.M.; Hendrickson, C.L.; Qian, K.N.; Quinn, J.P.; Marshall, A.G., High-resolution field desorption/ionization Fourier transform ion cyclotron resonance mass analysis of nonpolar molecules, *Anal. Chem.*, **2003**, 75, 2172–2176.

133. Rodgers, R.P.; Schaub, T.M.; Marshall, A.G., Petroleomics: MS returns to its roots, *Anal. Chem.*, **2005**, 77, 20A–27A.

134. Marshall, A.G.; Rodgers, R.P., Petroleomics: the next grand challenge for chemical analysis, *Acc. Chem. Res.*, **2004**, 37, 53–59.

135. Schaub, T.M.; Linden, H.B.; Hendrickson, C.L.; Marshall, A.G., Continuous-flow sample introduction for field desorption/ionization mass spectrometry, *Rapid Commun. Mass Spectrom.*, **2004**, 18, 1641–1644.

136. Rodgers, R.P.; Klein, G.C.; Marshall, A.G., Petroleomics: ESI FT-ICR MS identification of hydrotreatment resistant neutral and acidic nitrogen species in crude oil, *Abstr. Papers Am. Chem. Soc.*, **2003**, 226, U535.

137. Cerda, B.A.; Horn, D.M.; Breuker, K.; McLafferty, F.W., Sequencing of specific copolymer oligomers by electron-capture-dissociation mass spectrometry, *J. Am. Chem. Soc.*, **2002**, 124, 9287–9291.

138. de Koster, C.G.; Duursma, M.C.; van Rooij, G.J.; Heeren, R.M.A.; Boon, J.J., Endgroup analysis of polyethylene glycol polymers by matrix-assisted laser desorption/ionization Fourier-transform ion cyclotron resonance mass spectrometry, *Rapid Commun. Mass Spectrom.*, **1995**, 9, 957–962.

139. van Rooij, G.J.; Duursma, M.C.; Heeren, R.M.A.; Boon, J.J.; de Koster, C.J., High resolution end group determination of low molecular weight polymers by matrix-assisted laser desorption ionization on an external ion source Fourier transform ion cyclotron resonance mass spectrometer, *J. Am. Soc. Mass Spectrom.*, **1996**, 7, 449–457.

140. Guo, X.H.; Grutzmacher, H.F.; Nibbering, N.M.M., Reactivity and structures of hydrogenated carbon cluster ions CnHx+ (n = 8,20; x = 4–12) derived from polycyclic aromatic hydrocarbons toward benzene: ion/molecule reactions as a probe for ion structures, *Eur. J. Mass Spectrom.*, **2000**, 6, 357–367.

141. Kim, S.; Stanford, L.A.; Rodgers, R.P.; Marshall, A.G.; Walters, C.C.; Qian, K.; Wenger, L.M.; Mankiewicz, P., Microbial alteration of the acidic and neutral polar NSO compounds revealed by Fourier transform ion cyclotron resonance mass spectrometry, *Org. Geochem.*, **2005**, 36, 1117–1134.

142. Wu, Z.G.; Rodgers, R.P.; Marshall, A.G., Comparative compositional analysis of untreated and hydrotreated oil by electrospray ionization Fourier transform ion cyclotron resonance mass spectrometry, *Energy Fuels*, **2005**, 19, 1072–1077.

143. Miyabayashi, K.; Naito, Y.; Miyake, M.; Tsujimoto, K., Quantitative capability of electrospray ionization Fourier transform ion cyclotron resonance mass spectrometry for a complex mixture, *Eur. J. Mass Spectrom.*, **2000**, 6, 251–258.

144. Barrow, M.P.; McDonnell, L.A.; Feng, X.D.; Walker, J.; Derrick, P.J., Determination of the nature of naphthenic acids present in crude oils using nanospray Fourier transform ion cyclotron resonance mass spectrometry: the continued battle against corrosion, *Anal. Chem.*, **2003**, 75, 860–866.

145. Qian, K.; Rodgers, R.P.; Hendrickson, C.L.; Emmett, M.R.; Marshall, A.G., Reading chemical fine print: resolution and identification of 3000 nitrogen-containing aromatic compounds from a single electrospray ionization Fourier transform ion cyclotron resonance mass spectrum of heavy petroleum crude oil, *Energy and Fuels*, **2001**, 15, 492–498.

146. Wu, Z.; Jernstroem, S.; Hughey, C.A.; Rodgers, R.P.; Marshall, A.G., Resolution of 10000 compositionally distinct components in polar coal extracts by negative-ion electrospray ionization Fourier transform ion cyclotron resonance mass spectrometry, *Energy and Fuels*, **2003**, 17, 946–953.

147. Stenson, A.C.; Landing, W.M.; Marshall, A.G.; Cooper, W.T., Ionization and fragmentation of humic substances in electrospray ionization Fourier transform-ion cyclotron resonance mass spectrometry, *Anal. Chem.*, **2002**, 74, 4397–4409.

148. Sleno, L.; Volmer, D.A.; Marshall, A.G., Assigning product ions from complex MS/MS spectra: the importance of mass uncertainty and resolving power, *J. Am. Soc. Mass Spectrom.*, **2005**, 16, 183–198.

149. Schmatloch, S.; Schubert, U.S., Techniques and instrumentation for combinatorial and high-throughput polymer research: recent developments, *Macromol. Rapid Commun.*, **2004**, 25, 69–76.

150. Dar, Y.L., High-throughput experimentation: a powerful enabling technology for the chemicals and materials industry, *Macromol. Rapid Commun.*, **2004**, 25, 34–47.

151. Schmid, D.G.; Grosche, P.; Bandel, H.; Jung, G., FTICR-mass spectrometry for high-resolution analysis in combinatorial chemistry, *Biotechnol. Bioeng.*, **2000**, 71, 149–161.

152. Marshall, A.G.; Hendrickson, C.L.; Shi, S.D.H., Scaling MS plateaus with high-resolution FT-ICRMS, *Anal. Chem.*, **2002**, 74, 253A–259A.

153. Ling, C.; Cottrell, C.E.; Marshall, A.G., Effect of signal-to-noise ratio and number of data points upon precision in measurement of peak amplitude, position and width in Fourier-transform spectrometry, *Chemom. Intell. Lab. Syst.*, **1986**, 1, 51–58.

154. Wang, T.C.L.; Ricca, T.L.; Marshall, A.G., Extension of dynamic range in Fourier-transform ion-cyclotron resonance mass-spectrometry via stored wave-form inverse Fourier-transform excitation, *Anal. Chem.*, **1986**, 58, 2935–2938.

155. Francl, T.J.; Sherman, M.G.; Hunter, R.L.; Locke, M.J.; Bowers, W.D.; McIver, R.T., Experimental determination of the effects of space charge on ion cyclotron resonance frequencies, *Int. J. Mass Spectrom. Ion Phys.*, **1983**, 54, 189–199.

156. Ledford, E.B.; Rempel, D.L.; Gross, M.L., Space-charge effects in Fourier-transform mass-spectrometry—mass calibration, *Anal. Chem.*, **1984**, 56, 2744–2748.

157. Li, Y.Z.; McIver, R.T.; Hunter, R.L., High-accuracy molecular-mass determination for peptides and proteins by Fourier-transform mass-spectrometry, *Anal. Chem.*, **1994**, 66, 2077–2083.

158. Dunbar, R.C.; Asamoto, B., *FT-ICR/MS* (Asamoto, B., editor). Camebridge: VCH Publisher, **1991**, pp. 29–81.

159. Schweikhard, L.; Marshall, A.G., Excitation modes for Fourier transform-ion cyclotron-resonance mass-spectrometry, *J. Am. Soc. Mass Spectrom.*, **1993**, 4, 433–452.

160. Dunbar, R.C.; Chen, J.H.; Hays, J.D., Magnetron motion of ions in the cubical ICR cell, *Int. J. Mass Spectrom. Ion Proc.*, **1984**, 57, 39–56.

161. Comisarow, M.B., Cubic trapped-ion cell for ion-cyclotron resonance, *Int. J. Mass Spectrom. Ion Proc.*, **1981**, 37, 251–257.

162. Elkind, J.L.; Weiss, F.D.; Alford, J.M.; Laaksonen, R.T.; Smalley, R.E., Fourier-transform ion-cyclotron resonance studies of H-2 chemisorption on niobium cluster cations, *Journal of Chemical Physics*, **1988**, 88, 5215–5224.

163. Comisarow, M.B.; Marshall, A.G., Fourier transform ion cyclotron resonance spectroscopy method and apparatus, USA Patent No. 3,937,955, issued February 14, **1976**.

164. Easterling, M.L.; Pitsenberger, C.C.; Kulkarni, S.S.; Taylor, P.K.; Amster, I.J., A 4.7 Tesla internal MALDI-FTICR instrument for high mass studies: performance and methods, *Int. J. Mass Spectrom. Ion Proc.*, **1996**, 157/158, 97–113.

165. Vartanian, V.H.; Laude, D.A., Optimization of a fixed-volume open geometry trapped ion cell for Fourier-transform ion-cyclotron mass-spectrometry, *Int. J. Mass Spectrom. Ion Proc.*, **1995**, 141, 189–200.

166. Gabrielse, G.; Haarsma, L.; Rolston, S.L., Open-endcap penning traps for high-precision experiments, *Int. J. Mass Spectrom. Ion Proc.*, **1989**, 88, 319–332.

167. Beu, S.C.; Laude, D.A., Open trapped ion cell geometries for Fourier-transform ion-cyclotron resonance mass-spectrometry, *Int. J. Mass Spectrom. Ion Proc.*, **1992**, 112, 215–230.

168. Beu, S.C.; Laude, D.A., Elimination of axial ejection during excitation with a capacitively coupled open trapped-ion cell for Fourier-transform ion-cyclotron resonance mass-spectrometry, *Anal. Chem.*, **1992**, 64, 177–180.

169. Guo, X.H.; Duursma, M.; Al-Khalili, A.; McDonnell, L.A.; Heeren, R.M.A., Design and performance of a new FT-ICR cell operating at a temperature range of 77-438 K, *Int. J. Mass Spectrom.*, **2004**, 231, 37–45.

170. Caravatti, P.; Alleman, M., The "infinity cell": a new trapped-ion cell with radiofrequency covered trapping electrodes for Fourier transform ion-cyclotron resonance mass spectrometry, *Org. Mass Spectrom.*, **1991**, 26, 514–518.

171. Skelton, R.; Dubois, F.; Zenobi, R., A MALDI sample preparation method suitable for insoluble polymers, *Anal. Chem.*, **2000**, 72, 1707–1710.

CHAPTER 5

TANDEM MASS SPECTROMETRY AND POLYMER ION DISSOCIATION

Michael J. Polce[1] *and Chrys Wesdemiotis*[2]

[1]Lubrizol Advanced Materials, Cleveland, and [2]Department of Chemistry, The University of Akron, Akron, Ohio

5.1 INTRODUCTION AND BASIC PRINCIPLES OF TANDEM MASS SPECTROMETRY (MS/MS)

5.1.1 Introduction

Synthetic polymers are generally produced with a distribution of molecular weights (MWs), which may be further convoluted by a variety of end groups and architectures. With copolymers, additional distributions of comonomers and comonomer block lengths are possible [1–5]. Fortunately, most components of this complex mixture differ in mass, and hence can be dispersed and analyzed by mass spectrometry (MS). Indeed, MS has been increasingly used in polymer research, as is evident from the considerable number of relevant reviews and books that have appeared in the last decade [6–30]. Today, MS is widely used to characterize the MW, molecular weight distribution (MWD), oligomer composition, and end groups of synthetic polymers [6–30]. As a dispersive method, MS is ideally suited for the identification of the minor synthetic products and degradation products of polymer synthesis, which would be difficult to detect by integrative analytical methods, such as IR or NMR spectroscopy. Minor and/or degradation products help to unveil the origin of unknown samples, and hence are particularly useful for the characterization of competitors' formulations in industrial laboratories.

Single-dimensional MS (one-stage MS) furnishes the mass of a synthetic product, based on which its elemental composition may be determined, depending on the molecular mass of the compound and mass measurement accuracy. If the mechanism of the synthetic route is well established, this compositional information can be translated into structural insight [31, 32]. With new polymerization methods and/or new types of materials, the compositional information alone is not sufficient to derive the structure, that is, connectivity, of a polymer. In these cases, two- or

MALDI Mass Spectrometry for Synthetic Polymer Analysis, Edited by Liang Li
Copyright © 2010 John Wiley & Sons, Inc.

multidimensional mass spectrometry (MS^2 or MS^n), generically referred to as tandem mass spectrometry (MS/MS), can be employed to decipher how the polymer's constituents are connected to each other [33]. Recent literature has documented that MS/MS is a promising method for the characterization of the individual end groups of polymer chains [32, 34], the analysis of copolymer sequences [35, 36], and the differentiation of isobaric or isomeric oligomers and polymer architectures [37, 38]. These capabilities will be demonstrated in this chapter for several classes of polymeric materials, including important industrial polymers.

5.1.2 Basic Principle of MS/MS

Every MS experiment starts with the generation of gas-phase ions from the analyte [39]. The most widely used ionization methods for synthetic polymers are matrix-assisted laser desorption ionization (MALDI) [40, 41] and electrospray ionization (ESI) [42]. Depending on the chemical properties of the substance under analysis and the sample preparation procedure, MALDI and ESI mainly yield ions by protonation, deprotonation, or metal ion addition, resulting in $[M + H]^+$, $[M - H]^-$, or $[M + X]^+$ (X = metal) quasi-molecular ions, respectively, from each oligomer in the polymeric analyte. In a single-stage MS experiment, the ions formed this way are separated by their mass-to-charge ratio (m/z) and their relative abundances are displayed versus the corresponding m/z values [33]. In an MS/MS experiment, the $[M + H]^+$, $[M - H]^-$, or $[M + X]^+$ quasi-molecular ion of one individual oligomer is mass-selected and energetically activated to decompose into structurally indicative fragments, which are subsequently dispersed by a second stage of mass analysis to yield the MS/MS spectrum of the selected oligomer [39]. Based on the fragments present in this spectrum, specific structural features can be assigned to the oligomer. In triple-stage mass spectrometry (MS^3), this procedure is repeated by selecting one of the mentioned fragments (first-generation or primary fragment) and inducing its further decomposition into second-generation (or secondary) fragments. Further selection/activation steps can follow if the instrumentation used allows it and if signal intensity is sufficient [39].

This chapter describes the instrumentation used in MS/MS studies of synthetic polymers. Fragmentation is a central step in an MS/MS experiment, as it is this event that yields structural information about the analyte. The fragmentations of synthetic polymer ions will be classified according to their mechanisms and discussed with specific examples from the literature or from current studies in the Akron laboratory.

5.2 MS/MS INSTRUMENTATION FOR POLYMER ANALYSIS

Mass separation devices can be divided into two broad categories, viz. beam and trap instruments; ions traverse through the former, but are stored in the latter. Beam mass dispersers include the quadrupole mass filter (Q), time-of-flight (ToF) drift tube, and electric (E) or magnetic (B) sectors. Examples of trap mass analyzers are the quadrupole ion trap (QIT; 3D, 2D or linear, Orbitrap) and the magnetic or

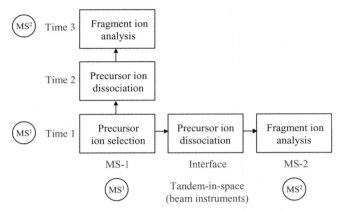

Scheme 5.1.   Differentiation between tandem-in-space and tandem-in-time mass spectrometry. Adapted from Reference 39 with permission.

ion cyclotron resonance (ICR) trap, which store ions using electric or magnetic fields, respectively. The simplest MS/MS experiment (MS^2) encompasses two stages of mass analysis, the first involving the selection of a specific oligomer ion (precursor ion) and the second the dispersion of the fragments formed from this ion (Scheme 5.1). In tandem-in-space experiments, the two mass separating events take place in two different mass analyzers (MS-1 and MS-2 in Scheme 5.1), which are separated by a reaction region in which the precursor ion is energetically activated to dissociate. Any of the above-mentioned devices can be used at the MS-1 and MS-2 stage, with the most common combinations listed in Table 5.1.

Alternatively, trapping analyzers permit tandem-in-time experiments, in which the mass selection, activation, and fragment dispersion steps are performed in time sequence within the same device, that is, the trap (times 1–3 in Scheme 5.1). Further stages of MS (e.g., MS^3, MS^4) are easily accomplished by repeating the steps mass selection → activation → mass dispersion. In sharp contrast, each added MS stage in tandem-in-space MS requires an additional mass-separating device. At the present time, most commercially available tandem-in-space mass spectrometers consist of two physically distinct mass analyzers, which may include tandem-in-time devices (see Table 5.1).

Over the last 20 years, several types of instruments have been used for MS/MS studies on synthetic polymers. A number of initial studies were performed on multisector and sector-quadrupole or sector-ToF hybrid mass spectrometers equipped with field desorption [43–45], fast atom bombardment (FAB) [46–48], MALDI [49–52], and more recently, direct probe chemical ionization [53–55]. Both MALDI and ESI, the currently major ionization methods for synthetic polymers [29, 30], do not perform well with sector instruments. As a pulsed ionization method, MALDI is not easily adapted to scanning instruments, such as sector, triple-quadrupole, or

TABLE 5.1.　Mass Analyzer Combinations in Tandem Mass Spectrometry Studies of Synthetic Polymers[a]

Stage 1 (Precursor Ion Selection)		Collision	Stage 2 (Fragment ion Analysis)		Ionization
Analyzer	Resolution[b]	Energy[c]	Analyzer	Resolution	Method[j]
Tandem-in-space					
Reflectron ToF	Low	PSD[d]	[h]	Low	MALDI
ToF	Intermediate	High[e]	ToF	High	MALDI
B	*Low*	*High*[e]	*E*	*Low*	*CI, FAB, FD, ESI*
EB or BE	*High*	*High*[e]	*EB or BE*	*High*	*CI, FAB, FD, ESI*
EB or BE	*High*	*High*[e]	*oa-ToF*	*High*	*ESI, MALDI*
EB or BE	*High*	*Low*[f]	*Q*	*Intermediate*	*CI, FAB, FD, ESI*
Q	Intermediate	Low[f]	Q	Intermediate	ESI
Q	Intermediate	Low[f]	QIT (2D)	Intermediate	ESI
Q	Intermediate	Low[f]	oa-ToF	high	ESI, MALDI
QIT (3D)	Intermediate	Low[g]	ToF	High	MALDI
QIT (2D)	Intermediate	Low[g]	Orbitrap	High	ESI
Tandem-in-time					
QIT (2D or 3D)	Intermediate	Low[g]	[i]	Intermediate	ESI, MALDI
FTICR	High	Variable[g]	[i]	High	ESI, MALDI

[a] Italicized entries refer to outdated, presently decommissioned, or rarely used instrument configurations.
[b] Low: a window of several mass-to-charge ratios is transmitted; intermediate: unit or near unit mass resolution; high: isobaric ions are individually selectable.
[c] In collisionally activated dissociation (CAD); high and low refer to laboratory-frame collision energies in the kiloelectron volt and electron volt regime, respectively.
[d] Post-source decay.
[e] CAD in a collision cell within a field-free region.
[f] CAD in an rf-only quadrupole collision cell.
[g] CAD inside the trap.
[h] Different parts of the same instrument are used for mass selection of the precursor ion and fragment ion analysis (see PSD section).
[i] The same mass analyzer is used for mass selection of the precursor ion and fragment ion analysis.
[j] Typical ionization methods interfaced with these configurations; CI, FD, and FAB denote chemical ionization, field desorption, and fast atom bombardment (also known as liquid secondary ion mass spectrometry), respectively.

sector-quadrupole tandem mass spectrometers; as a result, commercial MALDI MS instruments with such mass analyzers either do not exist or are gradually being phased out. ESI-equipped sector instruments are also slowly being discontinued due to the difficulty of interfacing an ESI source, which requires potentials of a few kilovolt for ion formation [39, 42], with the high acceleration potentials (3–10 keV) necessary for sector operation. MS/MS studies nowadays mainly employ ToF, quadrupole, or trap mass analyzers for precursor ion selection and fragment ion separation. All of these mass analyzers are readily connected to MALDI and ESI sources; moreover, all have substantially higher transmission efficiencies than electric or magnetic sectors and allow for fast, repetitive scanning, ion accumulation before scanning, and/or simultaneous detection of a wide range of mass-to-charge ratios without scanning, providing the enhanced sensitivity needed for MS/MS with pulsed ion sources or MS/MS of minor components in a mixture [39]. The instrument configurations most widely employed in current polymer MS/MS studies will be discussed in more detail after a brief description of the activation methods used to cause polymer ion fragmentation.

5.2.1 Precursor Ion Activation Methods

Collisionally Activated Dissociation (CAD)   Most commonly, the mass-selected oligomer ion is energetically excited to fragment via CAD [56, 57]. In CAD, an ion that has been accelerated or decelerated to a given kinetic energy (laboratory-frame kinetic energy, E_{lab}) undergoes collisions with a gaseous target, for example, Ar or air. This process converts part of the ion kinetic energy into internal energy, which is redistributed over all degrees of freedom of the ion and can cause fragmentation anywhere in the ionized molecule, independent of which part of it collided with the gaseous target molecules (ergodic dissociation). The types of dissociations taking place depend on the amount of rovibrational energy quanta deposited upon collision, the rate constants of the possible dissociations, and the time available for dissociation. Since the collisions transfer a distribution of internal energies, many competing reactions can occur and the ultimately observed spectrum contains all fragments formed at the various rovibrational energies accessed over the course of the experiment.

The maximum amount of internal energy that can be deposited in a collision is given by the center-of-mass collision energy, E_{cm}, which is related to the ion kinetic energy (E_{lab}) via the equation $E_{cm} = [m/(m + M)] \times E_{lab}$, where m and M are the masses of the gaseous target and selected oligomer ion, respectively. Collisions in the electron volt kinetic energy range involve direct rovibrational excitation of the precursor ion and can deposit internal energies up to E_{cm} per collision [56, 57]. Collisions involving kiloelectron volt ions may involve direct rovibrational excitation or initial electronic excitation that is subsequently converted to rovibrational excitation; the former is more likely for larger ions, such as those obtained from synthetic polymers [57]. CAD experiments generally involve multiple collisions (stepwise excitation), and dissociation occurs after enough internal energy has been accumulated to enable fragmentation within the time the precursor ions spend in the activation region. Due to their size, polymer ions are associated with substantial kinetic and competitive shifts; that is, the energies needed for fragmentation exceed the corresponding dissociation thresholds, the extra amounts being required to cause dissociation within the available observation time (kinetic shift) or to make an energetically less favorable reaction competitive (competitive shift) [58].

The internal energy distributions resulting after CAD on polymer ions have not been systematically evaluated. Based on reported spectra, the average internal energy transferred depends on the type of instrumentation used (Table 5.1), increasing in the order: CAD in QITs < CAD in radiofrequency (rf) only collision cells (in all instruments with at least one quadrupole mass analyzer) < CAD in time-of-flight (ToF)/ToF and sector/ToF configurations. The energy requirements of the different dissociation channels can be qualitatively assessed by measuring tandem mass spectra as a function of collision energy (E_{lab}); the relative intensities of the fragmentations proceeding over higher barriers generally increase with collision energy.

Other Activation Methods   In surface-induced dissociation (SID) [59], a mass-selected precursor ion with low kinetic energy collides with a solid surface and, as a result, part of its kinetic energy is converted to internal energy, causing dissociation. SID proceeds under strictly single-collision conditions and, as shown with ions from organic and biological substances, it deposits a narrower internal energy distribution than CAD, thereby permitting the acquisition of energy-resolved tandem mass spectra at a significantly higher-energy resolution [59–62]. SID can be performed with tandem quadrupole, sector/quadrupole, ToF/ToF, and trap instruments [57]. Since this method does not require introduction of a gas, it is particularly useful for Fourier transform ion cyclotron resonance (FTICR) instruments, in which a very low pressure ($<10^{-8}$ mbar) is essential for obtaining tandem mass spectra with high resolution [61, 62]. SID has not been applied to synthetic polymer samples.

Another method that does not necessitate the introduction of a collision gas is infrared multiphoton dissociation (IRMPD) [57]. Here, a low-power (<100 W) continuous-wave CO_2 laser, emitting at 10.6 µm, is used for precursor ion activation. Nonresonant multiphoton absorption takes place to produce energetically exited precursor ions, which later decompose, after randomization of their internal energy over all degrees of freedom, as with CAD. Photon absorption over a long time (tens to hundreds of millisecond) is necessary to deposit enough energy for dissociation. Consequently, IRMPD has primarily been used with trap instruments, in which ions can be stored for long time intervals [63, 64]. IRMPD is ideally suitable for FTICR mass spectrometers, whose very low background pressure reduces the probability of deactivating collisions during the photon absorption process. In QITs, deactivation is minimized by reducing the pressure of the He bath gas in the trap region. Many recent applications of IRMPD to peptide and other biomolecular ions have been reported [57, 62]. Although synthetic polymers have not been studied yet by this method, pertinent applications, especially using FTICR MS, are anticipated.

The methods discussed so far can be used with MALDI- or ESI-generated precursor ions of any charge state. In contrast, electron capture dissociation (ECD) is possible only with multiply charged precursors, which are readily available only with ESI [65]. In ECD, multiply charged cations capture electrons with near thermal energies (<0.2 eV) and are reduced to radical cations that dissociate rapidly by radical site reactions. Dissociation of the radical ions is believed to occur without randomization of their internal energy (i.e., nonergodically). Nonergodic dissociation mainly involves direct bond scissions and eliminates or minimizes misleading rearrangements, which facilitates structural determinations [58]. ECD is performed in FTICR traps, where thermal electrons can be easily produced, for example, using a heated tungsten filament [57, 65–67]. With negative precursor ions, higher-energy electrons ($\sim$20 eV) are used to cause fragmentation via electron detachment dissociation (EDD) [66]. ECD and EDD are impossible in QITs, because their strong rf fields prevent storage of low-energy electrons for the times needed to produce detectable amounts of fragments (up to minutes). A closely related variant, electron transfer dissociation (ETD) [68, 69], can, however, be performed in QITs. ETD involves ion–ion reactions [70]. Multiply charged cations, formed by ESI, are mixed in the trap with anthracene anions (or other reagent ions), formed by negative chemical ionization (CI) (or another ionization method) in a different external ion source. Cation/anion collisions lead to proton as well as electron transfer. The radical cations emerging from electron transfer (reduction) undergo very similar fragmentations to those observed upon ECD. With negatively charged precursor ions, the reagent ion is changed to Xe^+ to oxidize the anions to radical anions, which subsequently decompose at the newly created radical sites [71]. ECD applications to polymer ions have recently been reported [72–76].

5.2.2 Post-Source Decay (PSD) MS

A reflectron ToF mass spectrometer, like the one shown in Figure 5.1 [39], can be used to acquire MS/MS spectra of precursor ions formed by MALDI [77, 78]. At a high laser intensity, a fraction of the ions is formed with enough internal energy to decompose metastably in the drift region between the ion source and the reflector entrance. The resulting fragment ions are said to arise from PSD [79]. Precursor ion and fragments have essentially the same velocities but different kinetic energies because of the principles of conservation of momentum and energy. Because of their identical velocities, precursor and fragment ions reach the reflector at the same time but are dispersed inside the reflector according to their kinetic energies. Compared to the precursor, a fragment ion has smaller mass and kinetic energy, and thus, it penetrates less into the reflector, spending less time inside the reflecting lenses, and arriving sooner at the detector in reflectron mode [79, 80]. At the typical acceleration voltages used in MALDI-ToF MS (~20 keV), not all fragment ions can be focused onto the detector simultaneously. Different m/z ranges are brought into focus by gradually decreasing the reflecting lens potentials. The m/z segments recorded at each voltage interval are subsequently pasted together to provide the PSD spectrum; the pasting and mass calibration are completed automatically by the data system [79, 80]. Usually, individual segments are obtained from different locations of the sample spot, and the fragment ion intensities within each segment vary; as a result, the

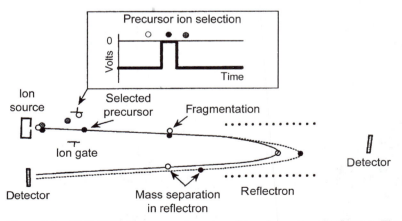

Figure 5.1.   MALDI-ToF mass spectrometer with an ion gate and reflectron. Three ions are shown to exit the ion source, one of which passes the ion gate and fragments en route to the reflectron (see text). Adapted from Reference 39 with permission.

reproducibility of relative abundances is low, which makes quantitative analyses by PSD impractical.

A pulsed ion deflector ("ion gate"), located after the ion source, is used to select a specific precursor ion [39, 79, 80]. The ion gate opens at a preprogrammed time that allows only a narrow m/z range to pass into the drift tube. The selection resolution generally is ~50; that is, a ~20-Da window is transmitted with a precursor ion of m/z 1000 [39]. This resolution can be varied to lower or higher values on some instruments. For the best mass accuracy (0.1 Da), the precursor ion should be centered within the transmitted m/z window [81]. Fragment ion resolution is low and comparable to that obtained using other kinetic energy dispersers (e.g., electric sectors) for mass analysis of metastable fragments, but it can be improved by careful optimization of the reflector potentials [81]. The fragmentation extent in PSD spectra can be increased by adding a collision cell in the beam path after the ion source. The abundances of low-mass products have been shown to increase significantly by switching from PSD to CAD mode [82].

The PSD method is exemplified in Figure 5.2 with the spectrum obtained from a lithiated poly(methyl methacrylate) (PMMA) oligomer. Radical ions are observed at lower masses, while the intermediate and higher m/z range of the spectrum is dominated by four homologous series of closed-shell fragments that contain either the initiating (a- and b-type) or terminating chain end (y- and z-type). The nomenclature and fragmentation pathways leading to the observed fragments will be discussed in more detail later. Numerous PSD studies on synthetic polymers have been reported [33, 35, 49, 81–99]. Qualitatively, PSD spectra look very similar to CAD spectra measured on other types of instruments (vide infra); however, spectral resolution and mass accuracy are poorer compared to most modern commercial tandem mass spectrometers.

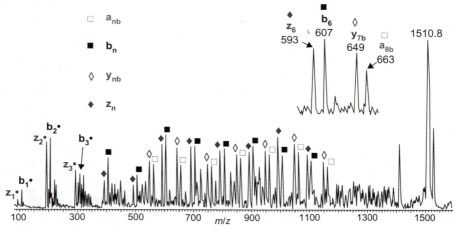

Figure 5.2. PSD mass spectrum of the lithiated 15-mer from a PMMA standard with the connectivity H–[CH₂–C(CH₃)CO₂CH₃]ₙ–H (m/z 1510.8). See section on mechanisms for nomenclature and structures of the fragments.

5.2.3   Sector Hybrid Tandem Mass Spectrometers

Up to a few years ago, a significant number of MS/MS studies on synthetic polymers utilized all-sector or sector hybrid tandem mass spectrometers of various designs; today, such instruments are used rarely, because their interfacing with MALDI or ESI is problematic (vide supra). Precursor ion selection has been achieved by a magnetic sector (B) or a combination of magnetic and electric (E) sectors, which allow for ion isolation at low and high resolution, respectively [45–55, 82, 83, 100–108]. After CAD, the fragment ion dispersion has been effected with one [48, 82, 102, 105] or more [49, 100] sectors, a quadrupole mass filter [46, 47, 53–55], or a ToF device [45, 49–52, 101–104, 106–109].

MS/MS studies on small oligomers (<~1000 Da) ionized by CI or FAB have been conducted with BE/Q or EBE/Q hybrid mass spectrometers, primarily by Lattimer et al. [46, 47, 53–55]. In such instruments, precursor ions are formed at high kinetic energy (5–8 keV), mass-selected with the sectors (MS-1), and decelerated to <200 eV for low-energy CAD in an rf-only octapole or hexapole collision cell located between BE and Q. The rf field focuses the fragment ions near the center of the collision cell, minimizing scattering losses and maximizing the transmission efficiency to the quadrupole mass filter, which is scanned for fragment mass analysis (MS-2).

Quadrupoles can achieve unit mass resolution up to ~1000 m/z units (i.e., $m/\Delta m < 1000$). Improved sensitivity and a higher MS-2 resolutions ($m/\Delta m \approx 1500$) can be obtained by attaching an orthogonal acceleration (oa) ToF drift tube at the end of sectors [50]. Commercial EBE-oa-ToF tandem mass spectrometers equipped with MALDI or ESI (Figure 5.3) [110] were briefly available in the 1990s and have been used in polymer MS/MS studies, in particular by Scrivens, Jackson, and coworkers who reported applications to several types of polymers with the MALDI version [49–51, 83, 101, 103, 104, 108, 109]. Oligomers ionized by MALDI were accelerated to 8 keV, mass-selected by EBE, and decelerated to 800 eV before CAD

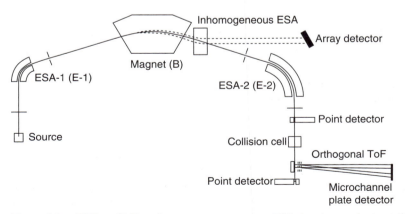

Figure 5.3.   EBE-oa-ToF tandem mass spectrometer. ESA denotes an electrostatic analyzer. Reproduced from Reference 110 with permission.

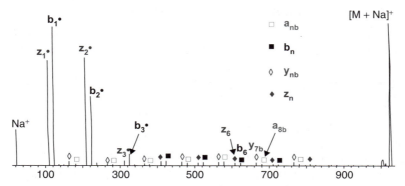

Figure 5.4.   MALDI-CAD mass spectrum of the sodiated 10-mer from a PMMA standard with the connectivity H–[CH$_2$–C(CH$_3$)CO$_2$CH$_3$]$_n$–H (*m/z* 1026), acquired on a EBE-oa-ToF tandem mass spectrometer. Only the all-^{12}C isotope at *m/z* 1026 is transmitted thought EBE. Adapted from the *International Journal of Mass Spectrometry*, vol. 188, Gidden, J., et al., Gas phase conformations of synthetic polymers: poly(methyl methacrylate) oligomers cationized by sodium ions, pp. 121–130, copyright 1999 with permission from Elsevier. See section on mechanisms for nomenclature and structures of the fragments.

with Xe targets in a confined collision cell within the field-free region between the sectors and the ToF device (Figure 5.3). The fragments were reaccelerated to 8 keV for mass analysis based on their ToF. A representative CAD spectrum, obtained from a sodiated PMMA oligomer is shown in Figure 5.4. Note the better signal-to-noise ratio and higher dynamic range in comparison to the PSD spectrum of Figure 5.2.

EBE-oa-ToF mass spectrometers with a MALDI source have not performed well in MS mode, that is, for the measurement of single-stage mass spectra. This could be due to engineering problems in the construction of the inhomogeneous electric sector needed to direct the ions exiting the nonscanning magnet onto the array detector (see Figure 5.3). (The low duty cycle of a pulsed ionization method [MALDI] prevents the use of a scanning device for ion separation.) Because of this shortcoming and the high acquisition and maintenance costs of such instruments, sector-ToF hybrids are no longer commercially available and have been replaced by simpler, more powerful, Q/ToF or ToF/ToF combinations (vide infra).

5.2.4   Quadrupole/Time-of-Flight (Q/ToF) Hybrid Tandem Mass Spectrometers

Q/ToF tandem mass spectrometers, which were introduced approximately 10 years ago [111], employ a quadrupole mass filter for precursor ion selection (MS-1) and a ToF mass analyzer for fragment ion separation (MS-2). These devices are arranged orthogonally, as depicted in Figure 5.5 [112]. MALDI, ESI, atmospheric pressure chemical ionization (APCI), and most recently, desorption electrospray ionization (DESI) and desorption atmospheric pressure chemical ionization (DAPCI) ion sources have been interfaced with such instruments [111–114]. In MS/MS mode, the ions produced in the ion source are transferred by a combination of ion lenses

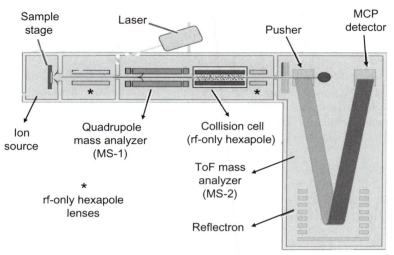

Figure 5.5. Waters® Q-Tof Ultima® MALDI Mass Spectrometer, a Q/ToF tandem mass spectrometer equipped with a MALDI source. Adapted from Reference 112 with permission, © Waters Corporation 2003. See color insert.

to a quadrupole mass analyzer, which is set to transmit only the desired oligomer ion. The selected ion continues to travel into a collision cell, an rf-only quadrupole or hexapole, where CAD takes place, usually with Ar gas. Laboratory-frame collision energies up to 200 eV can be adjusted by appropriate floating of the collision cell. The fragments generated upon CAD are accelerated orthogonally by ~10 kV into a ToF drift tube for mass analysis. Unlike sector/ToF hybrids, Q/ToF mass spectrometers can also be operated as single-stage mass analyzers; for this, the dispersing quadrupole is switched to rf-only mode, so that all ions formed in the source pass through MS-1 to be orthogonally accelerated into the ToF drift tube (MS-2) for mass analysis. Hence, in both MS and MS/MS operation, mass dispersion is performed at the ToF device, which provides significantly higher resolution and mass accuracy than a quadrupole mass filter [111–113].

Precursor ions up to ~3,000 m/z have been fragmented in Q/ToF instruments. The collision energy can be adjusted to maximize the abundances of fragment ions or varied to obtain breakdown graphs, which provide information about dissociation energetics and mechanisms. Mass resolving power and mass accuracy at the ToF analyzer are >10,000 and <10 ppm, respectively, and an even higher-mass resolving power (~20,000) can be achieved by dual reflection within the ToF tube (W-mode compared to the V-mode shown in Figure 5.5) [112]. Because of these advantages, Q/ToF instrumentation has essentially replaced sector hybrids and is increasingly used in polymer MS/MS studies in conjunction with both ESI [34, 114–118] and MALDI [36, 119].

The complete isotopic cluster or a single isotope of the precursor ion can be mass-selected with the quadrupole mass filter by proper adjustment of its direct current (DC) and rf potentials. Usually a window of masses covering all isotopes is transmitted for increased sensitivity and easy identification of fragments that contain elements with unique isotopic compositions. For example, Ag^+-containing fragments give rise to two all-^{12}C peaks, $2\,m/z$ units apart from each other and with an abundance ratio of ~1:1, resulting from the $^{107}Ag/^{109}Ag$ isotopes; similarly, Cu-containing fragments give all-^{12}C doublets in a ~3:1 abundance ratio due to the $^{63}Cu/^{65}Cu$ isotopes. If the precursor ion forms fragments that are similar in mass ($\Delta m < 4\,Da$), the corresponding isotopic clusters overlap, obscuring the identification of the less abundant fragment. In such cases, selection of the monoisotopic, all-^{12}C precursor ion substantially facilitates interpretation of the MS/MS spectrum. This advantage is demonstrated in Figure 5.6 with the CAD spectrum of a silverated poly(isoprene) oligomer.

Recent work in the authors' laboratory has shown that CAD of Ag^+-cationized poly(isoprene) oligomers causes homolytic cleavage at the allylic bonds, creating incipient radical ions which undergo hydrogen rearrangement(s) and/or β scission(s) to yield homologous series of fragment ions that contain either or none of the original end groups. Several of these fragment series differ in mass by 1–2 Da; the less abundant ones can be conclusively identified only if the monoisotopic precursor ion is selected (cf. Figure 5.6).

Although Q/ToF instrumentation allows for the measurement of single-stage mass spectra (vide supra), it may distort the MWD of a polymer because of the rf fields adjusted on the quadrupole and the hexapolar (or quadrupolar) lenses that the oligomer ions must pass en route to the orthogonally located ToF mass analyzer (see Figure 5.5). Axial extraction from the ion source without the use of such lenses, as available in MALDI-ToF mass spectrometers, bypasses this problem and is therefore

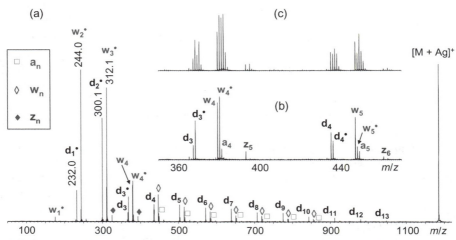

Figure 5.6. (a) MALDI-CAD mass spectrum of the monoisotopic (i.e., all-^{12}C) [15-mer + ^{107}Ag]$^+$ ion from a poly(isoprene) standard with the connectivity C_4H_9–[$CH_2CH=C(CH_3)CH_2$]$_n$–H, acquired on a Q/ToF tandem mass spectrometer with Ar collision gas at $E_{lab} = 60\,eV$. See section on mechanisms for nomenclature and structures of the fragments. (b) Expanded trace of the m/z 300–500 region. (c) Expanded trace of the same m/z region of a spectrum measured from the complete isotopic cluster of the same oligomer ion.

most suitable for the determination of MWDs. When upgraded to tandem ToF configurations, such instruments also permit MS/MS experiments with better control of the CAD conditions and with a higher resolution and fragment ion mass accuracy than is possible with the PSD approach described in Section 5.2.2.

5.2.5   ToF/ToF Tandem Mass Spectrometers

In the commercially available ToF/ToF instruments [120–122], a short linear ToF tube (MS-1) is connected axially with a reflectron ToF mass spectrometer (MS-2), as shown in Figure 5.7. Ions formed by MALDI, the only ionization method applied thus far to such instruments [120–123], are extracted into ToF-1 after an appropriate delay time with a potential of ~8 kV. The delay time is optimized to focus the desired precursor ion in the center of the timed ion selector (TIS), which consists of two deflection gates [120–122]. The opening and closing of these gates is synchronized so that only a very narrow m/z window passes through, typically the entire isotopic cluster of the precursor ion. Mass resolutions (m/Δm) of 400–1000 have been reported at m/z ~1000, with the higher values obtained at threshold laser power.

In the design shown in Figure 5.7 [121], the selected precursor ion is decelerated upon entering the collision cell to ~1–3 keV. CAD in the cell produces fragments, all of which have essentially the same velocity as the precursor ion. Hence, the packet of undissociated precursor and fragment ions enters simultaneously the second source, where it is reaccelerated to ~15–20 keV for mass analysis through ToF-2.

In a slightly different design, the collision cell is located closer to the ion source and in front of the TIS [122]. Now, TIS transmits the complete packet of precursor ion and its fragments, because all have the same velocity and reach the TIS gates at the same time [122]. Subsequently, the selected ion packet is reacceler-

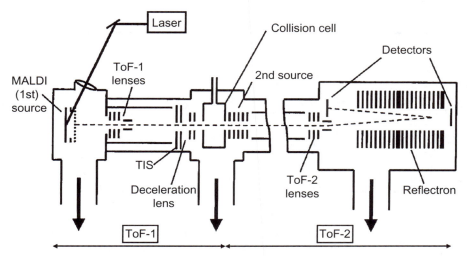

Figure 5.7.   MALDI-ToF/ToF tandem mass spectrometer. Reprinted by permission of Elsevier from De novo sequencing of peptides using MALDI-ToF/ToF, by Yergey, A.L., et al., *Journal of the American Society for Mass Spectrometry*, vol. 13, pp. 784–791, copyright 2002 by the American Society for Mass Spectrometry.

ated and mass-dispersed through ToF-2, as explained above. Either design can be used as a simple, reflectron MALDI-ToF mass spectrometer (MS mode) by grounding the TIS gates, deceleration lenses, and second source, and by using a higher acceleration potential at the first (i.e., MALDI) source. MALDI-ToF/ToF tandem mass spectrometers can provide mass resolutions of 15,000–25,000 or 1500–3000, in MS or MS/MS mode, respectively; the corresponding mass accuracies are <20 ppm (MS) and <100 ppm (MS/MS) [121, 124, 125].

MALDI-ToF/ToF instrumentation has been used to characterize poly(ethylene glycol) (PEG) [123], poly(ester amide)s [124], polyesters [125], and polystyrenes [126, 127]. The quality of MS/MS spectra obtained with such equipment is illustrated in Figure 5.8 with the CAD spectrum of a sodiated polyester 6-mer [125]. The seven major fragment ion series generated from this precursor ion have been marked on the spectrum by different labels. They have been rationalized by charge-remote fragmentations involving hydrogen rearrangements, β C–C bond scissions in the adipic acid moiety, and ester C–O bond cleavages [125]. More details on such reactions will be provided in the section concerning polymer ion fragmentation mechanisms (vide infra).

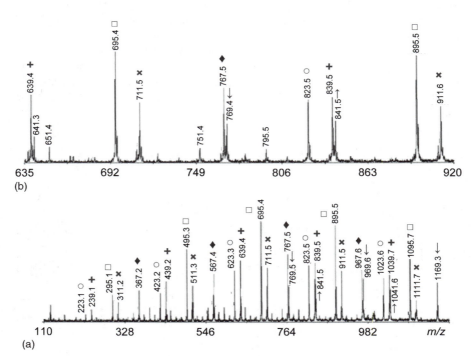

Figure 5.8.   (a) MALDI-CAD mass spectrum of the sodiated 6-mer from a poly(butylene adipate) with the connectivity H–[O(CH$_2$)$_4$OC(=O)(CH$_2$)$_4$C(=O)]$_n$–OH (m/z 1241.6), acquired on a ToF/ToF tandem mass spectrometer with Ar as collision gas at E_{lab} = 2 keV. (b) Expanded trace of the m/z 635–920 region. Reproduced from Reference 125 with permission.

Higher collision energies are used in ToF/ToF (1–8 keV) than in Q/ToF (<200 eV) tandem mass spectrometers. The time spent in the collision cell is, however, longer in Q/ToF instruments, where the ions move forward more slowly and the collision cell is longer than in ToF/ToF instruments, leading to a larger number of collisions in the Q/ToF vis à vis the ToF/ToF design. As a result, the internal energy distributions deposited in either design onto a given precursor ion and the resulting fragmentation patterns are very similar. Mass accuracy and resolution are overall higher with Q/ToF instruments; on the other hand, ToF/ToF instruments are more suitable for the determination of true MWDs when they are operated in the MS mode (vide supra).

5.2.6 Triple Quadrupole (QqQ) Tandem Mass Spectrometers

QqQ tandem mass spectrometers employ quadrupole mass filters for both precursor ion selection as well as fragment ion separation [39]. The first and third quadrupole serve as MS-1 and MS-2, respectively, while the center, rf-only quadrupole is the collision cell; an rf-only hexapole or octapole can also be used for the latter purpose. Triple quadrupole mass spectrometers are essential in metabolomics and metabonomics research, which requires quantitative analyses of small molecules (usually <500 Da) in complex mixtures [128, 129]. The ease of interfacing such instruments with chromatography, their fast and versatile scanning modes, and their high dynamic range make them ideally suitable for such applications. Because QqQ instruments have a relatively small mass range and are incompatible with pulsed ion sources, such as MALDI, their use in materials and polymer science has been limited. The few applications reported thus far concerned MS/MS studies of small oligomers (400–1000 Da) from polyamide [92], polyglycol [82], polyaniline [130], poly(vinyl acetate) [131], and cyclic polylactic acid [98] samples ionized by ESI.

5.2.7 QIT Mass Spectrometers

The QIT is an ion storage device consisting of a ring electrode and two end caps (Figure 5.9) [39, 132–141]. The end caps have small holes, through which ions can be injected, for example, from an external ESI or MALDI ion source [135, 136], or ejected for detection. An oscillating field of fixed frequency (0.75–1.2 MHz) is generated in the QIT by grounding the end caps and applying a high-voltage, variable rf potential to the ring electrode. The amplitude of the rf field (up to 30 kV) determines the m/z range that can be trapped. The kinetic energy gained by the ions upon injection in the trap and repulsive forces between the trapped ions increase their velocities and could lead to their ejection from the trap; this is prevented by pressurizing the ion trap with He bath gas (~10^{-3} Torr), which collisionally cools the ions, continually pushing them toward the trap center. Most modern traps have no DC supply [39, 141, 142]. The motion of ions in the trap is manipulated by an auxiliary rf voltage of variable frequency (0.01–1.2 MHz) and amplitude (V_{p-p} < 30 V) supplied to one end cap and used for ion isolation, fragmentation, and mass analysis (vide infra).

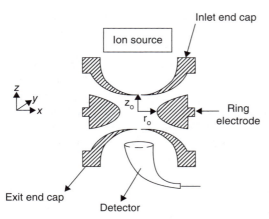

Figure 5.9.   Schematic view of the three-dimensional quadrupole ion trap (QIT). The ring electrode lies in the *x-y* plane. Ions are injected to and ejected from the trap along the *z*-axis. In an ideal trap, $r_o^2 = 2z_o^2$; generally, commercial traps deviate from this size.

Ion motion in the QIT is described by the Mathieu equation, which contains the dimensionless trapping parameter $q_z = 8zeV/m(r_o^2 + 2z_o^2)\Omega^2$, where V and Ω are the amplitude and angular frequency of the main (fundamental) rf field (on the ring electrode), *m* and *z* are the mass and charge of the ion, and r_o and z_o are the radial and axial dimensions of the trap, respectively (Figure 5.9). Each ion has a unique q_z value, which is inversely proportional to the corresponding mass-to-charge ratio (*m/z*) and reveals whether the ion can or cannot be trapped. Ions have stable trajectories in the trap, if $q_z < 0.908$. Such ions are held in the trap within a pseudo-potential well whose depth (D_z) is proportional to q_z ($D_z = q_zV/8$); ions with smaller mass-to-charge ratios oscillate in deeper wells than ions with larger *m/z* values, because q_z decreases with *m/z*. Ions with $q_z > 0.908$ cannot be trapped, as their trajectories are unstable; they are ejected through the end caps along the z axis, with the 50% moving toward the detector exit (cf. Figure 5.9) being measured. At a given rf potential (V), the *m/z* with $q_z = 0.908$ gives the low-mass cutoff. Similarly, inserting $q_z = 0.908$ and the maximum possible V in the above equation gives the maximum mass-to-charge ratio that can be trapped. The mass range can be increased by increasing V, decreasing Ω or the size of the trap (r_o, z_o), and causing ejection to occur at q_z values smaller than 0.908 [39, 141].

In MS mode, ions that have been accumulated in the trap for a certain period (few microseconds to ~1 s), depending on ion current produced upon ionization, are mass-analyzed by gradually raising the rf potential (V) to sequentially bring ions of increasing *m/z* into unstable trajectories (i.e., $q_z > 0.908$), so that they exit the trap and strike an external ion detector. This method is called mass-selective ion ejection at the stability limit. An alternative method involves mass analysis by resonant ejection. Trapped ions oscillate with a secular frequency, ω, which is lower than the main rf field frequency, Ω, and decreases with increasing *m/z*. Each *m/z* value has its own characteristic secular frequency (at a given V and Ω). If the auxiliary rf voltage (applied to the end caps) is tuned to the secular frequency of a trapped ion, the ion will absorb energy, and the amplitude of its oscillations will continually increase until it is ejected from the trap in the z direction. V and the frequency of the auxiliary field can be adjusted to cause ejection at $q_z < 0.908$. For example, the Bruker Esquire ion trap causes ejection at $q_z = 0.25$, which extends the mass range to m/z ~6000 [39, 142].

QIT electronics systems can generate auxiliary rf voltages with multiple frequencies that excite and eject a broad m/z range of ions. This feature is utilized in MS/MS experiments to mass-select a specific precursor ion; a broadband composite of the secular frequencies of all ions stored in the trap except the frequency of the precursor ion is synthesized and applied to isolate only a narrow mass-to-charge window around the precursor ion. The auxiliary field is subsequently used to increase the kinetic energy of the selected precursor ion via resonance excitation, so that it undergoes CAD with the He bath gas in the trap; this is effected with an rf voltage alternating at the precursor ion's secular frequency and having a relatively small amplitude (~1 V), in order to avoid ejection (vide infra). Typical excitation times are 20–60 ms. The fragments formed upon CAD are scanned by resonant ejection or ejection at the stability limit, as outlined above. Alternatively, a specific fragment ion may be selected for further CAD (MS^3), and this procedure can be repeated several times (MS^n) if necessary. The individual steps of an MS or MS^n experiment are preprogrammed into a scan function, which describes the time sequence of the events that will take place in the trap to yield the desired spectrum. Figure 5.10 includes a scan function for an MS^2 experiment with ions generated in an external ESI source [142].

CAD in a QIT involves multiple collisions, each depositing a certain amount of internal energy onto the precursor ion [57]. Dissociation occurs after enough internal energy has been gained to cause fragmentation within the time the activated precursor ion spends in the trap (millisecond time frame). Due to this stepwise excitation mechanism, the dominant fragments in QIT-CAD mass spectra result from dissociations with low critical energies. Since the fragment ions are not in

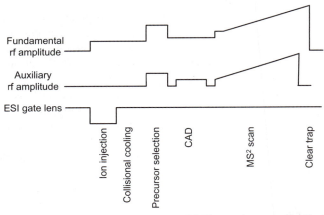

Figure 5.10. Scan function for an MS/MS experiment on the Bruker Esquire QIT equipped with an external ESI source.

resonance with the auxiliary rf field, they are not energetically excited and the probability of their consecutive fragmentation is low. This is in sharp contrast to CAD in an rf-only quadrupole collision cell (as in QqQ and Q/ToF instruments), where laboratory-frame collision energies are significantly larger and the first-generation fragment ions retain enough kinetic energy to be collisionally activated and undergo consecutive dissociations as they traverse the collision cell [57]. As a result of these differences, CAD spectra acquired with ion traps may contain overall fewer fragments, or a higher proportion of the energetically most favorable fragments, than spectra obtained with beam instruments. If desired, sequential fragmentation can be induced in the ion trap by adding further stages of activation/dissociation, that is, through MS^n experiments.

Quadrupole ion traps have been used in a significant number of MS/MS studies on a diverse array of synthetic polymers [37, 38, 90, 102, 143–158], including polyesters [37, 143, 144, 146, 156], polyethers [102, 147–149, 152, 153, 157, 158], polyamides [151, 154, 155], functionalized polyisobutylene [38], and poly(vinylidene fluoride) [150], as well as poly(propylene imine) dendrimers [90, 145]. The vast majority of these studies utilized ESI to ionize the polymers; in fact, QITs have been the most frequently used instruments for MS/MS experiments of oligomer ions produced by ESI (to this point). Since atmospheric pressure MALDI became available on QITs [136, 159, 160], MS/MS studies on MALDI-generated polyglycol ions have also been reported [152, 158].

The analytical power of multistage QIT-MS is demonstrated with MS^n spectra of an amine functionalized polyisobutylene (PIB) [38]. Due to the high intrinsic basicity of amine groups, such PIBs protonate readily. CAD (MS^2) of the protonated 6-mer produces one fragment via elimination of water (18 Da), cf. Figure 5.11(a); this reaction is characteristic for β-hydroxy amines, confirming the presence of such a substituent in the analyzed PIB [38]. Isolation of the water loss product (m/z 422.5) followed by CAD (MS^3) gives rise to a series of 56-Da losses, cf. Figure 5.11(b), in agreement with the PIB backbone of the oligomer (56 Da is the mass of the PIB repeat unit) [38]. Clearly, the combination of MS^2 and MS^3 permits a more complete structural characterization than simple, one-step MS/MS.

During CAD experiments in the QIT, the q_z value of the precursor ion is adjusted in the 0.2–0.3 range, to allow resonance excitation without exceeding the pseudo-potential well, which would cause ejection; for the same reason, as mentioned above, the amplitude of the auxiliary rf field is kept low (<~1 V) [39]. Setting the precursor ion q_z within 0.2–0.3 brings fragment ions with m/z values less than 22–33% of the precursor ion mass-to-charge ratio below the low-mass cutoff; as a consequence, the lower mass section of a CAD spectrum, typically the region below ~1/3 of the precursor ion m/z, is generally not measurable. This region may contain abundant and/or structurally significant fragments, as documented in Figure 5.12, which depicts the CAD spectra of a lithiated PMMA oligomer acquired using QIT and Q/ToF mass spectrometers [161].

From the tandem mass spectra of Figure 5.12, end-group information can be deduced based on the fragments in the middle- and high-mass end of the spectrum, which are detected in the QIT scan. The radical ions appearing at low masses, marked by a superscripted dot, provide important mechanistic information, however,

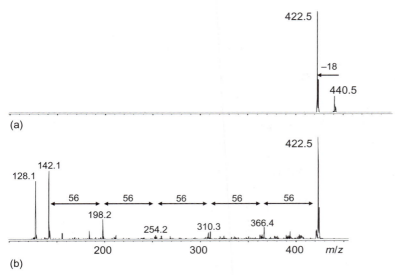

Figure 5.11.   Multistage ESI-CAD mass spectra of the protonated 6-mer from the terminally functionalized polyisobutylene H–[CH$_2$C(CH$_3$)$_2$]$_n$–CH$_2$CH(OH)(CH$_3$)CH$_2$–NHCH$_3$, acquired with a QIT mass spectrometer. (a) MS^2 spectrum of [M + H]$^+$ (m/z 440.5); (b) MS^3 spectrum of [M + H − H$_2$O]$^+$ (m/z 422.5). Adapted from Reference 38 with permission.

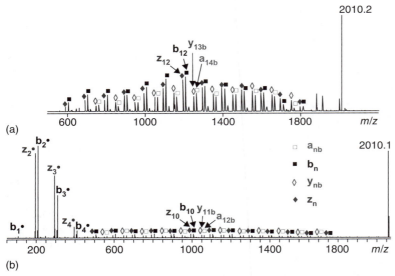

Figure 5.12.   CAD mass spectra of the lithiated 19-mer from a PMMA standard with the connectivity H–[CH$_2$–C(CH$_3$)CO$_2$CH$_3$]$_n$–H (m/z 2010), acquired using (a) a Bruker Esquire-LC QIT equipped with ESI and (b) a Waters Q/ToF Ultima mass spectrometer equipped with MALDI. The complete isotopic envelope was mass-selected in both cases. Essentially the same fragments with very similar relative abundances are observed with both methods in the m/z 600–2000 region. See section on mechanisms for nomenclature and structures of the fragments.

without which the underlying fragmentation mechanism cannot be elucidated unequivocally [161]. Moreover, the radical ions also reveal end-group information and, if detected, would provide this insight even in a spectrum with inferior signal/noise ratio, in which the middle- and high-mass ranges would be buried below noise level.

Low-mass discrimination is partly bypassed by using the pulsed Q CAD technique [162]; the precursor ion is activated at high q_z with a short (~0.1 ms), high-energy rf pulse and, after a ~0.1 ms delay time, q_z is lowered to trap the fragments. Similar improvement is observed with high amplitude short time excitation (HASTE) at more conventional q_z values (~0.25) [163]; for this, the auxiliary rf voltage is turned on at an increased amplitude for 1–2 ms (vs. the normal 20–40 ms), and afterward the main (trapping) rf voltage is lowered so that fragment ions below the low-mass cutoff remain trapped. Alternatively, fragmentation can be effected by photodissociation (PD) or ETD, which does not alter the precursor ion kinetic energy and, hence, does not cause the low-mass cutoff associated with CAD. The Q and HASTE CAD techniques as well as PD and ETD have not yet been applied to synthetic polymers.

Other shortcomings of QITs are their limited dynamic range, resolution, and mass accuracy, primarily resulting from space charge effects [140]. Linear (2-D) traps can store more ions than conventional 3-D traps, leading to a significantly increased dynamic range [164–166]. Moreover, the trapping efficiency is improved in the 2-D versus the 3-D configuration, also for lower-mass ions, which increases the sensitivity and lowers the detection limits [164–166]. On the other hand, mass resolution and accuracy are increased dramatically by extending a QIT into a QIT/ToF mass spectrometer [167–170]. In these, only recently commercialized hybrid instruments, the QIT is used for ion accumulation and storage, and as a reaction region for MS^n experiments, while the ToF mass spectrometer is used for mass analysis of the ions injected or formed in the trap. The QIT/ToF and Q/ToF designs are conceptually very similar, both having the high resolution and mass accuracy of a ToF mass analyzer (in MS as well as tandem MS mode), but only the former also is suitable for multistage MS studies. MALDI and ESI sources can be attached to QIT/ToF mass spectrometers [167–169], and a most recent application to PEG and poly(propylene glycol) oligomers [171] affirms their enhanced fragment ion resolution and mass accuracy.

Increased resolution and mass accuracy have also been achieved by interfacing a linear (2-D) ion trap with the Orbitrap mass analyzer, another newly introduced innovation [172, 173]. Also in this case, the trap is used for ion storage and/or MS^n experiments. The final product ions are ejected from the 2-D trap into a so-called C-trap, a device that converts beams into short pulses that are transferred orthogonally to the Orbitrap [174]. The Orbitrap is an electrostatic ion trap composed of coaxial inner and outer electrodes that resemble a spindle and a barrel, respectively. Ions trapped in the Orbitrap spin around the center electrode, inducing an image current on the outer electrode, which is measured over a certain time interval, amplified, and converted to a frequency spectrum via Fourier transformation. Since the frequency of oscillation of an ion along the center electrode (axial oscillation) is inversely proportional to $(m/z)^{0.5}$, frequency is easily converted to m/z to yield a mass spectrum. Ion detection in the Orbitrap is very similar to that in the magnetic FTICR trap (vide infra and Chapter 4). Resolutions up to 100,000 and mass accuracies better than 0.1 ppm have been demonstrated in analyses of peptides and metabolites in the 500–1000 Da range [172–174]. The (2-D) QIT/Orbitrap hybrid has not been used yet in polymer studies, but future applications are highly probable, given the high-mass accuracy and high resolution of this instrument and its lower acquisition and maintenance cost as compared with an FTICR mass spectrometer.

5.2.8   FTICR Mass Spectrometers

FTICR mass spectrometry (FTICR MS or FTMS) combines exceptionally high resolution and mass accuracy for precursor ion selection and fragment ion analysis with the capability to perform multiple tandem-in-time mass spectrometry (MS^n) experiments using a variety of activation methods, including CAD, IRMPD, SID, and ECD [39, 57, 62, 67]. Since both MALDI and ESI can be interfaced with FTMS, this method has been used in a number of MS/MS polymer characterizations and, because of the described advantages, increased use in the polymer science field is anticipated [8, 72–74, 175–182]. The FTMS methodology and its application to polymers are covered in considerable detail in Chapter 4.

5.3   FRAGMENT ION NOTATION FOR SPECTRAL PRESENTATION AND INTERPRETATION

The need for a consistent and systematic nomenclature for polymer fragmentation has recently been recognized by the polymer MS community [183]. Nomenclature schemes have been established for the fragment ions generated upon MS/MS of most biopolymer classes, inter alia peptides and proteins [184, 185], carbohydrates [186], nucleic acids [187], and lipids [188]. Such schemes have facilitated spectral interpretation, the derivation of dissociation mechanisms, and comparisons of data from different studies, and have also provided more clarity in the communication and discussion of experimental results. MS/MS of synthetic polymers would similarly benefit from a meaningful nomenclature that relates the acronyms used to specific structural features. Synthetic macromolecules span a diverse range of compounds, each characterized by a unique backbone connectivity, set of end groups, and (if applicable) branching architecture. Since the types of bonds that can be cleaved during an MS/MS experiment vary among different polymer classes, each one must have its own fragment nomenclature, as is also true with biopolymers. Expanding upon a recent proposal for a common fragment ion notation in the mass spectra of synthetic polymers [189], this chapter provides guidelines for naming the MS/MS fragments from linear polymers.

5.3.1   Linear Polymers with Defined Initiating (α) and Terminating (ω) End Groups

Figure 5.13 illustrates a straightforward nomenclature scheme for polymer chains with defined initiating (α) and terminating (ω) end groups [189]. Such polymers are

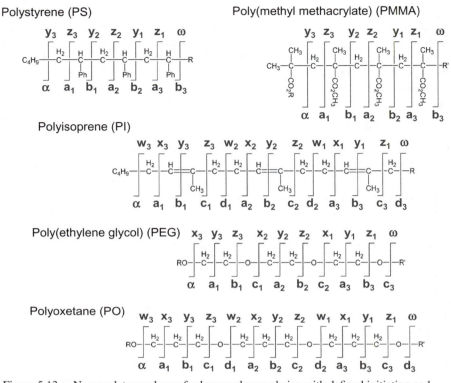

Figure 5.13.   Nomenclature scheme for homopolymer chains with defined initiating and terminating chain ends; α and ω designate ions containing the corresponding end groups only and no piece of the backbone. Most commonly R and R' are hydrogen atoms or alkyl groups, but can also be functionalized substituents. Ph abbreviates a phenyl (C_6H_5) substituent. In PMMAs with R = CH_3, the α chain end becomes identical to H + repeat unit.

usually produced by living polymerization techniques or by radical polymerizations. Fragments that retain the initiating chain end are designated by lowercase letters from the beginning of the alphabet, for example, a_n, b_n, c_n, and d_n, with the number of letters used being equal to the number of different bonds in the polymer chain. Conversely, fragments carrying the terminating (ω) chain end are designated by lowercase letters from the end of the alphabet, for example, z_n, y_n, x_n, and w_n. The subscripted $_n$ gives the number of complete or partial repeat units contained in the fragment; a_n, b_n, c_n, d_n, etc. are counted from the initiating chain end while z_n, y_n, x_n, w_n, etc. are counted from the terminating chain end, as shown in Figure 5.13 for polymers with two to four different backbone bonds. If a fragment ion contains only the α or ω end group (or pieces thereof), but no part of the polymer backbone, it is given the notation α or ω, respectively (Figure 5.13). The letter symbols include the ion charge, which can be a metal ion, the proton, or a deprotonated site, and whose specific identity is commonly provided in the MS/MS spectrum shown or the corresponding figure caption.

The described nomenclature is similar with the one used for fragment ions from ionized peptides [184, 185]. Peptides consistently have three different backbone bonds, however; as a result, letters a–c and x–z suffice to name their fragment ions from backbone cleavages, leaving other letters of the alphabet for the products of alternative fragmentations, for example, combined backbone and side chain cleavages (d, v, w). In synthetic polymers, the number of backbone bonds varies (Figure 5.13), requiring the use of more letters for backbone cleavages and of a different nomenclature for side chain eliminations (vide infra). As with peptides, dissociation of a polymer backbone bond creates fragments with one original and one new chain end. The structure of the newly generated chain end is indicated by appropriate subscripts or superscripts, added to the notation of Figure 5.13 according to the rules listed below. A major objective is to keep the notation as concise and informative as possible. The examples given to illustrate the nomenclature refer to the oligomers depicted in Figure 5.13 and assume ionization by Ag^+ for polystyrene (PS) and polyisoprene (PI), or Na^+ for PMMA, PEG, and polyoxetane (PO), unless noted otherwise:

(a) A fragment ion with an unpaired electron (radical ion, distonic ion) is designated by a superscripted radical. For example, $b_2{}^{\bullet}$ from the polystyrene shown in Figure 5.13 represents the radical ion $[C_4H_9–CH_2CH(Ph)–CH_2CH(Ph)]^{\bullet} + Ag]^+$; similarly, $z_2{}^{\bullet}$ from the polyoxetane is the radical ion $[^{\bullet}O–CH_2CH_2CH_2O–R' + Na]^+$.

(b) A fragment with one H atom less than the radical ions mentioned under (a) is designated by a single letter code with no further addition. These fragments have unsaturated chain ends and nominally arise by cleavage of a backbone bond followed by β scission of an $H^{\bullet}$ radical. For example, the b_2 and z_2 fragments from the PS and PO mentioned in (a) have the structures $[C_4H_9–CH_2CH(Ph)–CH=CH(Ph) + Ag]^+$ and $[O=CHCH_2CH_2O–R' + Na]^+$, respectively.

(c) A double prime is used to designate fragment ions that gained a saturated chain end, that is, if they have two more H atoms than the closed-shell ions mentioned under (b). Thus, $b_2{}''$ and $z_2{}''$ from the discussed PS and PO would be $[C_4H_9–CH_2CH(Ph)–CH_2CH_2(Ph) + Ag]^+$ and $[HO–CH_2CH_2CH_2O–R' + Na]^+$, respectively.

(d) If side chains (pendants) are present, unsaturated chain ends can also be formed by loss of a side chain. This is indicated by appropriate subscripts. For example, $[C_4H_9–CH_2C(CH_3)(Ph)–CH_2C(CH_3)(Ph)–CH_2{}^{\bullet} + Ag]^+$, which is the $a_3{}^{\bullet}$ radical ion from a silverated poly(α-methylstyrene), can lose $CH_3{}^{\bullet}$ or $Ph^{\bullet}$ radicals to yield a terminal double bond. The corresponding products, $[C_4H_9–CH_2C(CH_3)(Ph)–CH_2C(Ph)=CH_2 + Ag]^+$ and $[C_4H_9–CH_2(CH_3)(Ph)–CH_2C(CH_3)=CH_2 + Ag]^+$, are termed a_{3a} and a_{3b}, respectively, that is, the loss of the smaller side chain is indicated by the subscript $_a$ and that of the larger side chain by the subscript $_b$. If the smaller substituent is an H atom, the subscripted $_a$ is omitted and fragment ion notation falls under rule (b). For example, $a_3{}^{\bullet}$ from the PS shown in Figure 5.13, viz. $[C_4H_9–CH_2CH(Ph)–CH_2CH(Ph)–$

$CH_2^\bullet + Ag]^+$, can form a terminal double bond by loss of $H^\bullet$ or $Ph^\bullet$ radicals, which lead to the fragment ions $[C_4H_9-CH_2CH(Ph)-CH_2C(Ph)=CH_2 + Ag]^+$ (a_3) and $[C_4H_9-CH_2CH(Ph)-CH_2CH=CH_2 + Ag]^+$ (a_{3b}), respectively. It is worth noting at this point that $a_n^\bullet$ ions from silverated poly(α-methylstyrene), viz. $[C_4H_9-\{CH_2C(CH_3)(Ph)\}_{n-2}-CH_2C(CH_3)(Ph)-CH_2^\bullet + Ag]^+$, not only give rise to a_{na} ($CH_3^\bullet$ loss) and a_{nb} ($Ph^\bullet$ loss) fragments, but also (after rearrangement) to a_n fragments via nominal $H^\bullet$ losses (vide infra).

(e) The described notation provides information about the structure of a fragment ion, but not necessarily about the corresponding reaction mechanism. Thus, the PS fragment $[C_4H_9-CH_2CH(Ph)-CH_2C(Ph)=CH_2 + Ag]^+$ (a_3) can arise by β scission of an $H^\bullet$ radical from $a_3^\bullet$, as explained in (d), or from a larger $b_n^\bullet$ radical ion after a backbiting rearrangement (see section on mechanisms). Further, the rules outlined here aid the naming of a fragment ion, but do not specify the exact dissociation mechanism, that is, the losses discussed are nominal, but not automatically actual. Independent of the exact mechanism, the resulting fragment ion carries the same letter code as the corresponding radical ion generated from homolytic cleavage at the same backbone bond, as exemplified in (d) for a-type fragments from poly(α-methylstyrene).

(f) Consecutive fragmentations often lead to internal fragments, that is, fragments missing both original end groups. Internal fragment ions are designated by capital letters near the center of the alphabet (J, K, L, etc.). They often have 2–3 repeat units and can be radical ions (e.g., $J_2^\bullet$) or closed-shell species (e.g., K_3).

(g) If the initiating and terminating chain ends are indistinguishable and the backbone has a symmetrical structure, the number of different backbone fragment ions is reduced. This situation is encountered frequently with polyethers. For the PEG shown in Figure 5.13, $a_n = y_n$, $b_n = x_n$, and $c_n = z_{n+1}$ if R = R'. Analogously, for the PO in Figure 5.13, $a_n = y_n$, $b_n = x_n$, $c_n = w_n$, and $d_n = z_{n+1}$ if R = R'. Due to this duplication, only one set of fragments (a_n, b_n, c_n, d_n, etc.) needs to be considered.

(h) A superscripted $^+$ is added to fragment ions produced from metalated precursor ions, which do not contain the metal ion. For example, b_1^+ from a silverated PS with the structure given in Figure 5.13 corresponds to the carbocation $C_4H_9-CH_2CH(Ph)^+$.

5.3.2 Linear Polymers with Undefined End Groups

The situation is more complex with linear polymers with undefined initiating and terminating substituents, such as condensation polymers. Two industrially important classes of polymers, polyesters and polyamides, usually fall under this category. Depending on the synthetic procedure and molar amounts of reagents used, these macromolecules may bear identical or different chain ends; for example, polyesters can be prepared with diacid, acid/alcohol, or dialcohol end groups. CAD of condensation polymers mainly leads to linear fragments with the same or new end group combinations and to cyclic fragments with different repeat unit content. Linear and cyclic fragments have been designated by the acronyms I_n^{AB} and C_n^{AB}, respectively, where l and c differentiate between the linear and cyclic architecture, the subscripted $_n$ gives the number of repeat units in the fragment, and the superscripted ^{AB} gives the chain ends in a linear or the combining ends in a cyclic fragment [156].

5.4 POLYMER FRAGMENTATION MECHANISMS

The vast majority of polymer ions are closed-shell species, that is, they bear no unpaired electrons. Such ions can dissociate via homolytic cleavages, which create radical intermediates, or via rearrangements, which involve concomitant bond cleavage and bond formation and lead to closed-shell products. The radical intermediates generated in homolytic cleavages may yield fragments that retain the radical site, or expel radicals to form closed-shell fragments. The kinetics of these competitive processes, which depend on the structure of the precursor ion, and the internal energy available, determine which reaction pathways predominate [58].

CAD encompasses charge-induced and charge-remote reactions. In charge-induced or charge-directed reactions, bond cleavage and bond formation take place at the charge site, that is, the charge (atom or group of atoms carrying the charge) is directly involved in the fragmentation process. In contrast, charge-remote reactions involve bond cleavage and bond formation at locations that are not directly bonded to the charge site. Numerous studies on peptide and fatty acid ions have shown that protonated species mainly undergo charge-induced dissociations, whereas deprotonated and metalated species decompose through a combination of charge-induced and charge-remote processes, the proportion of which is dictated by the structure of the precursor ion and its internal energy distribution [47, 190, 191]. These principles also apply to synthetic polymers. Examples of these types of fragmentations will be discussed in this section.

5.4.1 Charge-Induced Fragmentations

Figure 5.14 shows the ESI-CAD mass spectrum of the lithiated 9-mer from a PEG standard with HO/H end groups (m/z 421 from PEG 600). The spectrum was acquired using a QIT, in which the average internal energies deposited on CAD are relatively low (see Section 5.2.7) [192]. Two homologous series of fragments, marked c_n'' and b_n, are clearly discerned. The more prominent series (c_n'') arises by nominal losses of 44n Da from the precursor ion, and the less abundant one by nominal losses of 44n + 18 Da.

The fragments observed in this case (Figure 5.14) are readily rationalized by charge-directed dissociation, promoted by the Li$^+$ Lewis acid, as depicted in Scheme 5.2. Li$^+$ coordination at the oxygens weakens the adjacent C–O bonds, facilitating their dissociation. Cleavage of a C–O bond, followed by an energetically favorable 1,2-hydride shift (rH^-) [58], produces an ion–dipole complex between an alkoxyethyl cation and a lithium alkoxylate. Proton transfer within this complex, from the positively charged ethyl group to the basic alkoxylate (rH^+) [193], followed by

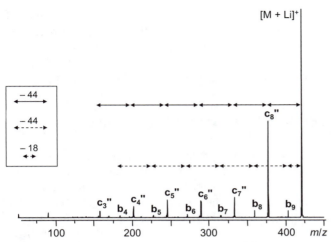

Figure 5.14. ESI-CAD mass spectrum of the Li⁺ adduct of HO–(CH₂CH₂O)₉–H (m/z 421), acquired on a quadrupole ion trap from PEG 600. Courtesy of Panthida Thomya [192].

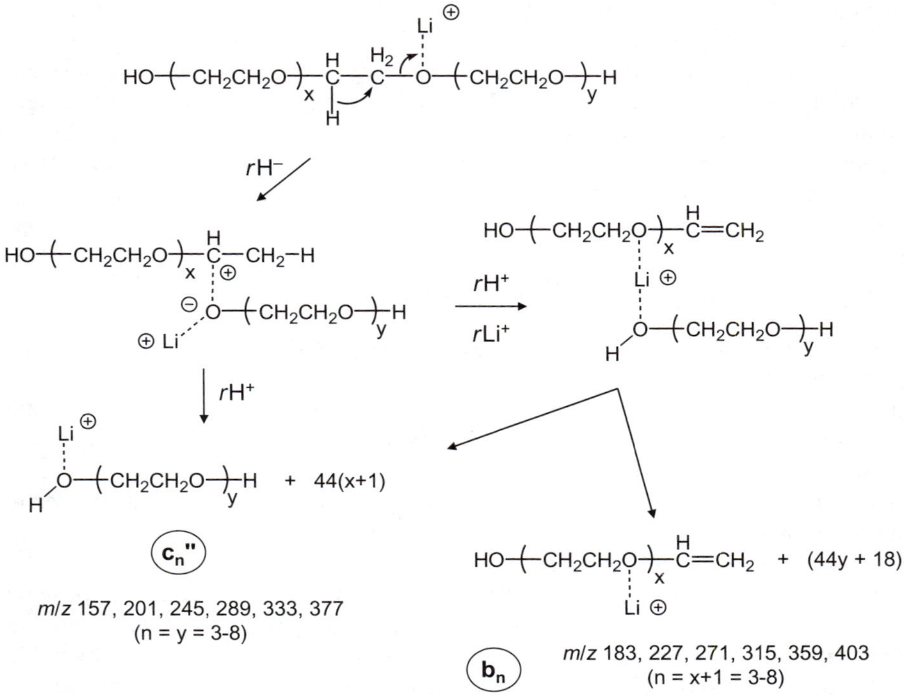

Scheme 5.2. Charge-induced fragmentation pathway of $[HO–(CH_2CH_2O)_n–H + Li]^+$ to truncated linear fragments with HO/H (c_n'') and HO/vinyl (b_n) end groups.

dissociation, gives rise to the major fragment ion series, c_n'', composed of truncated PEG oligomers with HO/H chain ends. Alternatively, proton transfer may be accompanied by rearrangement of the ligands (rLi$^+$) to form a Li$^+$-bound complex between two truncated PEG chains with HO/vinyl and HO/H end groups. The latter complex can dissociate to yield either the c_n'' or the b_n series. Since the PEG ligands in the Li$^+$-bound dimer have comparable structures, they should have similar Li$^+$ affinities, leading to approximately equal abundances for the c_n'' and b_n ions generated through this channel. The much higher relative abundances of the c_n'' ions point out that the principal route of the lithiated PEG 9-mer to these fragments must be the direct dissociation of the initially formed ion–dipole complex (left part of Scheme 5.2). A particularly high abundance is noticed for the largest c_n'' fragment, which arises by the nominal loss of one monomer unit (44 Da). The mechanism of this reaction is still under investigation; possibly, it is catalyzed by the terminal OH proton.

Charge-induced reactions are generally associated with lower critical energies than charge-remote reactions [194]. Consequently, they proceed most efficiently at CAD conditions that deposit low internal energies, which are prevalent with QITs (vide supra). Charge-induced dissociations have been observed with both cations and anions. The charge can be viewed as an electrophile in the former and a nucleophile in the latter, promoting Lewis acid chemistry and nucleophilic displacements, respectively.

5.4.2   Charge-Remote Rearrangements

A charge-remote rearrangement describes a reaction that involves concerted bond making and bond breaking without the intermediacy of radical species. In principle, such a reaction can take place in the absence of the charge and may also occur upon the thermal or photochemical activation of the substance under study. Commonly observed charge-remote rearrangements are intramolecular transesterifications and related displacements, as well as processes falling under the category of pericyclic reactions, which include retro-cyclcoadditions, sigmatropic rearrangements, and atom/group transfer reactions (e.g., retro-ene reactions and β eliminations via cyclic transition states) [195].

The elimination of isobutylene from the side chains of poly(t-butyl methacrylate), Figure 5.15, provides an illustrative example of a charge-remote rearrangement [196]. This reaction can be accounted for by a β elimination involving hydrogen transfer to a carbonyl oxygen via an energetically favorable six-membered ring transition state, as shown in Scheme 5.3. Sequential i-C$_4$H$_8$ losses are the only significant dissociations observed. The lack of backbone cleavages, which proceed efficiently from other collisionally activated methacrylates (see Figures 5.2 and 5.4), provides evidence that the consecutive charge-remote H rearrangements proceed over substantially lower barriers than any competitive dissociations.

Most polymer ion MS/MS studies reported thus far have invoked charge-remote rearrangements to rationalize the observed fragmentation patterns. Such mechanisms have been proposed for the dissociations of, inter alia, polyglycols [47, 48], polyesters [75, 125, 156, 177], polyamides [124], polystyrenes [51], poly(methyl methacrylate)s [50], and poly(dimethyl siloxane)s [192, 197, 198].

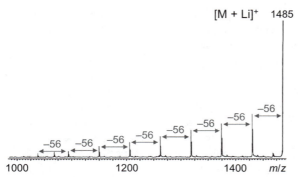

[M + Li]⁺ 1485

Figure 5.15.   MALDI-CAD mass spectrum of the lithiated 10-mer from a poly(t-butyl methacrylate) with the connectivity $C_4H_9-[CH_2C(CH_3)(COO-tC_4H_9)]_n-H$ (m/z 1485), acquired using a Q/ToF tandem mass spectrometer and Ar as the collision gas at $E_{lab} = 60$ eV. The losses of 9 and 10 i-C_4H_8 molecules are also observed with abundances near noise level. Courtesy of Kittisak Chaicharoen [196].

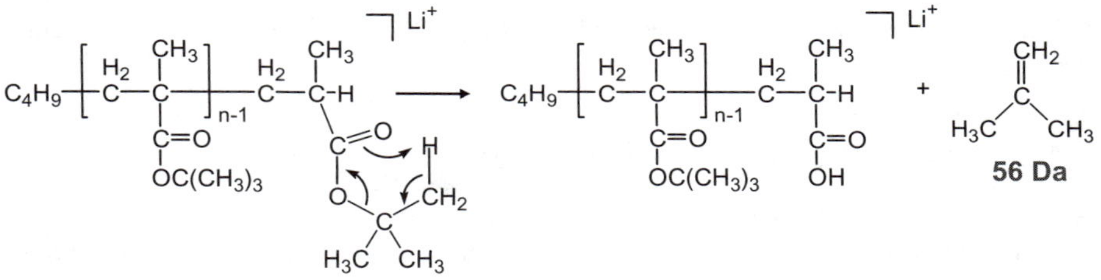

Scheme 5.3.   Elimination of isobutylene from the side chains of poly(t-butyl methacrylate) via charge-remote hydrogen rearrangement.

The rearrangement shown in Scheme 5.3 is localized in the side chains. In other types of metalated polymers, nominal H rearrangements from the backbone to other sites within the backbone or at the side chains are possible, and these reactions can lead to breakup of the polymer frame [47, 48, 50, 51, 125]. Commonly, two smaller oligomers are created this way, either of which may maintain the charge. In a number of studies, it has been found that one of the expected fragments is either not formed or present with minuscule abundance [50, 51, 125]. Such a result is usually attributed to large differences in the metal ion affinities of the two pieces resulting from the rearrangement. That the metal ion affinities of oligomers with many common functional groups and differing only in their chain ends are so different that one yields a dominant fragment and the other none or a minor one, is however improbable and calls for a revision in the proposed mechanism. For example, charge-remote rearrangements cannot explain adequately the fragmentation behavior of substituted polyethylenes (PS, PMMA, etc.). The following section discusses alternative pathways via radical intermediates, which have been considered much less frequently thus far, but can adequately rationalize the MS/MS characteristics of several polymer classes.

5.4.3 Charge-Remote Fragmentations via Radical Intermediates

The CAD spectra of metalated polyether, polydiene, polystyrene, and polyacrylate oligomers acquired with sector/ToF, Q/ToF, and ToF/ToF instrumentation contain sizable peaks at low masses corresponding to radical ions, as documented in Figures 5.4, 5.6, 5.12, and 5.16 [49–51, 161, 199–204]. The appearance of such products clearly demonstrates that homolytic bond cleavages take place in collisionally excited polymer ions.

CAD induces fragmentation by increasing the rovibrational energy of the system, in complete analogy to thermal degradation which, with the mentioned polymer classes, is known to proceed through radical intermediates [205–208]. Indeed, a number of abundant fragments in the CAD spectra of polystyrene and polyacrylate oligomers have been observed in pyrolyses of such compounds [205–208], consistent with charge-remote fragmentation based on principles of radical chemistry in the mass spectrometer. The dissociations of silverated PS oligomers with the structure $C_4H_9-[CH_2CH(Ph)]_n-H$ will be discussed in detail to illustrate the types of radical reactions occurring at the internal energies and time scale accessible in MS/MS experiments [201–203]. As will be shown, all CAD fragments (Figure 5.16) can be rationalized by random homolytic cleavages followed by β scissions and backbiting rearrangements, which are typical radical site reactions.

Random homolyses along the PS backbone lead to four types of incipient radical ions, which are depicted in Figure 5.17. Two contain the initiating chain end

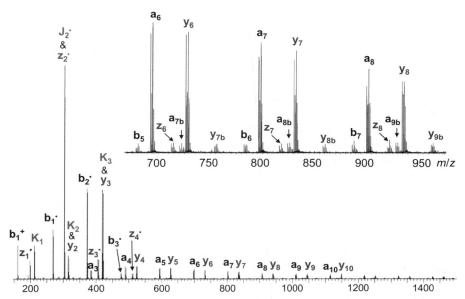

Figure 5.16.   MALDI-CAD spectrum of the silverated 18-mer from a polystyrene standard with the connectivity $C_4H_9-[CH_2CH(Ph)]_n-H$ (m/z 2038), acquired with a Q/ToF tandem mass spectrometer and Ar collision gas at $E_{lab} = 130\,eV$. Reproduced from Reference 202 with permission from the American Chemical Society via the Copyright Clearance Center.

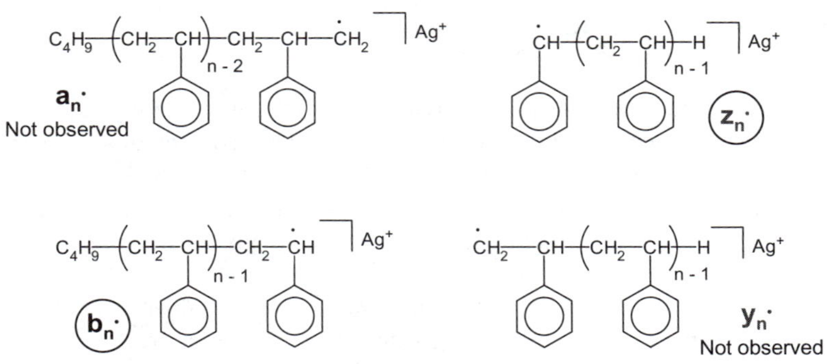

Figure 5.17. Charged radicals produced via homolytic cleavages in the backbone of a polystyrene with the connectivity $C_4H_9-[CH_2CH(Ph)]_n-H$. Adapted from Reference 202 with permission from the American Chemical Society via the Copyright Clearance Center.

$(a_n{}^\bullet, b_n{}^\bullet)$ and two the terminating chain end $(y_n{}^\bullet, z_n{}^\bullet)$; further, two of these radicals contain benzylic radical sites $(b_n{}^\bullet, z_n{}^\bullet)$, while the other two have primary $CH_2{}^\bullet$ termini $(a_n{}^\bullet, y_n{}^\bullet)$. Only small $(n = 1–4)$ benzylic radical ions are detected in the CAD spectrum (Figure 5.16), indicating that all primary radicals as well as the larger benzylic radicals readily lose the radical center via consecutive dissociations. The main fragmentation channels of radicals are bond scissions in β position to the radical site and 1,5-H$^\bullet$ rearrangements via six-membered ring transition states ("backbiting") [209]; β cleavage of a hydrogen atom would yield closed-shell a_n, b_n, y_n, and z_n fragments, whereas β cleavage of a phenyl group, which is possible from $a_n{}^\bullet$ and $y_n{}^\bullet$ only, would yield a_{nb} and y_{nb} fragments, all of which are observed (cf. inset of Figure 5.16). Note, however, that fragments a_n and y_n, which dominate the medium- and high-mass range of the CAD spectrum, are considerably more abundant than the remaining β scission products (viz. b_n, z_n, a_{nb}, and y_{nb}). The much higher proportion of a_n and y_n is the result of backbiting in the incipient $b_n{}^\bullet$ and $z_n{}^\bullet$ radical ions, where a benzylic H atom can be abstracted by the original radical site over a six-membered ring to give more stable, tertiary benzylic radicals. The backbiting process is illustrated for the $a_n{}^\bullet$ radicals in Scheme 5.4 [201–203].

Backbiting moves the radical site into an interior position, where it can promote β C–C scissions on either side of the PS chain (see Scheme 5.4). One of these scissions generates the internal radical ion $J_2{}^\bullet$ $(m/z$ 302) and an a_n fragment; the other scission generates the internal trimer K_3 $(m/z$ 419) and a smaller benzylic radical $(b_{n-3}{}^\bullet)$. Similar backbiting on the $z_n{}^\bullet$ radical (not shown) gives rise to the same internal fragments $J_2{}^\bullet$ and K_3, to a y_n fragment, and to a smaller z-type radical $(z_{n-3}{}^\bullet)$. The smaller benzylic radicals can undergo renewed backbiting, until their size is reduced to the point that a further step becomes impossible or energetically not feasible; this, in turn, explains the observation of solely small benzylic radicals $(n = 1–4)$ with quite high relative abundances.

Because polystyrene $C_4H_9-[CH_2CH(Ph)]_n-H$ does not possess a unique ω end group, the internal ions $J_2{}^\bullet$ $(m/z$ 302) and K_3 $(m/z$ 419) are identical with the terminal ions $z_2{}^\bullet$ and y_3, respectively. The overlapping fragment ions become separable if the ω chain end is functionalized; in the MALDI-CAD spectrum of silverated

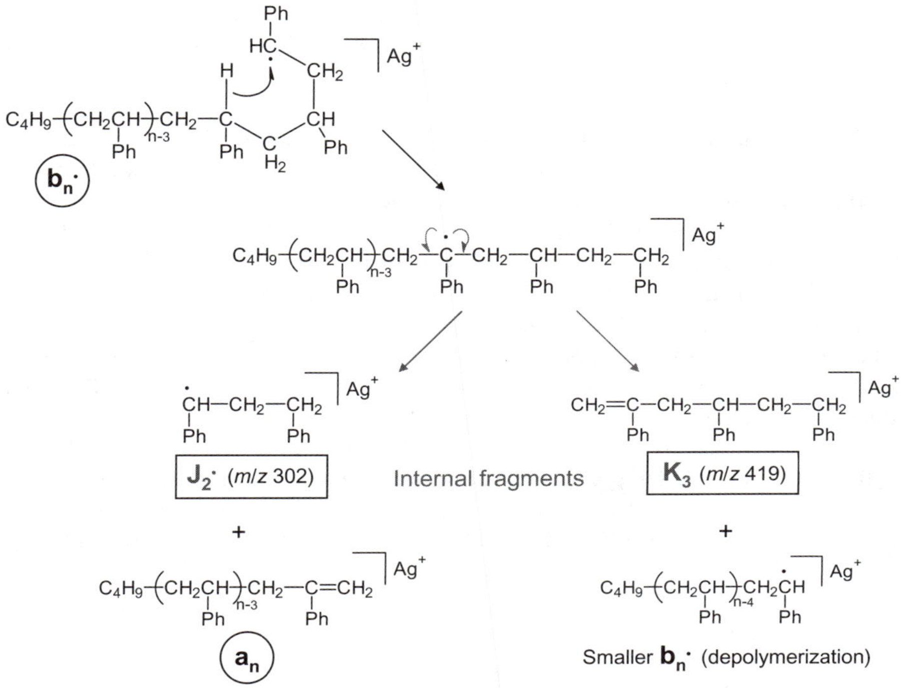

Scheme 5.4. Backbiting in the incipient benzylic radical ions, arising from random homolytic cleavages of a polystyrene chain with C_4H_9/H end groups. Reproduced from Reference 202 with permission from the American Chemical Society via the Copyright Clearance Center.

C_4H_9–$[CH_2CH(Ph)]_n$–CH_2CH_2OH, the internal $J_2{}^{\bullet}$ and K_3 fragments remain at m/z 302 and 419, respectively, while the terminal fragments $z_2{}^{\bullet}$ and y_3 move $44\, m/z$ units higher, now appearing at m/z 346 and 463, respectively. With the ω hydroxylated polystyrene, $J_2{}^{\bullet}$ is more abundant than $z_2{}^{\bullet}$ (by ~1.5 times) and K_3 is more abundant than y_3 (by ~7 times) [201, 202]. These trends suggest that the predominant components of the composite peaks in the CAD spectrum of silverated C_4H_9–$[CH_2CH(Ph)]_n$–H (Figure 5.16) are the internal fragments $J_2{}^{\bullet}$ and K_3. This is not surprising considering that $J_2{}^{\bullet}$ and K_3 could be produced in *every* backbiting cycle, if charge is retained on these fragments.

In addition to K_3 (silverated styrene trimer), the CAD spectrum also contains silverated styrene monomer (K_1) and dimer (K_2). The monomer ion, C_8H_8–Ag^+ (m/z 211), presumably originates from styrene molecules released by depolymerization of the initially formed radical ions $a_n{}^{\bullet}$, $b_n{}^{\bullet}$, $y_n{}^{\bullet}$, and $z_n{}^{\bullet}$. K_2 can be rationalized by two consecutive backbiting events, involving successive 1,7- and 7,3-H rearrangements (7,3- is a reverse 1,5-H rearrangement). These reactions transform the terminal benzylic radicals $b_n{}^{\bullet}/z_n{}^{\bullet}$ to the isomeric benzylic radicals C_4H_9–$[CH_2CH(Ph)]_{n-2}$–$CH_2C^{\bullet}(Ph)$–$CH_2CH_2(Ph)/(Ph)CH_2$–$CH_2C^{\bullet}(Ph)$–$[CH_2CH(Ph)]_{n-2}$–H, respectively, which can decompose via β C–C bond scission to K_2, $[CH_2{=}C(Ph)CH_2CH_2(Ph) + Ag]^+$ (m/z 315) [202].

The backbiting process outlined in Scheme 5.4 is disabled in the poly(α-methylstyrene) frame, C_4H_9–$[CH_2C(CH_3)(Ph)]_n$–H (PαMeS), which lacks the benzylic hydrogens necessary for this rearrangement. As a result, the backbiting signature ions, viz. the internal fragments $J_2^{\bullet}$ and K_3 (see Scheme 5.4), are not observed in the MALDI-CAD spectrum of the corresponding Ag^+ adduct (Figure 5.18). Small, terminal radical ions ($b_1^{\bullet}$, $b_2^{\bullet}$, $z_1^{\bullet}$, $z_2^{\bullet}$) are now the dominant fragments. Surprisingly, the medium- and high-mass range of the spectrum mainly shows a_n and y_n peaks, similar to PS, where these ions were largely created by backbiting (vide supra). These a_n and y_n fragments cannot be formed by β-H$^{\bullet}$ loss from the respective radical ions $a_n^{\bullet}$ and $y_n^{\bullet}$ because the structures of the latter, C_4H_9–$[CH_2C(CH_3)(Ph)]_{n-1}$–$CH_2^{\bullet}$ ($a_n^{\bullet}$) and $^{\bullet}CH_2C(CH_3)(Ph)$–$[CH_2C(CH_3)(Ph)]_{n-1}$–H ($y_n^{\bullet}$), do not carry an H atom in β position to the radical site. The a_n and y_n fragments can, however, be reconciled if the primary radicals $a_n^{\bullet}$ and $y_n^{\bullet}$ undergo the 1,2-phenyl shift shown in Scheme 5.5 [201–203, 210, 211].

The fragmentation characteristics of silverated PS vs. PαMeS are duplicated with lithiated or sodiated poly(methyl acrylate) (PMA) versus PMMA, pointing out that free radical chemistry could be the preferred dissociation avenues for many other substituted polyethylenes. Free radical reactions can account for MS/MS frag-

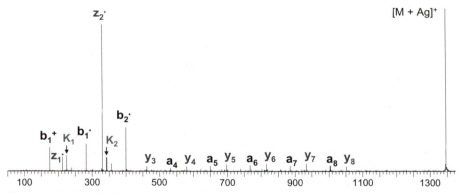

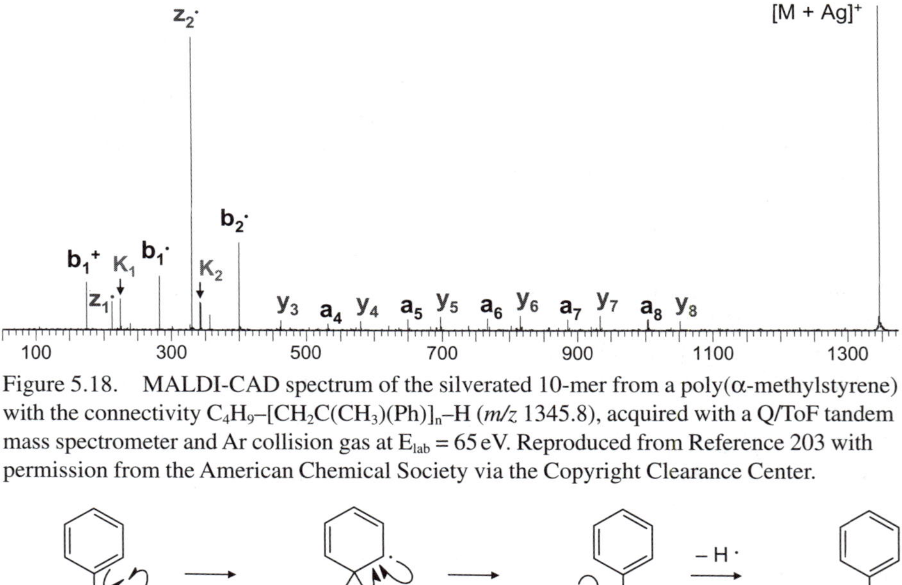

Figure 5.18. MALDI-CAD spectrum of the silverated 10-mer from a poly(α-methylstyrene) with the connectivity C_4H_9–$[CH_2C(CH_3)(Ph)]_n$–H (m/z 1345.8), acquired with a Q/ToF tandem mass spectrometer and Ar collision gas at $E_{lab} = 65$ eV. Reproduced from Reference 203 with permission from the American Chemical Society via the Copyright Clearance Center.

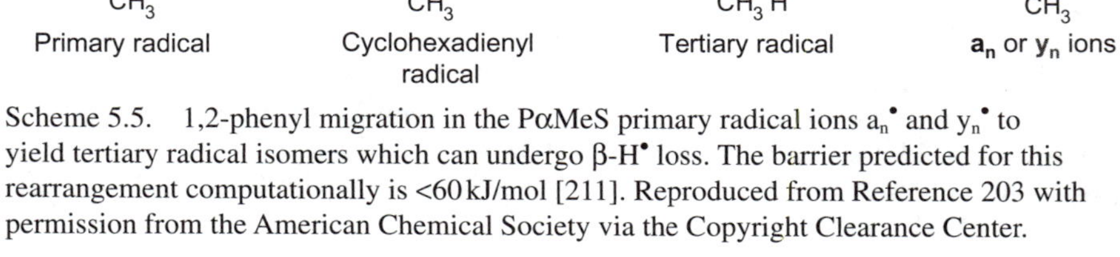

Scheme 5.5. 1,2-phenyl migration in the PαMeS primary radical ions $a_n^{\bullet}$ and $y_n^{\bullet}$ to yield tertiary radical isomers which can undergo β-H$^{\bullet}$ loss. The barrier predicted for this rearrangement computationally is <60 kJ/mol [211]. Reproduced from Reference 203 with permission from the American Chemical Society via the Copyright Clearance Center.

mentation of metalated polydienes [200] and fluorinated polyethers [36] as well. Based on research published so far, it appears that dissociations via radical intermediates are associated with considerable barriers, which may suppress such reactions if the structure of the polymer under study allows for alternative charge-induced processes or charge-remote rearrangements. For example, PMMA oligomers decompose via radical intermediates upon CAD, as noted above, but poly(t-butyl methacrylate) oligomers do not (see Section 5.4.2), as their structure permits energetically more favorable rearrangement pathways. When free radical chemistry is in competition with other decomposition mechanisms, it is most pronounced at higher internal energies; this is the case with metalated PEGs, which mainly fragment via ion–dipole complexes at the lowest internal energies (see Section 5.4.1), but increasingly via radical intermediates as the internal energy is raised.

5.5   OUTLOOK

Instruments of many different configurations are now available for MS/MS studies on synthetic polymers ionized by MALDI or ESI. The experiments reported thus far have focused on surveying the fragments generated from polymer ions and assessing (primarily) the end group information that they provide. The exact fragmentation mechanisms of many classes of polymer ions still remain largely unknown and the determinants for preference of particular dissociation pathways are not yet well understood. Fortunately, new insight on these latter issues is emerging, as they are essential for choosing the proper MS/MS method and conditions, and for correctly interpreting the resulting spectra in order to deduce structural information. The currently available instrumentation allows for a variety of experiments to ascertain the mechanisms, kinetics, and energetics of polymer ion fragmentation, so that MS/MS can be applied with more confidence to the analysis of unknown polymer architectures and copolymer sequences, which are still formidable tasks to perform.

MS/MS studies via CAD have so far been limited to precursor ions below m/z 3000. Singly or multiply charged ions of larger mass-to-charge ratios do not yield detectable fragments due to poor fragmentation efficiencies. With increasing size, the energy deposited by collisional activation is redistributed over a larger number of degrees of freedom (vide supra), leading to unfavorable dissociation kinetics within the time window of MS analyses. Higher internal energies may be transferred by SID, which involves only one collision event and results in efficient activation of the precursor ion within picoseconds [62]. Another promising activation method is ECD and its closely related ETD variant (vide supra); these methods are capable of transferring large amounts of internal energy locally causing rapid, nonergodic dissociation; that is, fragmentation occurs before the deposited energy is randomized over the available degrees of freedom [65, 67]. Indeed, multiply protonated third-generation polyamidoamine (PAMAM) dendrimers with a mass of 7076.5 Da have been successfully fragmented by ECD [76]. Since ECD and ETD have become available on a variety of commercial instruments, future work in this area is expected and will certainly extend the applicability of MS/MS to larger and more complex polymers.

ACKNOWLEDGMENTS

The authors thank the National Science Foundation (CHE-0517909, CHE-0833087, and DMR-0821313) for generous financial support and David E. Dabney for helpful discussions. Research projects have also been partly supported by Lubrizol, Inc. and Omnova Solutions, Inc.

REFERENCES

1. Cowie, J.M.G., *Polymers: Chemistry & Physics of Modern Materials*. 2nd edition. London: Blackie Academic & Professional, **1991**.
2. Kennedy, J.P.; Iván, B., *Designed Polymers by Carbocationic Macromolecular Engineering*. Munich: Hanser Publishers, **1992**.
3. Hsieh, H.L.; Quirk, R.P., *Anionic Polymerization: Principles and Practical Applications*. New York: Marcel Dekker, **1996**.
4. Carraher, C.E., Jr., *Seymour/Carraher's Polymer Chemistry*. 5th edition. New York: Marcel Dekker, **2000**.
5. Rodriguez, F.; Cohen, C.; Ober, C.; Archer, L.A., *Principles of Polymer Systems*. 5th edition. New York: Taylor & Francis, **2003**.
6. Sheng, L.-S.; Shew, S.L.; Winger, B.E.; Campana, J.E., A new generation of mass spectrometry for characterizing polymers and related materials. In *Hyphenated Techniques in Polymer Chemistry: Thermal, Spectroscopic and Other Methods* (Provder, T.; Urban, M.W.; Barth, H.G., editors). Washington, DC: ACS, **1994**, pp. 55–72.
7. Scrivens, J.H., The characterization of synthetic polymer systems, *Adv. Mass Spectrom.*, **1995**, 13, 447–464.
8. Jackson, C.A.; Simonsick, W.J., Jr., Application of mass spectrometry to the characterization of polymers, *Curr. Opin. Solid State Mater. Sci.*, **1997**, 2, 661–667.
9. Saf, R.; Mirtl, C.; Hummel, K., Electrospray ionization mass spectrometry as an analytical tool for non-biological monomers, oligomers and polymers, *Acta Polym.*, **1997**, 48, 513–526.
10. Wu, K.J.; Odom, R.W., Characterizing synthetic polymers by MALDI MS, *Anal. Chem.*, **1998**, 70, 456A–461A.
11. Räder, H.J.; Schrepp, W., MALDI-TOF mass spectrometry in the analysis of synthetic polymers. *Acta Polym.*, **1998**, 49, 272–293.
12. Nielen, M.W.F., MALDI time-of-flight mass spectrometry of synthetic polymers, *Mass Spec. Rev.*, **1999**, 18, 309–344.
13. Pasch, H.; Ghahary, R., Analysis of complex polymers by MALDI-TOF mass spectrometry, *Macromol. Symp.*, **2000**, 152, 267–278.
14. Scrivens, J.H.; Jackson, A.T., Characterisation of synthetic polymer systems, *Int. J. Mass Spectrom.*, **2000**, 200, 261–276.
15. Haddleton, D.M., Mass spectrometry of polymers. In *Emerging Themes in Polymer Science* (Ryan, A.J., editor). Cambridge, UK: Royal Society of Chemistry, **2001**, pp. ••–••.
16. Hanton, S.D., Mass spectrometry of polymers and polymer surfaces, *Chem. Rev.*, **2001**, 101, 527–569.
17. Montaudo, G.; Lattimer, R.P. (editors), *Mass Spectrometry of Polymers*. Boca Raton, FL: CRC Press, **2002**.
18. McEwen, C.N.; Peacock, P.M., Mass spectrometry of chemical polymers, *Anal. Chem.*, **2002**, 74, 2743–2748.
19. Murgasova, R.; Hercules, D.M., Polymer characterization by combining liquid chromatography with MALDI and ESI mass spectrometry, *Anal. Bioanal. Chem.*, **2002**, 373, 481–489.
20. Macha, S.F.; Limbach, P.A., Matrix-assisted laser desorption/ionization (MALDI) mass spectrometry of polymers, *Curr. Opin. Solid State Mater. Sci.*, **2002**, 6, 213–220.

21. Pasch, H.; Schrepp, W., *MALDI-TOF Mass Spectrometry of Synthetic Polymers*. Berlin: Springer, **2003**.

22. Murgasova, R.; Hercules, D.M., MALDI of synthetic polymers—an update. *Int. J. Mass Spectrom.*, **2003**, 226, 151–162.

23. McEwen, C.N., Recent developments in polymer characterization using mass spectrometry, *Adv. Mass Spectrom.*, **2004**, 16, 215–227.

24. Peacock, P.M.; McEwen, C.N., Mass spectrometry of synthetic polymers. *Anal. Chem.* **2004**, 76, 3417–3428.

25. Montaudo, G.; Carroccio, S.; Montaudo, M.S.; Puglisi, C.; Samperi, F., Recent advances in MALDI mass spectrometry of polymers, *Macromol. Symp.*, **2004**, 218, 101–112.

26. Zhang, L.K.; Rempel, D.; Pramanik, B.N.; Gross, M.L., Accurate mass measurements by Fourier transform mass spectrometry, *Mass Spectrom. Rev*, **2005**, 24, 286–309.

27. Montaudo, G.; Samperi, F.; Montaudo, M.S.; Carroccio, S.; Puglisi, C., Current trends in matrix-assisted laser desorption/ionization of polymeric materials, *Eur. J. Mass Spectrom.*, **2005**, 11, 1–14.

28. Jagtap, R.N.; Ambre, A.H., Overview literature on matrix-assisted laser desorption ionization mass spectrometry (MALDI MS): basics and its applications in characterizing polymeric materials, *Bull. Mater. Sci.*, **2005**, 28, 515–528.

29. Montaudo, G.; Samperi, F.; Montaudo, M.S., Characterization of synthetic polymers by MALDI-MS, *Prog. Polym. Sci.*, **2006**, 31, 277–357.

30. Peacock, P.M.; McEwen, C.N., Mass spectrometry of synthetic polymers, *Anal. Chem.*, **2006**, 78, 3957–3964.

31. Wesdemiotis, C.; Arnould, M.A.; Lee, Y.; Quirk, R.P., MALDI TOF mass spectrometry of the products from novel anionic polymerizations, *Polym. Prepr.*, **2000**, 41(1), 629–630.

32. Arnould, M.A.; Polce, M.J.; Quirk, R.P.; Wesdemiotis, C., Probing chain-end functionalization reactions in living anionic polymerization via matrix-assisted laser desorption ionization time-of-flight mass spectrometry, *Int. J. Mass Spectrom.*, **2004**, 238, 245–255.

33. Polce, M.J.; Wesdemiotis, C., Introduction to mass spectrometry of polymers. In *Polymer Mass Spectrometry* (Lattimer, R.P.; Montaudo, G., editors). Boca Raton, FL, **2002**, pp. 1–29.

34. Jackson, A.T.; Slade, S.E.; Scrivens, J.H., Characterisation of poly(alkyl methacrylate)s by means of electrospray ionisation-tandem mass spectrometry (ESI-MS/MS), *Int. J. Mass Spectrom.*, **2004**, 283, 265–277.

35. Arnould, M.A.; Wesdemiotis, C.; Geiger, R.J.; Park, M.A.; Buehner, R.W.; Vandervorst, D., Structural characterization of polyester copolymers by MALDI mass spectrometry, *Prog. Org. Coat.*, **2002**, 45, 305–312.

36. Wesdemiotis, C.; Pingitore, F.; Polce, M.J.; Russell, V.M.; Kim, Y.; Kausch, C.M.; Connors, T.H.; Medsker, R.E.; Thomas, R.R., Characterization of a poly(fluorooxetane) and poly(fluorooxetane-co-THF) by MALDI mass spectrometry, size exclusion chromatography, and NMR spectroscopy, *Macromolecules*, **2006**, 39, 8369–8378.

37. Adamus, G.; Montaudo, M.S.; Montaudo, G.; Kowalczuk, M., Molecular architecture of poly[(R,S)-3-hydroxybutyrate-co-6-hydroxyhexanoate] and poly[(R,S)-3-hydroxybutyrate-co-(R,S)-2-hydroxyhexanoate] oligomers investigated by electrospray ionization ion-trap multistage mass spectrometry, *Rapid Commun. Mass Spectrom.*, **2004**, 18, 1436–1446.

38. Wollyung, K.M.; Wesdemiotis, C.; Nagy, A.; Kennedy, J.P., Synthesis and mass spectrometry characterization of centrally and terminally amine-functionalized polyisobutylenes, *J. Polym. Sci. Part A: Polym. Chem.*, **2005**, 43, 946–958.

39. De Hoffmann, E.; Stroobant, V., *Mass Spectrometry. Principles and Applications*. 2nd edition. Chichester: John Wiley & Sons, **2001**.

40. Tanaka, K.; Waki, H.; Ido, S.; Akita, Y.; Yoshida, Y.; Yoshida, T., Protein and polymer analyses up to m/z 100,000 by laser ionization time-of-flight mass spectrometry, *Rapid Commun. Mass Spectrom.*, **1988**, 2, 151–153.

41. Karas, M.; Hillenkamp, F., Laser desorption ionization of proteins with molecular masses exceeding 10,000 daltons, *Anal. Chem.*, **1988**, 60, 2299–2301.

42. Fenn, J.B.; Mann, M.; Meng, C.K.; Wong, S.F.; Whitehouse, C.M., Electrospray ionization for mass spectrometry of large biomolecules, *Science*, **1989**, 246, 64–71.

43. Craig, A.G.; Derrick, P.J., Spontaneous fragmentation of cationic polystyrene chains, *J. Am. Chem. Soc.*, **1985**, 107, 6707–6708.

44. Craig, A.G.; Derrick, P.J., Production and characterization of beams of polystyrene ions, *Aust. J. Chem.*, **1986**, 39, 1421–1434.

45. Jackson, A.T.; Jennings, R.C.K.; Scrivens, J.H.; Green, M.R.; Bateman, R.H., The characterization of complex mixtures by field desorption-tandem mass spectrometry, *Rapid Commun. Mass Spectrom.*, **1998**, 12, 1914–1924.

46. Lattimer, R.P.; Münster, H.; Budzikiewicz, H., Tandem mass spectrometry of poly glycols, *Int. J. Mass Spectrom. Ion Proc.*, **1989**, 90, 119–129.

47. Lattimer, R.P., Tandem mass spectrometry of poly(ethylene glycol) lithium-attachment ions, *J. Am. Soc. Mass Spectrom.*, **1994**, 5, 1072–1080.

48. Selby, T.L.; Wesdemiotis, C.; Lattimer, R.P., Dissociation characteristics of $[M + X]^+$ ions (X = H, Li, Na, K) from linear and cyclic polyglycols, *J. Am. Soc. Mass Spectrom.*, **1994**, 5, 1081–1092.

49. Jackson, A.T.; Yates, H.T.; Scrivens, J.H.; Critchley, G.; Brown, J.; Green, M.R.; Bateman, R.H., The application of matrix-assisted laser desorption/ionization combined with collision-induced dissociation to the analysis of synthetic polymers, *Rapid Commun. Mass Spectrom.*, **1996**, 10, 1668–1674.

50. Jackson, A.T.; Yates, H.T.; Scrivens, J.H.; Green, M.R.; Bateman, R.H., Utilizing matrix-assisted laser desorption/ionization-collision induced dissociation for the generation of structural information from poly(alkyl methacrylate)s, *J. Am. Soc. Mass Spectrom.*, **1997**, 8, 1206–1213.

51. Jackson, A.T.; Yates, H.T.; Scrivens, J.H.; Green, M.R.; Bateman, R.H., Matrix-assisted laser desorption/ionization-collision induced dissociation of poly(styrene), *J. Am. Soc. Mass Spectrom.*, **1998**, 9, 269–274.

52. Bottrill, A.R.; Giannakopulos, A.; Waterson, C.; Haddleton, D.M.; Lee, K.S.; Derrick, P.J., Determination of end groups of synthetic polymers by matrix-assisted laser desorption/ionization: high-energy collision-induced dissociation, *Anal. Chem.*, **1999**, 71, 3637–3641.

53. Lattimer, R.P., Mass spectral analysis of low-temperature pyrolysis products from poly(ethylene glycol), *J. Anal. Appl. Pyr.*, **2000**, 56, 61–78.

54. Lattimer, R.P.; Williams, R.C., Low-temperature pyrolysis products from a polyether-based urethane, *J. Anal. Appl. Pyr.*, **2002**, 63, 85–104.

55. Lattimer, R.P., Pyrolysis mass spectrometry of acrylic acid polymers, *J. Anal. Appl. Pyr.*, **2003**, 68–69, 3–14.

56. Busch, K.L.; Glish, G.L.; McLuckey, S.A., *Mass Spectrometry/Mass Spectrometry. Techniques and Applications of Tandem Mass Spectrometry*. New York: VCH Publishers, **1988**.

57. Sleno, L.; Volmer, D.A., Ion activation methods for tandem mass spectrometry, *J. Mass Spectrom.*, **2004**, 39, 1091–1112.

58. McLafferty, F.W.; Tureček, F., *Interpretation of Mass Spectra*. 4th edition. Mill Valley, CA: University Science Books, **1993**.

59. Cooks, R.G.; Ast, T.; Mabud, M.A., Collisions of polyatomic ions with surfaces, *Int. J. Mass Spectrom. Ion Proc.*, **1990**, 100, 209–265.

60. Dongré, A.R.; Somógyi, Á.; Wysocki, V.H., Surface-induced dissociation: an effective tool to probe structure, energetics, and fragmentation mechanisms of protonated peptides, *J. Mass Spectrom.*, **1996**, 31, 339–350.

61. Laskin, J.; Denisov, E.V.; Shukla, A.K.; Barlow, S.E.; Futrell, J.H., Surface-induced dissociation in a Fourier transform ion cyclotron resonance mass spectrometer: instrument design and evaluation, *Anal. Chem.*, **2002**, 74, 3255–3261.

62. Laskin, J.; Futrell, J.H., Activation of large ions in FT-ICR mass spectrometry, *Mass Spectrom. Rev.*, **2005**, 24, 135–167.

63. Xie, Y.; Lebrilla, C.B., Infrared multiphoton dissociation of alkali metal-coordinated oligosaccharides, *Anal. Chem.*, **2003**, 75, 1590–1598.

64. Wilson, J.J.; Brodbelt, J.S., Infrared multiphoton dissociation for enhanced de novo sequence interpretation of N-terminal sulfonated peptides in a quadrupole ion trap, *Anal. Chem.*, **2006**, 78, 6855–6862.

65. Zubarev, R.A.; Kelleher, N.L.; McLafferty, F.W., Electron capture dissociation of multiply charged protein cations: a nonergodic process, *J. Am. Chem. Soc.*, **1998**, 120, 3265–3266.

66. Zubarev, R.A., Reactions of polypeptide ions with electrons in the gas phase, *Mass Spectrom. Rev.*, **2003**, 22, 57–77.

67. Cooper, H.J.; Håkansson, K.; Marshall, A.G., The role of electron capture dissociation in biomolecular analysis, *Mass Spectrom. Rev.*, **2005**, 24, 201–222.

68. Syka, J.E.P.; Coon, J.J.; Schroeder, M.J.; Shabanowitz, J.; Hunt, D.F., Peptide and protein sequence analysis by electron transfer dissociation mass spectrometry, *Proc. Natl. Acad. Sci.*, **2004**, 101, 9528–9533.

69. Pitteri, S.J.; Chrisman, P.A.; McLuckey, S.A., Electron transfer ion/ion reactions of doubly protonated peptides: effect of elevated bath gas temperature, *Anal. Chem.*, **2005**, 77, 5662–5669.

70. Pitteri, S.; McLuckey, S.A., Recent developments in the ion/ion chemistry of high-mass multiply charged ions, *Mass Spectrom. Rev.*, **2005**, 24, 931–958.

71. Coon, J.J.; Shabanowitz, J.; Hunt, D.F.; Syka, J.E.P., Electron transfer dissociation of peptide anions, *J. Am. Soc. Mass Spectrom.*, **2005**, 16, 880–882.

72. Cerda, B.A.; Horn, D.M.; Breuker, K.; Carpenter, B.K.; McLafferty, F.W., Electron capture dissociation of multiply-charged oxygenated cations: a nonergodic process, *Eur. Mass Spectrom.*, **1999**, 5, 335–338.

73. Cerda, B.A.; Breuker, K.; Horn, D.M.; McLafferty, F.W., Charge/radical site initiation versus Coulombic repulsion for cleavage of multiply charged ions: charge solvation in poly(alkene glycol) ions, *J. Am. Soc. Mass Spectrom.*, **2001**, 12, 565–570.

74. Cerda, B.A.; Horn, D.M.; Breuker, K.; McLafferty, F.W., Sequencing of specific copolymer oligomers by electron-capture dissociation mass spectrometry, *J. Am. Chem. Soc.*, **2002**, 124, 9287–9291.

75. Koster, S.; Duursma, M.C.; Boon, J.J.; Heeren, R.M.A.; Ingemann, S.; van Benthem, R.A.T.M.; de Koster, C.G., Electron capture and collisionally activated dissociation mass spectrometry of doubly charged hyperbranched polyesteramides, *J. Am. Soc. Mass Spectrom.*, **2003**, 14, 332–341.

76. Lee, S.; Han, S.Y.; Lee, T.G.; Chung, G.; Lee, D.; Oh, H.B., Observation of pronounced b•, y cleavages in the electron capture dissociation mass spectrometry of polyamidoamine (PAMAM) dendrimer ions with amide functionalities, *J. Am. Soc. Mass Spectrom.*, **2006**, 17, 536–543.

77. Tang, X.; Ens, W.; Mayer, F.; Standing, K.G.; Westmore, J.B., Measurement of unimolecular decay in peptides of masses greater than 1200 units by a reflecting time-of-flight mass spectrometer, *Rapid Commun. Mass Spectrom.*, **1989**, 3, 443–448.

78. Spengler, B.; Kirsch, D.; Kaufmann, R., Metastable decay of peptides and proteins in matrix-assisted laser-desorption mass spectrometry, *Rapid Commun. Mass Spectrom.*, **1991**, 5, 198–202.

79. Spengler, B., Post-source decay analysis in matrix-assisted laser desorption/ionization mass spectrometry of biomolecules, *J. Mass Spectrom.*, **1997**, 32, 1019–1036.

80. Cotter, R.J., *Time-of-Flight Mass Spectrometry: Instrumentation and Applications in Biological Research*. Washington, DC: ACS Professional Reference Books, American Chemical Society, **1997**.

81. Hoteling, A.J.; Owens, K.G., Improved PSD and CID on a MALDI TOFMS, *J. Am. Soc. Mass Spectrom.*, **2004**, 15, 523–535.

82. Hoteling, A.J.; Kawaoka, K.; Goodberlet, M.C.; Yu, W.-M.; Owens, K.G., Optimization of matrix-assisted laser desorption/ionization time-of-flight collision-induced dissociation using poly(ethylene glycol), *Rapid Commun. Mass Spectrom.*, **2003**, 17, 1671–1676.

83. Scrivens, J.H.; Jackson, A.T.; Yates, H.T.; Green, M.R.; Critchley, G.; Brown, J.; Bateman, R.H.; Bowers, M.T.; Gidden, J., The effect of the variation of cation in the matrix-assisted laser desorption/ionization-collision induced dissociation (MALDI-CID) spectra of oligomeric systems, *Int. J. Mass Spectrom. Ion Proc.*, **1997**, 165/166, 363–375.

84. Przybilla, L.; Räder, H.-J.; Müllen, K., Post-source decay fragment ion analysis of polycarbonates by matrix-assisted laser desorption/ionization time-of-flight mass spectrometry, *Eur. Mass Spectrom.*, **1999**, 5, 133–143.

85. Muscat, D.; Henderickx, H.; Kwakkenbos, G.; van Benthem, R.; de Koster, C.G.; Fokkens, R.; Nibbering, N.M.M., In-source decay of hyperbranched polyesteramides in matrix-assisted laser desorption-ionization time-of flight mass spectrometry, *J. Am. Soc. Mass Spectrom.*, **2000**, 11, 218–227.

86. Goldschmidt, R.J.; Wetzel, S.J.; Blair W.R.; Guttman, C.M., Post-source decay in the analysis of polystyrene by matrix-assisted laser desorption/ionization time-of-flight mass spectrometry, *J. Am. Soc. Mass Spectrom.*, **2000**, 11, 1095–1106.

87. Przybilla, L.; Francke, V.; Räder, H.J.; Müllen, K., Block length determination of a poly(ethylene oxide)-*b*-poly(*p*-phenylene ethynylene) diblock copolymer by means of MALDI-TOF mass spectrometry combined with fragment-ion analysis, *Macromolecules*, **2001**, 34, 4401–4405.

88. Kéki, S.; Bodnár, I.; Borda, J.; Deák, G.; Batta, G.; Zsuga, M., Matrix-assisted laser desorption/ionization mass spectrometric study of the oligomers formed from lactic acid and diphenylmethane diisocyanate, *Macromolecules*, **2001**, 34, 7288–7293.

89. Murgasova, R.; Hercules, D.M., Matrix-assisted laser desorption/ionization collision-induced dissociation of linear single oligomers of nylon-6, *J. Mass Spectrom.*, **2001**, 36, 1098–1107.

90. Adhiya, A.; Wesdemiotis, C., Poly(propylene imine) dendrimer conformations in the gas phase: a tandem mass spectrometry study, *Int. J. Mass Spectrom.*, **2002**, 214, 75–88.

91. Laine, O.; Laitinen,T.; Vainiotalo, P., Characterization of polyesters prepared from three different phthalic acid isomers by CID-ESI-FT-ICR and PSD-MALDI-TOF mass spectrometry, *Anal. Chem.*, **2002**, 74, 4250–4258.

92. Fournier, I.; Marie, A.; Lesage, D.; Bolbach, G.; Fournier, F.; Tabet, J.C., Post-source decay time-of-flight study of fragmentation mechanisms of protonated synthetic polymers under matrix-assisted laser desorption/ionization conditions, *Rapid Commun. Mass Spectrom.*, **2002**, 16, 696–704.

93. Laine, O.; Trimpin, S.; Räder, H.J.; Müllen, K., Changes in post-source decay fragmentation behavior of poly(methyl methacrylate) polymers with increasing molecular weight studied by matrix-assisted laser desorption/ionization time-of-flight mass spectrometry, *Eur. J. Mass Spectrom.*, **2003**, 9, 195–201.

94. Neubert, H.; Knights, K.A.; de Miguel, Y.R.; Cowan, D.A., MALDI TOF post-source decay investigation of alkali metal adducts of apolar polypentylresorcinol dendrimers, *Macromolecules*, **2003**, 36, 8297–8303.

95. Ameduri, B.; Ladavière, C.; Delolme, F.; Boutevin, B., First MALDI-TOF mass spectrometry of vinylidene fluoride telomers endowed with low defect chaining, *Macromolecules*, **2004**, 37, 7602–7609.

96. Hanton, S.D.; Parees, D.M.; Owens, K.G., MALDI PSD of low molecular weight ethoxylated polymers, *Int. J. Mass Spectrom.*, **2004**, 238, 257–264.

97. Berndt, U.E.Ch.; Zhou, T.; Hider, R.C.; Liu, Z.D.; Neubert, H., Structural characterization of chelator-terminated dendrimers and their synthetic intermediates by mass spectrometry, *J. Mass Spectrom.*, **2005**, 40, 1203–1214.

98. Osaka, I.; Watanabe, M.; Takama, M.; Murakami, M.; Arakawa, R., Characterization of linear and cyclic polylactic acids and their solvolysis products by electrospray ionization mass spectrometry, *J. Mass Spectrom.*, **2006**, 41, 1369–1377.

99. Vergne, M.J.; Li, H.; Murgasova, R.; Hercules, D.M., Synthesis and mass spectral characterization of poly(amic methylester) oligomers, *Macromolecules*, **2006**, 39, 6928–6935.

100. Jackson, A.T.; Jennings, K.R.; Scrivens, J.H., Generation of average mass values and end group information of polymers by means of a combination of matrix-assisted laser desorption/ionization-mass spectrometry and liquid secondary ion-tandem mass spectrometry, *J. Am. Soc. Mass Spectrom.*, **1997**, 8, 76–85.

101. Gidden, J.; Jackson, A.T.; Scrivens, J.H.; Bowers, M.T., Gas phase conformations of synthetic polymers: poly(methyl methacrylate) oligomers cationized by sodium ions, *Int. J. Mass Spectrom.*, **1999**, 188, 121–130.

102. Stolarzewicz, A.; Neugebauer, D.; Silberring, J., Electrospray ionization tandem mass spectrometry for poly(propylene oxide) starting and end group analysis, *Rapid Commun. Mass Spectrom.*, **1999**, 13, 2469–2473.

103. Borman, C.D.; Jackson, A.T.; Bunn, A.; Cutter, A.L.; Irvine, D.J., Evidence for the low thermal stability of poly(methyl methacrylate) polymer produced by atom transfer radical polymerisation, *Polymer*, **2000**, 41, 6015–6020.

104. Jackson, A.T.; Bunn, A.; Hutchings, L.R.; Kiff, F.T.; Richards, R.W.; Williams, J.; Green, M.R.; Bateman, R.H., The generation of end group information from poly(styrene)s by means of matrix-assisted laser desorption/ionisation-collision induced dissociation, *Polymer*, **2000**, 41, 7437–7450.

105. Wu, J.; Polce, M.J.; Wesdemiotis, C., Unimolecular chemistry of Li$^+$- and Na$^+$-coordinated polyglycol radicals, a new class of distonic radical ions, *J. Am. Chem. Soc.*, **2000**, 122, 12786–12794.

106. Yalcin, T.; Gabryelski, W.; Li, L., Structural analysis of polymer end groups by electrospray ionization high-energy collision-induced dissociation tandem mass spectrometry, *Anal. Chem.*, **2000**, 72, 3847–3852.

107. Botrill, A.R.; Giannakopulos, A.; Millichope, A.; Lee, K.S.; Derrick, P.J., Combination of time-of-flight mass analysers with magnetic-sector instruments: in-line and perpendicular arrangements. Applications to poly(ethylene glycol) with long-chain end groups, *Eur. J. Mass Spectrom.*, **2000**, 6, 225–232.

108. Jackson, A.T.; Bunn, A.; Priestnall, I.M.; Borman, C.D.; Irvine, D.J., Molecular spectroscopic characterisation of poly(methyl methacrylate) generated by means of atom transfer radical polymerisation (ATRP), *Polymer*, **2006**, 47, 1044–1054.

109. Jackson, A.T.; Green, M.R.; Bateman, R.H., Generation and end-group information from polyethers by matrix-assisted laser desorption/ionisation collision-induced dissociation mass spectrometry, *Rapid Commun. Mass Spectrom.*, **2006**, 20, 3542–3550.

110. Bateman, R.H.; Green, M.R.; Scott, G., A combined magnetic sector-time-of-flight mass spectrometer for structural determination studies by tandem mass spectrometry, *Rapid Commun. Mass Spectrom.*, **1995**, 9, 1227–1233.

111. Morris, H.R.; Paxton, T.; Dell, A.; Langhorne, J.; Berg, M.; Bordoli, R.S.; Hoyes, J.; Bateman, R.H., High sensitivity collisionally activated decomposition tandem mass spectrometry on a novel quadrupole/orthogonal-acceleration time-of-flight mass spectrometer, *Rapid Commun. Mass Spectrom.*, **1996**, 10, 889–896.

112. *Q-ToF Ultima MALDI User's Guide*. Atlas Park, Manchester: Micromass UK Limited M22 5PP. Available at http://www.waters.com

113. Harvey, D.J.; Bateman, R.H.; Bordoli, R.S.; Tyldesley, R., Ionisation and fragmentation of complex glycans with a quadrupole time-of-flight mass spectrometer fitted with a matrix-assisted laser desorption/ionisation ion source, *Rapid Commun. Mass Spectrom.*, **2000**, 14, 2135–2142.

114. Jackson, A.T.; Williams, J.P.; Scrivens, J.H., Desorption electrospray ionization mass spectrometry of low molecular weight synthetic polymers, *Rapid Commun. Mass Spectrom.*, **2006**, 20, 2717–2727.

115. Collins, S.; Rimmer, S., Tandem (quadrupole-time-of-flight) electrospray mass spectrometry of oligo(vinyl acetates) with isopropylol or (1-hydroxyethyl)-2-oxyisopropanyl end groups, *Rapid Commun. Mass Spectrom.*, **2004**, 18, 3075–3078.

116. Jackson, A.T.; Scrivens, J.H.; Williams, J.P.; Baker, E.S.; Gidden, J.; Bowers, M.T., Microstructural and conformational studies of polyether copolymers, *Int. J. Mass Spectrom.*, **2004**, 238, 287–297.

117. Terrier, P.; Buchmann, W.; Desmazières, B.; Tortajada, J., Block lengths and block sequence of linear triblock and glycerol derivative diblock copolyethers by electrospray ionization–collision-induced dissociation mass spectrometry, *Anal. Chem.*, **2006**, 78, 1801–1806.

118. Jiang, X.; Schoenmakers, P.J.; van Dongen, J.L.J.; Lou, X.; Lima, V.; Brokken-Zijp, J., Mass spectrometric characterization of functional poly(methyl methacrylate) in combination with critical liquid chromatography, *Anal. Chem.*, **2003**, 75, 5517–5524.

119. Warburton, K.E.; Clench, M.R.; Ford, M.J.; White, J.; Rimmer, D.A.; Carolan, V.A., Characterisation of derivatised monomeric and prepolymeric isocyanates by matrix-assisted laser desorption/ionization time-of-flight mass spectrometry and structural elucidation by tandem mass spectrometry, *Eur. J. Mass Spectrom.*, **2005**, 11, 565–574.

120. Medzihradszky, K.F.; Campbell, J.M.; Baldwin, M.A.; Falick, A.M.; Juhasz, P.; Vestal, M.L.; Burlingame, A.L., The characteristics of peptide collision-induced dissociation using a high-performance MALDI-TOF/TOF tandem mass spectrometer, *Anal. Chem.*, **2000**, 72, 552–558.

121. Yergey, A.L.; Coorssen, J.R.; Backlund, P.S., Jr.; Blank, P.S.; Humphrey, G.A.; Zimmerberg, J.; Campbell, J.M.; Vestal, M.L., De novo sequencing of peptides using MALDI-TOF/TOF, *J. Am. Soc. Mass Spectrom.*, **2002**, 13, 784–791.

122. Suckau, D.; Resemann, A.; Schuerenberg, M.; Hufnagel, P.; Franzen, J.; Holle, A., A novel MALDI LIFT-TOF/TOF mass spectrometer for proteomics, *Anal. Bioanal. Chem.*, **2003**, 376, 952–965.

123. Giannakopulos, A.E.; Thomas, B.; Colburn, A.W.; Reynolds, D.J.; Raptakis, E.N.; Makarov, A.A.; Derrick, P.J., Tandem time-of-flight mass spectrometer (TOF-TOF) with a quadratic-field ion mirror, *Rev. Sci. Instrum.*, **2002**, 73, 2115–2123.

124. Rizzarelli, P.; Puglisi, C.; Montaudo, G., Sequence determination in aliphatic poly(ester amide)s by matrix-assisted laser desorption/ionization time-of-flight and time-of-flight/time-of-flight tandem mass spectrometry, *Rapid Commun. Mass Spectrom.*, **2005**, 19, 2407–2418.

125. Rizzarelli, P.; Puglisi, C.; Montaudo, G., Matrix-assisted laser desorption/ionization time-of-flight/ time-of-flight tandem mass spectra of poly(butylene adipate), *Rapid Commun. Mass Spectrom.*, **2006**, 20, 1683–1694.

126. Vergne, M.J., Mass Spectrometry of Functionalized Polymers. PhD Dissertation, Nashville, TN: Vanderbilt University, **2004**.

127. Arnould, M.A. (Xerox Corporation), unpublished results.

128. Kamel, A.; Prakash, C., High performance liquid chromatography/atmospheric pressure ionization/ tandem mass spectrometry (HPLC/API/MS/MS) in drug metabolism and toxicology, *Current Drug Metabolism*, **2006**, 7, 837–852.

129. Smyth, W.F., Electrospray ionization-mass spectrometry (ESI-MS) and liquid chromatography-electrospray ionization-mass spectrometry (LC-ESI-MS) of selected pharmaceuticals, *Current Pharmaceutical Analysis*, **2006**, 2, 299–311.

130. Dolan, A.R.; Wood, T.D., Synthesis and characterization of low molecular weight oligomers of soluble polyalanine by electrospray ionization mass spectrometry, *Synth. Met.*, **2004**, 143, 243–250.

131. Giguère, M.-S.; Mayer, P.M., Climbing the internal energy ladder: the unimolecular decomposition of ionized poly(vinyl acetate), *Int. J. Mass Spectrom.*, **2004**, 231, 59–68.

132. Cooks, R.G.; Kaiser, R.E., Jr., Quadrupole ion trap mass spectrometry, *Acc. Chem. Res.*, **1990**, 23, 213–219.

133. Todd, J.F.J.; Penman, A.D., The recent evolution of the quadrupole ion trap mass spectrometer—an overview, *Int. J. Mass Spectrom. Ion Proc.*, **1991**, 106, 1–20.

134. March, R.E., A musing on the present state of the ion trap and prospects for future applications, *Org. Mass Spectrom.*, **1991**, 26, 627–632.

135. McLuckey, A.A.; Van Berkel, G.J.; Goeringer, D.E.; Glish, G.L., Ion trap mass spectrometry of externally generated ions, *Anal. Chem.*, **1994**, 66, 689A–696A.

136. Qin, J.; Chait, B.T., Identification and characterization of posttranslational modifications of proteins by MALDI ion trap mass spectrometry, *Anal. Chem.*, **1997**, 69, 4002–4009.

137. Jonscher, K.R.; Yates, J.R., III., The quadrupole ion trap mass spectrometer—a small solution to a big problem, *Anal. Biochem.*, **1997**, 244, 1–15.

138. March, R.E., An introduction to quadrupole ion trap mass spectrometry, *J. Mass Spectrom.*, **1997**, 32, 351–369.

139. McLuckey, S.A.; Wells, J.M., Mass analysis at the advent of the 21st century, *Chem. Rev.*, **2001**, 101, 571–606.

140. March, R.E., Quadrupole ion trap mass spectrometry: a view at the turn of the century, *Int. J. Mass Spectrom.*, **2000**, 200, 285–312.

141. March, R.E.; Todd, J.F.J., *Quadrupole Ion Trap Mass Spectrometry*. 2nd edition. Hoboken, NJ: Wiley-Interscience, **2005**.

142. *Esquire-LC Operations Manual*, Vol. 1, version 3.1. Bremen, Germany: Bruker Daltonik GmbH, **1999**.

143. Jedli ski, Z.; Adamus, G.; Kowalczuk, M.; Schubert, R.; Szewczuk, Z.; Stefanowicz, P., Electrospray tandem mass spectrometry of poly(3-hydroxybutanioc acid) end groups analysis and fragmentation mechanism, *Rapid Commun. Mass Spectrom.*, **1998**, 12, 357–360.

144. Arslan, H.; Adamus, G.; Hazer, B.; Kowalczuk, M., Electrospray ionisation tandem mass spectrometry of poly [(R,S)-3-hydroxybutanoic acid] telechelics containing primary hydroxy end groups, *Rapid Commun. Mass Spectrom.*, **1999**, 13, 2433–2438.

145. McLuckey, S.A.; Asano, K.G.; Gregory Schaaff, T.; Stephenson, J.L., Jr., Ion trap collisional activation of protonated poly(propylene imine) dendrimers: generations 1–5. *Int. J. Mass Spectrom.*, **2000**, 195/196, 419–437.

146. Adamus, G.; Kowalczuk, M., Electrospray multistep ion trap mass spectrometry for the structural characterization of poly[(R,S)-3-hydroxybutanoic acid] containing a β-lactam end group, *Rapid Commun. Mass Spectrom.*, **2000**, 14, 195–202.

147. Chen, R.; Tseng, A.M.; Uhing, M.; Li, L., Application of an integrated matrix-assisted laser desorption/ionization time-of-flight, electrospray ionization mass spectrometry and tandem mass spectrometry approach to characterizing complex polyol mixtures. *J. Am. Soc. Mass Spectrom.*, **2001**, 12, 55–60.

148. Chen, R.; Li, L., Lithium and transition metal ions enable low energy collision-induced dissociation of polyglycols in electrospray ionization mass spectrometry, *J. Am. Soc. Mass Spectrom.*, **2001**, 12, 832–839.

149. Bogan, M.J.; Agnes, G.R., Poly(ethylene glycol) doubly and singly cationized by different alkali metal ions: relative cation ion affinities and cation-dependent resolution in a quadrupole ion trap mass spectrometer, *J. Am. Soc. Mass Spectrom.*, **2002**, 13, 177–186.

150. Marie, A.; Fournier, F.; Tabet, J.C.; Améduri, B.; Walker, J., Collision-induced dissociation studies of poly(vinylidene) fluoride telomers in an electrospray-ion trap mass spectrometer, *Anal. Chem.*, **2002**, 74, 3213–3220.

151. Yuan, G.; Tang, F.; Zhu, C.J.; Liu, Y.; Zhao, Y.F., Fragmentation mechanisms of polyamides containing *N*-methylimidazole by electrospray ionization tandem mass spectrometry, *Rapid Commun. Mass Spectrom.*, **2003**, 17, 2015–2018.

152. Creaser, C.S.; Reynolds, J.C.; Hoteling, A.J.; Nichols, W.F.; Owens, K.G., Atmospheric pressure matrix-assisted laser desorption/ionisation ion trap mass spectrometry of synthetic polymers: a comparison with vacuum matrix-assisted laser desorption/ionization time-of-flight mass spectrometry, *Eur. J. Mass Spectrom.*, **2003**, 9, 33–44.

153. Okuno, S.; Ohmoto, M.; Arakawa, R., Analysis of polypropyleneglycols using electrospray ionization mass spectrometry: effects of cationizing agents on the mass spectra, *Eur. J. Mass Spectrom.*, **2003**, 9, 97–103.

154. Yuan, G.; Zhou, J., Electrospray ionization mass spectral fragmentation mechanisms of DNA-recognizing polyamides containing *N*-methylpyrrole and *N*-methylimidazole, *Rapid Commun. Mass Spectrom.*, **2004**, 18, 1397–1402.

155. Wang, J.; Zhou, J.; Yuan, G., Electrospray negative ion mass spectrometry of polyamides containing *N*-methylpyrrole and *N*-methylimidazole, *J. Mass Spectrom.*, **2005**, 40, 688–689.

156. Arnould, M.A.; Vargas, R.; Buehner, R.W.; Wesdemiotis, C., Tandem mass spectrometry characteristics of polyester anions and cations formed by electrospray ionization, *Eur. J. Mass Spectrom.*, **2005**, 11, 243–256.

157. Zhou, G.-J.; Xu, P.-X.; Ye, Y.; Zhao, Y.-F., Electrospray ionization tandem mass spectrometry investigation of the polymerization of tetrahydrofuran initiated by phosphorus oxychloride, *Eur. J. Mass Spectrom.*, **2005**, 11, 319–324.

158. Hanton, S.D.; Parees, D.M.; Zweigenbaum, J., The fragmentation of ethoxylated surfactants by AP-MALDI-QIT, *J. Am. Soc. Mass Spectrom.*, **2006**, 17, 453–458.

159. Laiko, V.V.; Baldwin, M.A.; Burlingame, A.L., Atmospheric pressure matrix-assisted laser desorption/ionization mass spectrometry, *Anal. Chem.*, **2000**, 72, 652–657.

160. Laiko, V.V.; Moyer, S.C.; Cotter, R.J., Atmospheric pressure MALDI/ion trap mass spectrometry, *Anal. Chem.*, **2000**, 72, 5239–5243.

161. Chaicharoen, K.; Polce, M.J.; Wesdemiotis, C., MP11-231, tandem mass spectrometry of poly(methyl methacrylate)s. *Proceedings of the 54th ASMS Conference on Mass Spectrometry and Allied Topics*, Seattle, WA, May 28–June 1, 2006.

162. Schwartz, J.C.; Syka, J.E.P.; Quarmby, S.T., TODam 11:35, improving the fundamentals of MS^n on 2D linear ion traps: new ion activation and isolation techniques. *Proceedings of the 53rd ASMS Conference on Mass Spectrometry and Allied Topics*, San Antonio, TX, June 5–9, 2005.

163. Cunningham, C., Jr.; Glish, G.L.; Burinsky, D.J., High amplitude short time excitation: a method to form and detect low mass product ions in a quadrupole ion trap mass spectrometer, *J. Am. Soc. Mass Spectrom.*, **2006**, 17, 81–84.

164. Schwartz, J.C.; Senko, M.W.; Syka, J.E.P., A two-dimensional quadrupole ion trap mass spectrometer, *J. Am. Soc. Mass Spectrom.*, **2002**, 13, 659–669.

165. Hager, J.W., A new linear ion trap mass spectrometer, *Rapid Commun. Mass Spectrom.*, **2002**, 16, 512–526.

166. Riter, L.S.; Gooding, K.M.; Hodge, B.D.; Julian, R.K., Jr., Comparison of the Paul ion trap to the linear ion trap for use in global proteomics, *Proteomics*, **2006**, 6, 1735–1740.

167. Available at http://www.ssi.shimadzu.com/products/product.cfm?product=lcms_it_tof

168. Available at http://www.shimadzu-biotech.net/pages/products/1/aximaqit.php

169. Available at http://www.hitachi-hitec.com/global/science/ms/nanofrontier.html

170. Hashimoto, Y.; Hasegawa, H.; Satake, H.; Baba, T.; Waki, I., Duty cycle enhancement of an orthogonal acceleration TOF mass spectrometer using an axially resonant excitation linear ion trap, *J. Am. Soc. Mass Spectrom.*, **2006**, 17, 1669–1674.

171. Okuno, S.; Kiuchi, M.; Arakawa, R., Structural characterization of polyethers using matrix-assisted laser desorption/ionization quadrupole ion trap time-of-flight mass spectrometry, *Eur. J. Mass Spectrom.*, **2006**, 12, 181–187.

172. Hu, Q.; Noll, R.J.; Li, H.; Makarov, A.; Hardman, M.; Cooks, R.G., The Orbitrap: a new mass spectrometer, *J. Mass Spectrom.*, **2005**, 40, 430–443.

173. Available at http://www.thermo.com/com/cda/product/detail/0,,10121795,00.html?CA=orbitrap

174. Makarov, A.; Denisov, E.; Kholomeev, A.; Balschun, W.; Lange, O.; Strupat, K.; Horning, S., Performance evaluation of a hybrid linear ion trap/Orbitrap mass spectrometer, *Anal. Chem.*, **2006**, 78, 2113–2120.

175. Pastor, S.J.; Wilkins, C.L., Sustained off-resonance irradiation and collision-induced dissociation for structural analysis of polymers by MALDI-FTMS, *Int. J. Mass Spectrom. Ion Proc.*, **1998**, 175, 81–92.

176. Mariarz, E.P., III; Baker, G.A.; Wood, T.D., Capitalizing on the high mass accuracy of electrospray ionization Fourier transform mass spectrometry for synthetic polymer characterization: a detailed investigation of poly(dimethylsiloxane), *Macromolecules*, **1999**, 32, 4411–4418.

177. Koster, S.; Duursma, M.C.; Boon, J.J.; Nielen, M.W.F.; de Koster, C.G.; Heeren, R.M.A., Structural analysis of synthetic homo- and copolyesters by electrospray ionization on a Fourier transform ion cyclotron resonance mass spectrometer, *J. Mass Spectrom.*, **2000**, 35, 739–748.

178. Koster, S.; de Koster, C.G.; van Benthem, R.A.T.M.; Duursma, M.C.; Boon, J.J.; Heeren, R.M.A., Structural characterization of hyperbranched polyesteramides: MS^n and the origin of species, *Int. J. Mass Spectrom.*, **2001**, 210/211, 591–602.

179. Koster, S.; Duursma, M.C.; Guo, X.; van Benthem, R.A.T.M.; de Koster, C.G.; Boon, J.J.; Heeren, R.M.A., Isomer separation of hyperbranched polyesteramides with gas-phase H/D exchange and a novel MS^n approach: DoDIP, *J. Mass Spectrom.*, **2002**, 37, 792–802.

180. McDonnell, L.A.; Derrick, P.J.; Powell, B.B.; Double, P., Sustained off-resonance irradiation collision-induced dissociation of linear, substituted and cyclic polyesters using a 9.4 T Fourier transform ion cyclotron resonance mass spectrometer, *Eur. J. Mass Spectrom.*, **2003**, 9, 117–128.

181. Han, S.Y.; Lee, S.; Oh, H.B., Comparable electron capture efficiencies for various protonated sites on the 3rd generation poly(propylene imine) dendrimer ions: applications by SORI-CAD and electron capture dissociation mass spectrometry (ECD MS), *Bull. Korean Chem. Soc.*, **2005**, 26, 740–746.

182. Felder, T.; Schalley, C.A.; Fakhrnabavi, H.; Lukin, O., A combined ESI- and MALDI-MS(/MS) study of peripherally persulfonylated dendrimers: false negative results by MALDI-MS and analysis of defects, *Chem. Eur. J.*, **2005**, 11, 5625–5636.

183. Wallace, W.E., Report on the 4th annual NIST polymer mass spectrometry workshop: "Future directions in soft ionization of polymers," *J. Am. Soc. Mass Spectrom.*, **2006**, 17, 280–282.

184. Roepstorff, P.; Fohlman, J., Proposal for common nomenclature for sequence ions in mass spectra of peptides, *Biomed. Mass Spectrom.*, **1984**, 11, 601.

185. Biemann, K., Contributions of mass spectrometry to peptide and protein structure, *Biomed. Environ. Mass Spectrom.*, **1988**, 16, 99–111.

186. Domon, B.; Costello, C.E., A systematic nomenclature for carbohydrate fragmentations in FAB-MS/MS spectra of glycolconjugates, *Glycoconjugate J.*, **1988**, 5, 397–409.

187. McLuckey, S.A.; Van Berkel, G.J.; Glish, G.L., Tandem mass spectrometry of small, multiply charged oligonucleotides, *J. Am. Soc. Mass Spectrom.*, **1992**, 3, 60–70.

188. Claeys, M.; Nizigiyimana, L.; Van den Heuvel, H.; Derrick, P.J., Mechanistic aspects of charge-remote fragmentation in saturated and mono-unsaturated fatty acid derivatives: evidence for homolytic cleavage, *Rapid Commun. Mass Spectrom.*, **1996**, 10, 770–774.

189. Jackson, A.T.; Lattimer, R.P.; Price, P.C.; Wallace, W.E.; Polce, M.J.; Wesdemiotis, C., 2705, Proposal for a common nomenclature of fragment ions in the mass spectra of synthetic polymers.

Proceedings of the 54th ASMS Conference on Mass Spectrometry and Allied Topics, Seattle, WA, May 28–June 1, 2006.

190. Cheng, C.; Gross, M.L., Applications and mechanisms of charge-remote fragmentation, *Mass Spectrom. Rev.*, **2000**, 19, 398–420.

191. Gross, M.L., Charge-remote fragmentation: an account of research on mechanism and applications, *Int. J. Mass Spectrom.*, **2000**, 200, 611–624.

192. Thomya, P., Structural Characterization of Complex Polymer Systems by Degradation/Mass Spectrometry. PhD Dissertation, The University of Akron: Akron, Ohio, **2006**.

193. Available at http://webbook.nist.gov/chemistry/

194. Dongre, A.R.; Jones, J.L.; Somogyi, A.; Wysocki, V., Influence of peptide composition, gas-phase basicity, and chemical modification on fragmentation efficiency: evidence for the mobile proton model, *J. Am. Chem. Soc.*, **1996**, 118, 8365–8374.

195. Carey, F.A.; Sundberg, R.J., *Advanced Organic Chemistry. Part B: Reactions and Synthesis*. 4th edition. New York: Kluwer Academic/Plenum Publishers, **2001**.

196. Chaicharoen, K., Mass and Tandem Mass Spectrometric Studies on Synthetic Polymers. PhD Dissertation, The University of Akron: Akron, Ohio, **2008**.

197. Chen, H., End group-assisted siloxane bond cleavage in the gas phase, *J. Am. Soc. Mass Spectrom.*, **2003**, 14, 1039–1048.

198. Leigh, A.M.; Wang, P.; Polce, M.J.; Wesdemiotis, C., MP-340, tandem mass spectrometry of poly(dimethylsiloxane)s. *Proceedings of the 53rd ASMS Conference on Mass Spectrometry and Allied Topics*, San Antonio, TX, June 5–9, 2005.

199. Polce, M.J.; Wesdemiotis, C., TOEam 10:55, Fundamental aspects of polymer fragmentation. *Proceedings of the 51st ASMS Conference on Mass Spectrometry and Allied Topics*, Montréal, Quebec, Canada, June 8–12, 2003.

200. Polce, M.J.; Wesdemiotis, C., TPP 303, Tandem mass spectrometry of polyisoprene, polybutadiene, and poly(2-vinyl pyridine). *Proceedings of the 52nd ASMS Conference on Mass Spectrometry and Allied Topics*, Nashville, TN, May 23–27, 2004.

201. Polce, M.J.; Wesdemiotis, C., Proposed modifications to MALDI-CID fragmentation mechanisms of silver ionized polystyrene oligomers. *Proceedings of the 54th ASMS Conference on Mass Spectrometry and Allied Topics*, Seattle, WA, May 28–June 1, 2006.

202. Polce, M.J.; Ocampo, M.; Quirk, R.P.; Wesdemiotis, C., Tandem mass spectrometry characteristics of silver-cationized polystyrenes: backbone degradation via free radical chemistry, *Anal. Chem.*, **2008**, 80, 347–354.

203. Polce, M.J.; Ocampo, M.; Quirk, R.P.; Leigh, A.M.; Wesdemiotis, C., Tandem mass spectrometry characteristics of silver-cationized polystyrenes: internal energy, size, and chain end versus backbone substituent effects, *Anal. Chem.*, **2008**, 80, 355–362.

204. Gies, A.P.; Vergne, M.J.; Orndorff, R.L.; Hercules, D.M., MALDI-TOF/TOF CID study of polystyrene fragmentation reactions, *Macromolecules*, **2007**, 40, 7493–7504.

205. Cameron, G.C.; MacCallum, J.R., Thermal degradation of polystyrene, *J. Macromol. Sci., Rev. Macromol. Chem.*, **1967**, 1, 327–359.

206. Schroeder, U.K.O.; Ebert, K.H.; Hamielec, A.W., On the kinetics and mechanism of thermal degradation of polystyrene: 2. formation of volatile compounds, *Macromol. Chem.*, **1984**, 185, 991–1001.

207. Scott, G. (editor). *Mechanisms of Polymer Degradation and Stabilization*. London: Elsevier, **1990**.

208. Tsuge, S.; Ohtani, H., Pyrolysis gas chromatography/mass spectrometry (Py-GC/MS). In *Mass Spectrometry of Polymers* (Montaudo, G.; Lattimer, R.P., editors). Boca Raton, FL: CRC Press, **2002**, pp. 113–147.

209. Bauld, N.L., *Radicals, Radical Ions, and Triplets: The Spin Bearing Intermediates of Organic Chemistry*. New York: Wiley, **1997**.

210. Suebold, F.H., Jr., The rearrangement of the neophyl radical, *J. Am. Chem. Soc.*, **1953**, 75, 2532–2533.

211. Tokmakov, I.V.; Lin, M.C., Combined quantum chemical/RRKM-ME computational study of the phenyl + ethylene, vinyl + benzene, and H + styrene reactions, *J. Phys. Chem. A*, **2004**, 108, 9697–9714.

CHAPTER 6

CONVENTIONAL MALDI SAMPLE PREPARATION

Kevin G. Owens[1] *and Scott D. Hanton*[2]

[1]Department of Chemistry, Drexel University, Philadelphia, PA, and [2]Air Products & Chemicals, Inc., Allentown, PA

6.1 INTRODUCTION

Sample preparation for matrix-assisted laser desorption ionization (MALDI) is deceptively simple. It is a tribute to the MALDI technique that wide ranges of sample compositions successfully produce usable mass spectra—spectra that can be utilized to identify the analytes present in the sample. In the progression of an analytical technique, once the question of "what is there?" is successfully answered, the next question is generally "how much is there?" While acceptable *qualitative* results may be obtained with rather crude control of the sample preparation step, good *quantitative* analysis generally requires more careful control of the sample preparation variables. The polymer mass spectrometrist was the first to be concerned with quantitative MALDI analysis, as the calculation of the number-average and weight-average molecular mass (M_n and M_w, respectively) from the measured mass spectrum depends on the accurate quantitative measurement of the amount of each oligomer—which are separate chemical species—present in the sample. Reproducible measurement of the peak intensities (peak areas) from the MALDI spectrum is the key to these measurements. While quantitation of the different chain-length oligomers in a polymer sample has long been practiced, the quantitative analysis of different chemistry components in a sample mixture is still an analytical challenge for several reasons as will be described in this chapter.

The MALDI vernacular is replete with phrases relating to the quality of the MALDI sample preparation. Good samples are described as having "sweet spots" where the "analyte lights up" when illuminated with the laser beam. If no analyte-related signals are observed, the analyte is said "not to fly" or that the "sample is unMALDIable." If no analyte signals are observed from a given sample, the procedure is often to choose another matrix off the shelf, prepare a new sample, and repeat the measurement. This process has sometimes been referred to as "the black art of MALDI sample preparation." Great efforts have been expended to remove the "art"

MALDI Mass Spectrometry for Synthetic Polymer Analysis, Edited by Liang Li
Copyright © 2010 John Wiley & Sons, Inc.

and bring MALDI sample preparation into the realm of science. Our own involvement in investigating details of the MALDI sample preparation process was an evolution from crude trial and error—looking for the right matrix—to development of a fundamental understanding of the principles that allow us to drive to solutions more quickly, to move through the same phase space of sample preparation within a few hours rather than days or weeks. It is the goal of this chapter to address some of the key roles that the "matrix" plays in MALDI analysis, hopefully enabling a more sophisticated choice of MALDI sample composition. Note that we wish to emphasize here the critical concepts behind solution-phase MALDI sample preparation. While several detailed examples are provided, we specifically chose not to include a list or table of successful recipes for various polymer chemistries. The interested reader is referred to the extensive polymer MALDI Recipes Page maintained by the Polymers Division at the National Institute of Standards and Technology (http://www.nist.gov/maldi) for detailed sample preparation recipes and references.

6.2 LASERS AND MASS SPECTROMETRY

MALDI is not the first technique where lasers are used to create ions for analysis by mass spectrometry. Laser desorption (LD) [1] was developed for those molecules that lacked the volatility or thermal stability to be heated to produce a significant vapor pressure of analyte in the source region of a conventional electron impact mass spectrometer. In LD the analyte molecules are deposited "neat" on the sample probe; the desorption laser beam is then directed at the sample surface. LD (what now is termed by some as "matrix-less MALDI") was used for the analysis of a number of different classes of analytes, including peptides [2] and synthetic polymers [3]. It was observed quite early on that low-molecular-mass analytes could be desorbed intact; however, higher-molecular-mass species generally underwent extensive fragmentation. The upper limit in molecular mass was several thousand Daltons, but depended on the analyte class. It was generally understood that a series of complicated gas-phase reactions (involving proton and/or cation transfer) were responsible for the creation of the molecular ions that were observed in the mass spectrum. In a 1982 review of the LD technique [4], Franz Hillenkamp described a number of characteristics of the LD process. A final observation was that the LD mass spectra are qualitatively influenced by the composition of their immediate surroundings. It was the Hillenkamp group's work on the LD of single amino acids and simple binary mixtures [5] that led to the development of MALDI as it is generally practiced today; the first matrix may have been the amino acid tryptophan, which enabled the analysis of alanine at desorption laser intensities far below what was required to produce ions from alanine alone.

   Over the years, a large number of matrix molecules have been identified. It is unfortunately still the case that the exact molecular characteristics that define a matrix are not well understood. Structures of a number of known matrix molecules, including 2,5-dihydroxybenzoic acid (DHB), 3,5-dimethoxy-4-hydroxycinnamic acid (sinapinic acid), α-cyano-4-hydroxycinnamic acid (CHCA), trans-indoleacrylic

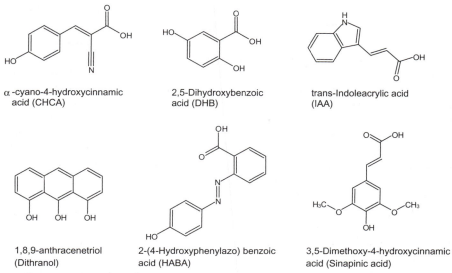

α-cyano-4-hydroxycinnamic
acid (CHCA)

2,5-Dihydroxybenzoic
acid (DHB)

trans-Indoleacrylic acid
(IAA)

1,8,9-anthracenetriol
(Dithranol)

2-(4-Hydroxyphenylazo) benzoic
acid (HABA)

3,5-Dimethoxy-4-hydroxycinnamic
acid (Sinapinic acid)

Figure 6.1.   Chemical structures of several common matrix compounds.

acid (IAA), 1,8,9-anthracenetriol (dithranol), and 2-(4-hydroxyphenylazo)benzoic acid (HABA) are shown in Figure 6.1. It is generally agreed that absorption of the laser wavelength chosen is critical, and that compounds should have a high molar absorptivity at the wavelength employed. In this chapter we will focus on MALDI performed in the UV region; MALDI can also be successfully accomplished using lasers in the IR [6, 7], but matrices for use in the IR region are different from those in use in the UV. It should be noted that since most MALDI instruments employ a nitrogen laser operating at 337 nm (due to its low initial purchase cost), molecules that have been identified as successful matrices in the literature are successful *due to the choice of that wavelength for analysis.* While MALDI can be practiced using a number of different UV wavelengths such as 355 nm produced by a Nd:YAG laser [8], little work has yet been expended in the search for new compounds with perhaps different and more interesting matrix properties using wavelengths other than 337 nm.

Aside from absorbing the laser light, a number of early papers suggested other important characteristics of successful matrix compounds. In an early review [9], Hillenkamp et al. noted that an important function of the matrix was to isolate the biopolymer analyte molecules from each other. Juhasz et al. [10] further suggested that *miscibility with the analyte in the solid phase* and adequate solubility in the solvent were key characteristics of a matrix compound. Recognizing that the matrix also played an important role in the ionization process, they also noted the matrix had to have a proper chemical composition to encourage ionization. It is not the goal of this chapter to construct the molecular template for a successful matrix compound. The goal is to recognize that the matrix plays several roles in the MALDI process, and that *each of these roles must be played well* to result in a successful MALDI experiment. It is our belief that an understanding of these roles is most important; while many researchers search for a *single molecule* that may be a good matrix for a given class of analytes, it may be possible that a *mixture of chemical compounds* may prove more effective if each component of the mixture is selected to optimize its role in the MALDI process. This separation of roles of the MALDI matrix was probably first evident in the analysis of synthetic polymer samples, as most analytes have a low proton affinity and require the addition of a cation to the mixture (even from adventitious sources) to observe any analyte signal in the mass spectrum. Acknowledgement of the separate roles of a matrix allows for the choice of sample components to fulfill each of the required roles independently. In this way, the best matrix for a given analysis may not be a single chemical compound but a combination of materials, each of which performs its own role optimally.

While this book focuses on the analysis of synthetic polymer samples, some of the work in our laboratories also involves the analysis of biological polymers, in particular peptides and proteins. We have generally found that the lessons learned in improving the analysis of one class of materials often help us in the analysis of another. Therefore, in some instances, examples involving the analysis of biological macromolecules may also be discussed.

6.3 THE MALDI PROCESS

In our view, four basic steps make up the MALDI process:

- sample preparation,
- analyte desorption,
- analyte ionization, and
- mass analysis.

It is our goal to focus on the sample preparation step, as other chapters in this book will cover ionization (Chapter 2) and mass analysis (Chapter 3) in much greater detail.

6.3.1 Sample Preparation

This is the only step conducted outside of the instrument; the analyte desorption and ionization steps occur in the instrument source within a few nanoseconds of the sample being illuminated by the laser beam. It should be noted, however, that choice of the MALDI sample composition in the *initial sample preparation step* also defines what will occur in each of the subsequent steps in the MALDI process. A chosen sample preparation may successfully isolate the analyte molecules from each other and enable them to desorb from the sample surface when illuminated by the laser beam; however, if the proper ionization reagent is not present, no analyte ions will be observed in the mass spectrum. In fact, the failure of the chosen *matrix sample mixture* in any of its required roles will result in the absence of analyte signal in the observed mass spectrum. When no analyte ions are observed in the acquired mass spectrum, it is unfortunately difficult to know where the MALDI process has failed. In our experience, when troubleshooting a problematic analysis, it is important to make use of a wide range of analytical techniques, rather than just MALDI itself, to probe the reasons for a failed analysis. While these auxiliary analytical techniques may not be available to all researchers, nor may they be used all the time even where they are available, results of their application demonstrate the importance of the key concepts presented here. Each of the four basic steps in the MALDI process listed above will be covered in more detail below, with the most extensive discussion focusing on details of the sample preparation step. Sample preparation is key to the MALDI process; some researchers claim that 80–90% of the success of a MALDI experiment is determined by it.

6.3.2 Analyte Desorption

As a defining difference between simple LD and MALDI, the matrix molecules play a key role in releasing *large* analyte molecules into the gas phase *intact*. While the desorption step is not covered explicitly as a chapter in this volume, a number of researchers have studied the MALDI desorption process [11]. Upon illumination with a short pulse of UV laser light, the matrix solid disintegrates, releasing the trapped analyte molecules into the gas-phase. In many cases the matrix undergoes rapid molecular dissociation; many matrices exhibit the loss of CO_2 or H_2O as the base peak in the matrix mass spectrum. All of this material (analyte molecules, intact matrix molecules, matrix fragments, matrix clusters, matrix-analyte clusters) is released into the gas phase; the high pressure immediately at the surface of the sample creates a supersonic expansion that carries the molecules away from the surface. The most important aspect of the supersonic expansion produced during the desorption step is that it creates a region of high probability for reaction. The relative velocity of the molecular species decreases as the desorbed molecules undergo a large number of collisions in the expanding plume. This was observed early on by many workers measuring the initial velocity of ions created in the MALDI process. For example, Beavis and Chait [12] found approximately equal initial velocity of 750 m/s for analyte ions of mass 1030, 5730, and 15,990. A number of traditional molecular beam studies have demonstrated that reaction cross sections generally increase as the relative kinetic energy of the reactants decreases [13–15]; the lower relative velocities allows more time for molecular reorientation to occur, ultimately leading to reaction. The desorbing MALDI plume can be viewed as a chemical reaction vessel where a variety of interesting chemical reactions may be observed. This is particularly important for the matrix in its role as ionization reagent.

6.3.3 Analyte Ionization

For preformed ions, the matrix only needs to isolate them from their surroundings and desorb them into the gas-phase. For most analyte molecules, however, the matrix also must create the ion that is observed. Several conditions must be fulfilled to observe analyte ions in the MALDI experiment: the appropriate ionization reagent must be present, the analyte molecules must have an affinity for the ionization reagent, and reaction conditions must be present that favor analyte ion production. As described above, the supersonic expansion within the desorbing MALDI plume creates the ideal reaction environment. In the case of most synthetic polymer analytes, the proton affinity is generally quite low. However, the alkali (or other cation) affinity is usually high. The molecule–molecule and ion–molecule reactions involved in the analyte ionization process are likely also related to the ion formation mechanisms operating in other "particle desorption" techniques such as fast atom bombardment (FAB), secondary ion mass spectrometry (SIMS), and plasma desorption (PD).

6.4 SAMPLE PREPARATION PROCESS

There are several obvious yet important steps to MALDI sample preparation. Before sample preparation proceeds, however, several important decisions need to be made. The fundamental decisions include:

- choice of the matrix,
- choice of the cationization agent,
- choice of the solvent(s), and
- decision of method of sample deposition.

For example, to analyze a low-molecular-mass polyethylene glycol (PEG) sample (e.g., PEG 1000), a small quantity of the PEG analyte can be added to one of a number of different matrices, in one of a number of solvents, at a variety of concentrations. A small drop of the solution is placed on the sample plate and allowed to dry under ambient conditions. No matter the choices, in most cases a representative mass spectrum can be obtained. The observed mass spectrum of PEG 1000 will generally show several ion series separated by 44 Da, the mass of an ethylene glycol monomer residue, the two main series being due to the sodium (major) and potassium (minor) cationized species. An example sample preparation method for PEG 1000 might be:

- create a 5 mg/mL solution of PEG 1000 in methanol,
- create a 0.25 M solution of DHB in methanol,
- mix the solutions in the volume ratio 10 : 100 PEG : DHB,
- spot 0.2 μL of the mixture on the target, and
- allow to air dry.

Unfortunately, not all analytes are as easy to analyze by MALDI as PEG 1000. In fact, PEG 1000 can be analyzed without a matrix in a straightforward LD experiment as long as some alkali is available for cationization. For the MALDI sample described above, alkali ions may not need to be explicitly added to the sample preparation. A sufficient quantity of sodium is present in most of the commercially available matrices (for many as a contaminant from the synthesis process); in other cases, contamination from the glass or plastic sample containers or the mass spectrometer

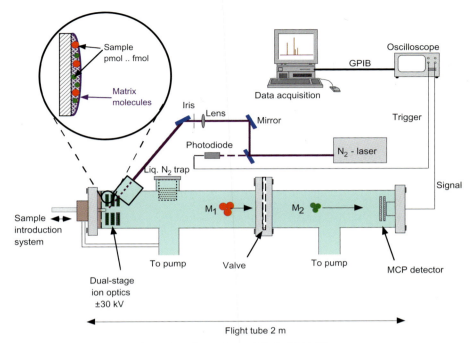

Figure 2.1. Schematic of a typical linear time-of-flight MALDI mass spectrometer. The valve separating the detector from the source region is optional but useful, for example, for maintenance or cleaning of the ion source.

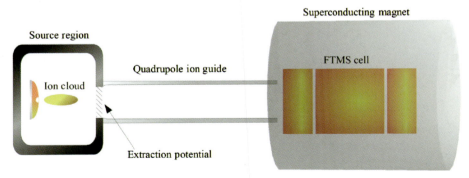

Figure 4.2. Schematic representation of a MALDI-FTMS instrument with external ionization.

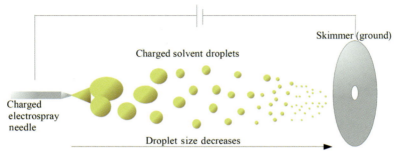

Figure 4.3.   Schematic representation of the ESI ionization process.

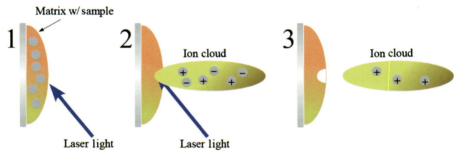

Figure 4.5.   Schematic representation of the MALDI ionization process.

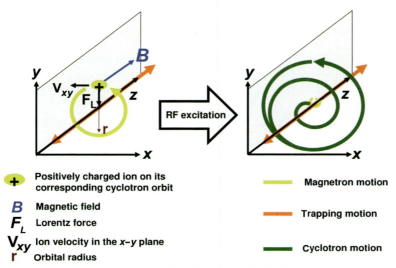

+ Positively charged ion on its corresponding cyclotron orbit

B Magnetic field

F_L Lorentz force

V_{xy} Ion velocity in the *x–y* plane

r Orbital radius

───── Magnetron motion

───── Trapping motion

───── Cyclotron motion

Figure 4.7. Cyclotron motion of ions in a magnetic field B caused by the balance between Lorentz force F_L and the centrifugal force F_Z. Schematic simplified representation of the ion motions in the ICR cell.

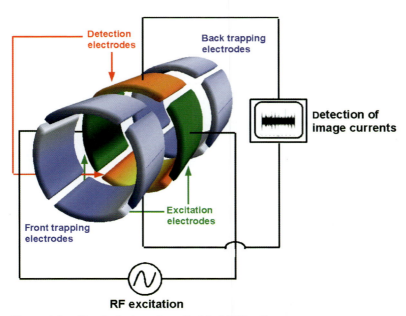

Figure 4.8. Simple design of a cylindrical ICR cell.

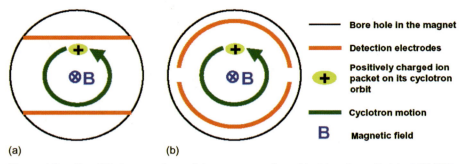

Figure 4.9. Simplified comparison of the geometry of a cubic (a) and a cylindrical (b) ICR cell within a bore hole of a superconducting magnet.

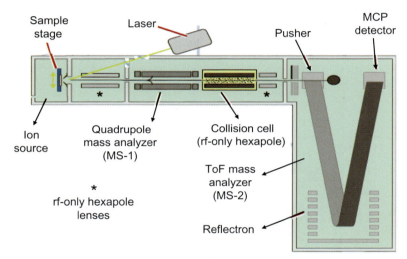

Figure 5.5. Waters® Q-Tof Ultima® MALDI Mass Spectrometer, a Q/ToF tandem mass spectrometer equipped with a MALDI source. Adapted from Reference 112 with permission, © Waters Corporation 2003.

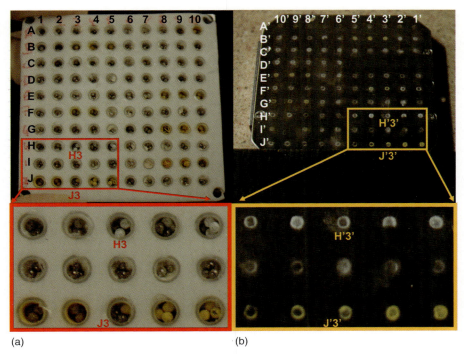

(a)   (b)

Figure 7.3.   Photographs of the loaded custom-made 100-sample plate (a) and the
sample-mirror imaged MALDI plate of the homogenized/transferred MALDI samples
(b) employing the vortex device for 5 min as well as their respective enlargements (bottom).

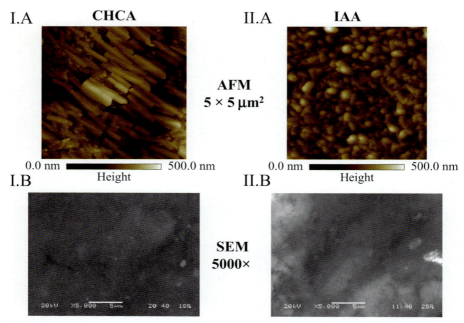

I.A **CHCA** II.A **IAA**

AFM
5 × 5 μm²

0.0 nm ▬▬▬▬ 500.0 nm 0.0 nm ▬▬▬▬ 500.0 nm
I.B   Height II.B   Height

SEM
5000×

Figure 7.5. Microscopy images of solvent-free MALDI samples prepared by the solvent-free vortex homogenization/spatula transfer method. Images I.A and B are from a sample prepared with PS/CHCA/Ag. The tapping mode AFM images show local, submicron morphology, rods for CHCA (I.A), and spheres for IAA (II.A). The SEM images show surprisingly homogeneous thin films for both samples (I.B and II.B) at high magnification (5000×).

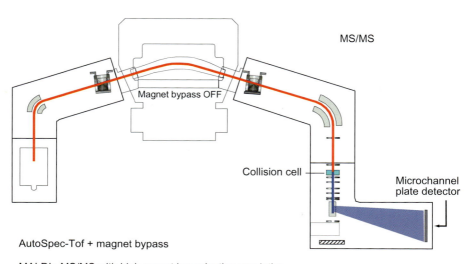

MS/MS

Magnet bypass OFF

Collision cell

Microchannel
plate detector

AutoSpec-Tof + magnet bypass

MALDI: MS/MS with high parent ion selection resolution

Figure 10.1. Schematic of MALDI-MS/MS instrument (Autospec-oa-TOF) used to generate most of the data shown in this chapter.

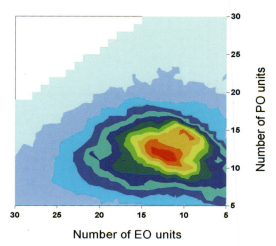

Number of EO units

Figure 11.12.   Two-dimensional plot of an EO-PO block copolymer combining the information of adsorption chromatography and of 14 MALDI-TOF spectra.

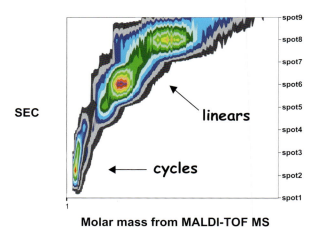

Molar mass from MALDI-TOF MS

Figure 11.14.   Two-dimensional plot created using SEC data (Figure 11.13a) and nine MALDI-TOF mass spectra (mass spectra of spot 8 is exemplarily shown in Figure 11.13b).

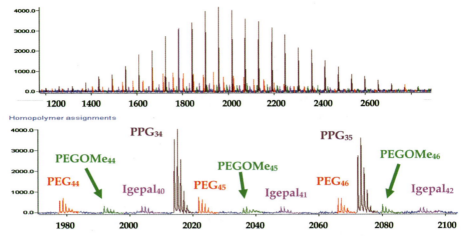

Figure 12.19. A MALDI mass spectrum from an equal mass blend of four common polymers with molecular mass of about 2000 Da: PEG, monomethyl PEG (PEGOMe), PPG, and an ethoxylated nonylphenol (Igepal Co890). The relative peak intensities show very different relative response factors for these polymers in this experiment in DHB.

sample plate itself is sufficient to produce ion series with adequate signal/noise (S/N) ratio. The distribution of the solid PEG 1000 sample on the probe may be somewhat irregular, but by averaging enough shots while the laser beam is rastered around on the surface of the sample a usable spectrum is obtained. As the analyte becomes more complex (e.g., by increasing the molecular mass or if the analyte is a mixture of different components) or when the chemical properties of the analyte are different from those of the matrix, closer attention needs to be paid to the sample preparation. Even in the case of PEG, higher-molecular-mass samples *do require* the presence of a matrix to obtain usable signal; in fact, as the molecular mass increases, the quantity of the matrix present, specifically the matrix-to-analyte (M/A) ratio, must increase as well.

While examples of facile analysis via MALDI abound, most analyses require closer consideration of the sample preparation details. For harder (usually what may be termed "real-world") samples, there are additional decisions that need to be made by the experimenter:

- choice of a co-matrix or other sample additive,
- choice of concentration of the analyte, matrix, co-matrix, and additive solutions,
- selection of M/A ratio,
- selection of co-matrix (or additive)-to-matrix ratio,
- selection of salt (i.e., cationization agent)-to-analyte ratio,
- selection of matrix-to-salt ratio,
- order of mixing of sample components,
- sample plate selection and preparation, and
- decision to use a single layer or multilayer sample preparation.

It may appear that not all MALDI sample preparations require these many decisions, that is, that not all MALDI sample protocols are this complicated. Why are many of the decisions listed above unnecessary in many experiments? It is largely dependent on the properties of the analyte and the required outcome of the experiment—and particularly whether qualitative or quantitative (i.e., molecular weight, rather than blend composition) information is desired. With the exquisite sensitivity of the MALDI process and modern TOFMS instruments, the failure to add a MALDI sample component or control one or more components' concentration may open up several reaction channels to the analyte. This may result in the collection of highly variable MALDI spectra from a single sample.

Once the decisions listed above are made, the process of sample preparation involves a number of discrete steps. As we are covering conventional MALDI sample preparation (Chapter 7 will cover the relatively newer technique of solvent-less sample preparation), the process starts with putting the various sample materials into solution:

- dissolution of the solid analyte,
- dissolution of the solid matrix,

- dissolution of the co-matrix or matrix additive (including the cationization reagent),
- production of a homogeneous sample solution,
- removal of the solvent or solvent mixture, and
- production of a homogeneous solid solution.

In our experience, the matrix, analyte and other sample additives must have adequate solubility in the solvents chosen. An analyte that is insoluble (which may be noted as a cloudy solution or a solution with solid residue on the bottom of the sample tube) generally produces poor or no MALDI results using the solution phase preparation techniques that we describe here. A significant problem is that many of the choices listed above are not experimentally independent, and their combined effect on the final MALDI mass spectrum is complicated. Further, it is difficult or sometimes impossible to design an experiment where the effect of a single choice or sample preparation step may be probed independently. Acknowledging these difficulties, we will discuss each of the choices and discrete steps in sample preparation listed above in more detail in the sections below.

6.4.1   Choice of the Matrix

The importance of the choice of the matrix is clear to even the most novice MALDI practitioner. Prior personal experience or successful results in the literature on similar analyte molecules is often used as a first guide to matrix selection. While prior experience is important, it is sometimes necessary to extend the MALDI analysis to previously unexplored sample chemistries. A good deal of the discussion below will focus on a means of rationally choosing a matrix for use with a particular analyte. In our experience, the incorporation of the analyte within the matrix solid is critical to the success of a MALDI experiment. It is this incorporation of analyte into the matrix that differentiates MALDI from the older technique of LD. In our coverage of this incorporation step we will invoke the concept of solubility or miscibility of the analyte with the solid matrix. As described in more detail below, it is not clear whether analyte molecules are truly separated (isolated) from each other on the molecular level, or whether they exist as small clusters or microdomains dispersed through the matrix solid. In our experience, good mixing (what could also be termed "intimate contact," i.e., close proximity) of the analyte, matrix, and other sample components leads to the most successful MALDI results.

   While it is clear that solubility of the analyte and the matrix in the chosen solvent is important, the key issue is the solubility (or miscibility) of the analyte *within the solid matrix sample*. The solubility of a solute within a solvent is governed by the intermolecular forces that are present in the mixture. These forces run from the very weak dispersion (i.e., London), through induction (dipole-induced dipole), orientation (dipole–dipole), donor–acceptor (hydrogen bonding or coordination, i.e., transition metal ions with pi systems) to the strongest Coulombic or ion–ion forces. Solubility is a key concept for a number of techniques employed routinely in chemistry. For example, in gas chromatography, the McReynolds phase constants are used to select the appropriate stationary phase for use with the analyte of interest. The best separation occurs when there is good interaction of the analyte with the stationary phase material. In its simplest form, the adage of "like dissolves like" is often invoked to choose a column for a particular separation. The Hildebrand solubility parameter [16] has been widely used to estimate the solubility of a solute within a solvent. While first developed only for use with hydrocarbons, the Hildebrand parameter has been extended for use with the stronger intermolecular interactions involved in hydrogen bonding solvents. There are a number of ways to measure the Hildebrand parameters, and lists for many solvents may be found in the literature. The concept of the Hildebrand parameter was extended by Hansen [17] by breaking it down into three components: dispersion, polarity, and hydrogen bonding; Hansen parameters are now widely used to estimate the solubility of one polymer within another.

Since the proper mixing (or good interaction of the analyte with the matrix) is a prerequisite for a good MALDI sample preparation, the concept of the solubility of the analyte within the matrix is key to a MALDI analysis. It should be noted that differential solubility in either the solution or solid state can be used as a means of separation or fractionation. This is well known to the biochemist, as large proteins may be precipitated from solution by the addition of a solvent such as cold acetone. Similarly for a polymer chemist, high-molecular-mass species can be preferentially precipitated by the addition of a nonsolvent to a solution of a high polydispersity polymer sample. In MALDI, if the solubility or miscibility of the analyte within the solid matrix is poor, then as the sample dries, it is expected that segregation of the analyte from the matrix will occur. As good interaction of the analyte and matrix is required to produce ions characteristic of the analyte, sample segregation will lead to poor MALDI results.

Incorporation of the analyte in the matrix can be studied in a number of ways. To isolate the incorporation step (since failure of any of the other steps in the MALDI process will lead to lack of signal in the mass spectrometer), other analytical techniques must be employed. The incorporation of analyte within the matrix solid has been studied by a number of groups using fluorescence microscopy. In these experiments, a fluorescently labeled analyte (a peptide or protein [18, 19, 20] or synthetic polymer [21]) is introduced into the MALDI sample preparation at the same concentration as the analyte would be normally. In our own work [18], fluorescently labeled analytes were found to be inhomogenously distributed through the solid matrix material. This experiment visually demonstrates that heterogeneous incorporation within the matrix solid is a prime cause of the variability of analyte ion signal for the commonly used dried droplet sample preparations. Dai et al.'s more recent confocal microscopy work [19] not only confirms these results, but also conclusively demonstrated that the fluorescently labeled analyte is indeed incorporated within the matrix crystals.

Further work in our laboratories has investigated the incorporation of synthetic polymer analytes within a number of different matrix materials using SIMS as a probe technique on a unique Physical Electronics, Inc., TRIFT II instrument. The instrument is equipped with both a nitrogen laser for MALDI work and a gallium liquid metal ion gun for traditional SIMS work. The laser and ion beams are aligned to strike the sample target at nearly the same location; a further advantage is that the instrument can be converted from MALDI to SIMS operation within minutes, allowing for a single sample to be interrogated by both the MALDI and SIMS techniques. In our experiments, the samples were prepared as for a normal MALDI analysis [22]. Since the SIMS technique only samples the first few atomic layers of a sample, if intact analyte ions are observed in the SIMS spectrum, this is an indication that the analyte is located on the surface of the sample. We interpret intense intact analyte ions as an indicator of a poor solubility of the analyte within the matrix. An example of these results is shown in Figure 6.2, where polystyrene 2900 is used as an analyte with both DHB and dithranol as a matrix. With DHB, intact analyte

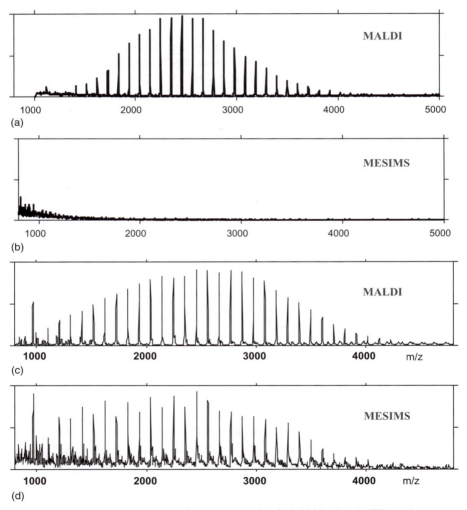

Figure 6.2.   Mass spectra obtained from the analysis of PS 2900 using AgTFA as the cationization reagent and dithranol as the matrix (panels a and b) or DHB as the matrix (panels c and d). Reproduced from Reference 22 with permission.

ions are observed in both the MALDI and SIMS experiments. The polystyrene (PS) is segregated out of the matrix solid, creating a high concentration of PS molecules in the surface layer. These results indicate that DHB is a poor matrix for PS. With dithranol as the matrix, a strong MALDI signal is obtained; the poor SIMS results indicate that the PS analyte is incorporated within (i.e., is soluble in) the dithranol matrix solid. These results also suggest that as a technique, MALDI has a much larger sampling depth than SIMS.

Our SIMS results [22] for a large number of polymer analyte/matrix combinations are summarized in Table 6.1. The IAA matrix is particularly illustrative. Note the series of "no" results in the middle of the table, indicating that no matrix-enhanced secondary ion mass spectrometry (MESIMS) signal is observed. At the left side of the table, both PEG and polypropylene oxide (PPO) are observed on the surface of the IAA—these analytes are *too polar* to be soluble within the IAA matrix. At the right side of the table, the polybutadiene (PBD) and polydimethylsiloxane (PDMS) are also found to be on the surface—these analytes are *too nonpolar* to be soluble within the IAA matrix. Hildebrand's solubility theory clearly states that in order for a solute to be soluble in a solvent, there must be a close match of the solubility parameters of the solute and the solvent. If there is too great of a mismatch—in either the polar or nonpolar direction—the result is a phase separation. An important related paper describing the miscibility of one polymer within another using Hansen parameters [23] further indicates that as the molecular mass of the polymer increases, there needs to be a closer match of the solute and solvent solubility parameters in order to form a true solid solution. This echoes the conventional MALDI wisdom that as the molecular mass of a polymer increases, the choice of the appropriate matrix becomes more crucial. A critical result of our SIMS work was the development of a "matrix solubility scale," where a number of commonly used matrices (and analytes) are listed in order from polar to nonpolar as shown in Figure 6.3. Choosing a matrix then involves judging the polarity of an analyte with respect to such a scale. As most researchers do not have access to this type of instrumentation, measurement of the solubility of both the analytes and matrix

TABLE 6.1. MESIMS Results for the Analysis of MALDI Samples Prepared from a Combination of the Polymer Analyte and the Matrix Specified. A "Yes" Entry Indicates Intact Cationized Analyte Peaks Were Observed in the MESIMS Spectrum

Polymer:	PEG	PPO	PEF	PVAc	PTMEG	PMMA	PS	PBD	PDMS
MW:	1000	2025	1500	1500	1800	2900	2450	1300	1500
DPBD				yes	no	**no**	yes	no	yes
Ret A		no*				no*	**no**	**no**	yes
Dith			yes	**no**	**no**	**no**	**no**	yes	no
IAA	yes	yes	**no**	**no**	**no**	**no**	**no**	yes	yes
FA	yes	yes	yes	yes	**no**	**no**	**no**	yes	yes
CHCA	yes	yes	yes	yes	yes	yes	yes	yes	yes
DHB	yes	yes	yes	yes	yes	yes	yes	no	yes
TU	yes		**no**	**no**	yes				

Matrix and polymer relative solubilities

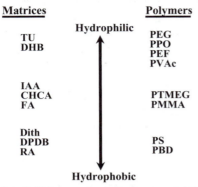

Figure 6.3. Relative matrix and polymer solubility scale developed from the MESIMS and liquid-phase solubility data. The more hydrophilic matrices and polymers are at the top of the figure and the more hydrophobic matrices and polymers are at the bottom of the figure. Abbreviations: DPBD = diphenylbutadiene, RA = all-trans retinoic acid, FA = ferulic acid, TU = thiourea, PPO = poly(propyleneoxide), PEF = poly(ethynylformamide), PVAc = poly(vinylacetate). Reproduced from Reference 22 with permission.

solids in a number of different solvents was suggested as a surrogate to the SIMS measurements.

In more recent work, we have investigated the compatibility of matrix and analyte compounds by using reverse-phase high-performance liquid chromatography (HPLC) [24]. It should be noted that HPLC is often used in chemistry to measure the hydrophobicity of compounds. Results for the HPLC analysis of a wide range of matrices are shown in Table 6.2. Here the absolute retention time is used as a measure of polarity—as we are using a reversed phase column, the most polar compounds elute first, followed by less polar materials. Note that the characteristics of the analytes can easily be measured using this HPLC technique. Following the HPLC analysis, matrix/analyte pairs were chosen by matching the measured retention times. It was demonstrated that the best MALDI results, in terms of analyte S/N ratio, were obtained by matching the retention time of the analyte to that of the matrix. This was demonstrated using a number of synthetic polymer analytes, including PEG, polymethylmethacrylate (PMMA), and PS, analytes that span the polarity range.

It is generally agreed that the matrix acts to isolate the analyte molecules from each other [9]. In this way the matrix acts as a *dispersant* to effectively eliminate analyte–analyte interactions. The concept of analyte/matrix solubility or miscibility is the direct result of this. One question that remains is whether the analyte must be dispersed at the molecular level within the matrix, or if intact molecular ions may be produced in MALDI from small clusters (dimers, trimers, etc.) of analyte. It is not completely clear from the evidence available whether the molecules must be isolated on the molecular level, or up to what size analyte clusters will still successfully produce ions in the MALDI experiment [25]. It may be that collisions experi-

TABLE 6.2. Reversed-Phase HPLC Retention Times for a Series of Matrix Compounds

Matrix Trivial Name	Monoisotopic Mass (Da)	HPLC RT (ELSD) (Min)	Compound Name
Nicotinic acid	123.032	0.73	Pyridine 3-carboxylic acid
DHBQ	140.011	0.88	2,5-dihydroxy-p-benzoquinone
DHB	154.027	1.64	2,5-dihydroxybenzoic acid
CHCA	189.043	3.11	Alpha-cyano-4-hydroxycinnamic acid
Nor-Harmane	168.069	3.39	9H-pyrido[3,4-b]indole
Sinapinic acid	224.068	4.12	3,5-dimethoxy-4-hydroxycinnamic acid
Ferulic acid	194.184	4.85	3-methoxy-4-hydroxycinnamic acid
IAA	187.063	6.14	Trans-indoleacrylic acid
HABA	242.069	6.27	2-(4-hydroxyphenylazo)benzoic acid
THAP	168.042	6.38	2′,4′,6′-trihydroxyacetophenone
MBT	166.986	6.81	2-mercaptobenzothiazole
CMBT	200.947	7.45	5-chloro-2-mercaptobenzothiazole
Dithranol	226.063	7.66	1,8,9-anthracentetriol
DCTB	250.147	8.71	2-[(2E)-3-(4-tert-butylphenyl)-2-methylprop-enylidene]malanonitrile
RA	300.209	8.73	All-trans retinoic acid

Source: Reproduced from Reference 24 with permission.
ELSD, evaporative light scattering detector.

enced by particles ejected from the solid during the desorption process may be energetic enough to dissociate small analyte clusters. It may also be the case, however, that due to the extreme sensitivity of the MALDI process, that only the small fraction of the analyte material that exists as isolated molecules in the matrix solid are observed as intact species in the MALDI experiment.

Recent work by Sheiko et al. on the dissociation of large molecules coated directly on metal surfaces may also shed light on an additional role of the matrix; in effect, the matrix protects the analyte from destruction by the surface [26]. These results would suggest that the choice of matrix surface is critical—as the sample surface in most instruments is part of the mass spectrometer source, they are generally constructed of metal so they define the electric field in the source region. Extending Sheiko's work, the MALDI matrix acts in the role of a dispersant to effectively eliminate the analyte–analyte *and* destructive *analyte–surface interactions* that decrease the probability of the desorption of intact analyte molecules.

6.4.2 Choice of Co-Matrices or Matrix Additives

As described above, incorporation of the analyte in the matrix is an important consideration for the production of useful MALDI data. Over the years, there has been significant work in the literature involving the use of matrix additives to improve the MALDI analysis, and in many cases, these additives enabled the analysis of materials that previously produced poor or no MALDI spectra. In a number of cases this is because matrices originally identified for the analysis of peptides and proteins were being used for samples with chemical properties (particularly solubility prop-erties) that differed significantly from that of peptides. For example, MALDI work on oligonucleotides proved difficult until fucose (a simple sugar) was used as a matrix additive [27]. In our view, the fucose can be viewed as a co-solvent, enabling the sugar-rich DNA strands to have better solubility in the relatively poorly matched matrix materials. Similarly, surfactants have been added to the MALDI preparation by the Linton group to improve the analysis of synthetic polymers [28], and more recently by Tummala and Limbach to improve the sequence coverage in peptide mass fingerprinting [29]. Matrix additives may also be included in the preparation to change the charge of the analyte, leading to better incorporation into the matrix solid. For example, attempts to analyze DNA gave rise to the inclusion of buffers to control the degree of ionization of the analyte in the MALDI preparation [30]. In fact, with strongly acidic matrices such as DHB and CHCA, analyte species with weakly acidic groups added to the sample as alkali salts will exist in solution (and the solid state) as protonated species—in effect the matrix can act as a "homogene-ous ion exchange resin." Note that the solubility of a protonated species in a par-ticular solvent is usually different from that of the corresponding alkali salt (e.g., the solubility of acetic acid vs. sodium acetate in water). In another peptide example, Juhasz and Biemann [31] found that the addition of basic peptides to the MALDI sample enabled the analysis of a highly acidic protein; the peptide protein complex was observed in the resulting mass spectrum. All of these examples of the use of matrix additives can be viewed as a means of impacting the incorporation of the analyte in the matrix solid solution.

In some cases, additives are instead added to the sample preparation to directly affect the desorption or ionization steps. Tsarbopoulos and co-workers introduced a mixture of 90% DHB with 10% 2,5-dihydroxybenzaldehyde (a mixture termed "super-DHB") [32]. It was proposed that the addition of the aldehyde weakened the bonding within the matrix crystal, lowering the energy required for desorption of the analyte from the matrix solid. Other researchers [33] (notably for the analysis of biological polymers such as oligonucleotides) have added materials such as crown-ethers (particularly 15-crown-5 and 18-crown-6) to bind with residual alkali contaminants (generally sodium or potassium) in the sample, increasing the proba-bility of observing protonated or deprotonated species over alkali-cationized species. Similarly, other researchers have added buffers such as diammonium hydrogen citrate (DAHC) [30] to the sample (usually DNA) to decrease the observation of alkali adducts of the analytes.

6.4.3 Choice of Cationization Reagent

Analytes with a high proton affinity (e.g., those containing a primary amine func-tionality) are generally protonated by most matrix molecules directly. For some analytes, however, the common (acidic) matrix molecules do not provide a usable ionization pathway. In these cases, a cationization reagent is required. The correct choice of ionization reagent depends upon the chemistry of the analyte. Oxygen-containing analytes without basic functionality such as polyethers and polyacrylates are best ionized using alkali ions. The adventitious sodium available in the sample from the matrix or analyte, or contamination from the sample containers or sample surfaces is often sufficient. While the alkali can be added simply in the form of a halide salt, better results are usually obtained using the alkali salt of the matrix [18], or a salt more soluble in organic solvents, such as the trifluoroacetate (TFA) salt. Presumably, this is because the solubility of the alkali salt in the matrix is higher, leading to closer association of the analyte, matrix, and cationization agent within the solid sample preparation. In this way, we extend the concept of solubility or miscibility in the matrix solid to the cationization agent as well. Choice of the alkali depends on the affinity of the analyte for the different alkali ions. In our own work, we have found that the alkali affinity can vary significantly for closely related analytes [34]. Figure 6.4 shows the results obtained for the analysis of an equimolar mixture of PEG and polytetramethylene glycol (PTMEG or polytetrahydrofuran).

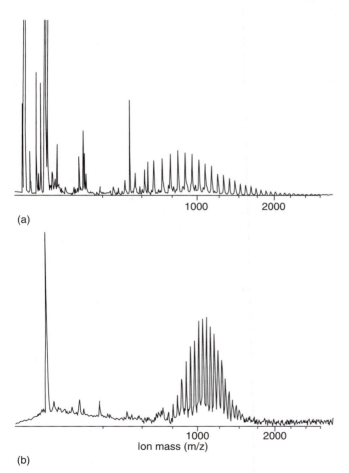

Figure 6.4. MALDI mass spectra of an equimolar mixture of PEG 1000 and PTMEG 1000 with the addition of (a) LiDHB and (b) KDHB as the cationization reagent. Reproduced from Reference 34 with permission.

Adding Li to the sample mixture (panel a) leads to observation of the PTMEG, while the addition of K (panel b) leads to the observation of the PEG. It was also observed that adduction with a specific alkali can change the determined molecular mass of the analyte, with higher-molecular-mass species appearing to have a preference for higher atomic number alkali ions [34]. From an analytical point of view, these results are a nightmare for a sample of unknown composition. Note, however, that various purification (e.g., ion exchange) or sample separation methods (e.g., chromatography) may be used to remove the high amounts of salts initially present in the sample; a suitable cationization reagent can then be added to enhance the specific adduct ion intensities in the MALDI spectrum.

Polymers that do not contain a heteroatom are usually more difficult to ionize. Unsaturated analytes such as PS and PBD are generally ionized using silver or copper ions, which can be added to the sample in the form of their TFA salts. These ions are known to interact strongly with the pi bonds present in the unsaturated carbon–carbon bonds, giving rise to the observed analyte–ion adducts. Analytes that lack both heteroatoms and unsaturations are the most difficult to ionize. In recent years, a significant effort has been made toward the analysis of polyethylene type materials due to their commercial significance [35, 36]. Ionization of these materials is acknowledged to be very difficult, and the most successful results are obtained by chemical modification of the analyte prior to MALDI analysis.

6.4.4 Choice of Solvent

In general, the best results are obtained using the same solvent for each of the components of the MALDI sample. If a binary (or higher order) solvent is used, the use of an azeotrope is the best choice. Considering the process of precipitation, the reason for selection of a solvent whose composition remains constant upon evaporation is clear. As the solvent evaporates, each of the components of the sample reach their saturation limit and start to precipitate from solution. If a nonazeotropic binary solvent is chosen, the composition of the solvent will change as the sample drop evaporates. The solubility of each of the matrix components and the analyte in each of the components of the binary solvent may be different, leading to significant segregation of one sample component from another as the sample dries.

6.4.5 Concentration of Sample Components

The concentrations of the various components of the MALDI sample are often chosen to set the M/A ratio, although little thought is usually given to the degree of saturation of the matrix, analyte, or cationization reagents in the final mixed solution. As the sample solution evaporates, if the analyte reaches its solubility limit first, it will begin to precipitate from solution, effectively forming small clusters or aggregates of analyte that are surrounded by solid matrix. As one role of the matrix is to "disperse" the analyte molecules from one another, this role will not be fulfilled, and the experiment is likely to fail. If the matrix reaches its solubility limit first, a significant quantity of the matrix may precipitate as fairly pure matrix crystals *before* the analyte reaches its solubility limit. This affects the amount of matrix available when the matrix and analyte do start to co-precipitate. Early work in the Li group [21, 37] on the analysis of polystyrene samples clearly demonstrated the effect of the relative solubility of analyte and matrix.

6.4.6   M/A Ratio

As the matrix plays several roles in the analysis (as dispersant, desorber, and ionization agent for protonation), it is likely that each analyte molecule requires the presence of many matrix molecules. In fact, the required M/A ratio generally ranges from 100:1 to 100,000:1, and tends to increase with the size of the analyte molecule. Owing to the usually large difference in the size of a matrix molecule compared to a synthetic or biological polymer analyte, it seems obvious that this ratio needs to be large, if an important role of the matrix is to separate one analyte molecule from another. Work in our laboratory using electrospray sample deposition (described further below) has been utilized to investigate the effect of M/A ratio on a MALDI analysis [38]. Individual standard solutions were prepared with a constant quantity of matrix; equal volumes of solutions having a different concentration of the bovine insulin analyte are added to each standard. Figure 6.5 shows the area under the insulin peak obtained as a function of M/A ratio. Note that the amount of analyte in the sample solution—and in the MALDI sample—increases monotonically in moving from *right to left* in the figure. We have colloquially referred to these plots as "Goldilocks curves." At a high M/A ratio, there is little analyte signal (too much matrix), at a low M/A ratio, there is also little analyte signal (too little matrix), and at some value of M/A ratio (approximately 5000 in this case), the analyte signal is maximized (the amount of matrix is "just right"). We interpret this as there being enough matrix molecules present in the sample so that they may fulfill, optimally, all of the roles that they are required to play (dispersant, desorber, and ionization agent). In our experience, the shape of the Goldilocks curves for peptide and protein analytes is general, although the width of the peak and position of the maximum is dependent upon the analyte and matrix combination. Importantly, it is the high reproducibility afforded by the electrospray deposition that allows these curves to be produced. A dramatic demonstration of the effect of M/A ratio on mass spectrum

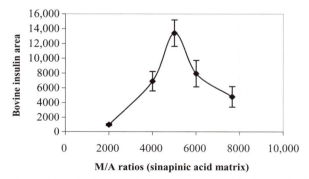

Figure 6.5.   Peak area of bovine insulin as a function of the M/A ratio using sinapinic acid as a matrix. Reproduced from Reference 38 with permission.

quality was presented by Gantt et al. in the analysis of bacterial extracts using sinapinic acid as the matrix [39]. Note that once the optimum M/A ratio is determined, it is constant; it is an experimental parameter that must be determined in the usual analytical method development process.

6.4.7   Co-Matrix (Additive)-to-Matrix Ratio

The choice of co-matrix (or other additive)-to-matrix ratio depends upon the reason the co-matrix is added. What must be remembered is that the "matrix" must both disperse the analyte and desorb it from the sample surface upon illumination by the laser beam. Anything added to the sample that does not contribute to the isolation or desorption process effectively must also be dispersed and desorbed itself by the remaining matrix materials. The matrix cannot tell analyte from co-matrix apart; if insufficient desorber is present, the overall analyte signal will be decreased. Care must be taken when using additives that quantities are controlled so that they do not interfere with the analyte signal.

6.4.8   Salt-to-Analyte Ratio and Salt-to-Matrix Ratio

The ratio of the number of moles of cationization reagent (colloquially referred to as the "salt") to analyte present is termed the "salt-to-analyte" or S/A ratio. Work in our labs has demonstrated that the cationization process is surprisingly efficient; titration of polymer analytes such as PEG and PMMA in common MALDI matrices have demonstrated that only near stoichiometric quantities of alkali cation are required to saturate the ion signal, indicating that the cationization process is nearly 100% efficient [33, 40, 41]. This is demonstrated in Figure 6.6, where a series of standards are prepared with constant quantities of matrix (DHB) and analyte (PMMA 6300); the amount of salt (NaTFA) increases in moving from left to right, effectively producing a titration curve. In this ideal case, the addition of additional salt (past the endpoint) leads to no change in the analyte signal. This work also showed that for some matrices, however, the presence of salt in excess of stoichiometric levels actually leads to a decrease in the analyte signal [40]. The exact cause of the decrease is still under investigation, but this work suggests that quantities of salt similar to the expected quantity of analyte are best for sample preparations.

In some cases, the added cationization agent may form clusters that interfere with the MALDI analysis. Macha et al. demonstrated that the addition of large quantities of Ag cation to certain matrices for the analysis of PS leads to the formation of silver clusters, which can interfere with the polymer signal when the data are collected at low resolution [42]. In a similar way, the salt-to-matrix (S/M) ratio is critical for those matrices that form very stable salt (particularly alkali ion) adducts. For example, CHCA forms very stable and predictable alkali clusters that have been well documented [43]. The formation of these undesired alkali–matrix clusters can interfere with the normal roles of the matrix within the sample preparation. In addition, the formation of matrix–alkali or cationization agent clusters leads to spectral congestion that may overlap with peaks of interest from the analyte [44].

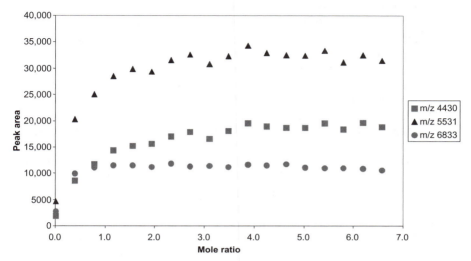

Figure 6.6. Peak area of the 4430 (squares) 5531 (triangles) and 6833 (circles) oligomers of PMMA 6300 as a function of the S/A ratio. Samples were prepared using DHB as a matrix (M/A = 1227) and NaTFA as the cationization reagent. Reproduced from Reference 40 with permission.

6.4.9 Order of Mixing of Sample Components

It is sometimes important what order the sample components are mixed if the final goal is to prepare a single sample solution. For example, low-molecular-mass PEG oligomers are readily analyzed using DHB or CHCA as a matrix. The oligomers are readily cationized by adventitious sodium ions. In some cases, however, cationization with a different alkali is preferred, and that alkali may be added to the sample in the form of the hydroxide. It is well known that PEG oligomers can be degraded by hydrolysis with high concentrations of base; to avoid that situation, the alkali hydroxide should be added to the acidic matrix solution *prior to* addition of the analyte. The reaction between the alkali hydroxide and matrix results in production of the alkali salt of the matrix and water. So that the extra water (a slow-drying solvent) does not affect the sample preparation, we prefer to prepare the alkali salt of the matrix in advance and dry down the material (using a rotary evaporator or Speed-Vac™, Thermo-Scientific, Inc.) prior to use. In this way, a known quantity of solid cationization reagent can be added to the sample preparation, enabling the S/A ratio to be controlled more effectively.

6.4.10 Sample Plate Selection and Preparation

An important part of the sample preparation process is selection and preparation of the sample plate. While some manufacturers offer a single choice of sample surface (usually stainless steel), others offer a range of plate materials, including gold, copper, and aluminum. For those interested in the analysis of biological samples, a number of plates with different, generally hydrophobic, surface coatings are also offered to help the deposited sample spot to dry in a smaller, defined location, enabling the sample spot to be more easily found and generally leading to more sensitive analyses. Early work on changing sample plate composition indicated there was little effect of the choice of surface; more recent work suggests choice of the surface may play a part in the sensitivity of the MALDI analysis [45]. In general, uncoated sample plates should be cleaned before use to remove any residual machining oil or other contaminants from the manufacturing process. In our lab, sample plates are cleaned by rinsing or ultrasonication with a succession of high-purity solvents of varying polarity: distilled water, methanol, tetrahydrofuran, hexanes, and then back to a final methanol rinse. If work will involve careful choice of a particular alkali ion for cationization, the surface is usually rinsed with an aqueous solution (50–100 mM) of diammonium hydrogen citrate, which is highly effective at removing alkali contaminants. Once the sample plate is used, previous samples may be removed by rinsing and wiping using a solvent in which both the analyte and matrix is highly soluble. For hard to remove samples, the sample plate can be polished using a finely divided alumina powder. When new, or after the plate has been cleaned using the polishing powder, it often needs to be "seasoned" with the matrix to be used. This was recognized early on by Fan and Beavis using DHB as a matrix [46], where it was suggested that deposition of the first sample leaves behind microscopic "seed crystals" which serve as nucleation sites for the creation of good matrix/ analyte crystals in subsequent sample applications. Alternatively, such deposition might also serve to bind the matrix to high energy sites on the sample plate surface that would otherwise irreversibly bind to the analyte molecules or lead to their destruction as described previously. To season the plate, a solution containing the matrix to be used can be applied (as for a dried droplet sample) and then removed by rinsing with the solvent in use.

6.4.11 Method of Sample Deposition

The final step in MALDI sample preparation is the removal of the solvent to produce the final solid matrix sample preparation. The goal of this step is the production of a thin uniform layer of solid sample containing analyte, matrix, other matrix additives, and the cationization reagent. It is during this final drying step that segregation of the sample components usually occurs, and there have been many methods suggested to improve the homogeneity of the final sample surface. A number of different "tricks" have been used in the preparation of the solid solution; in some cases, special sample preparation devices have even been developed. Heating of the sample (or sample plate before deposition), blowing dry nitrogen over the sample, depositing the sample droplet under vacuum, and the use of fast-drying solvents (e.g., the thin-layer method [47]) all decrease the sample drying time and promote the formation of "impure" crystals of the organic matrix compound. The use of room temperature drying at atmospheric pressure or slow-drying solvents (including water) encourage crystal purification, which results in exclusion of the analyte (and/or cationization agents) from the matrix solid as the sample dries. Overall, the goal of the sample deposition method is to produce, as closely as possible, a homogeneous solid solution. Having all components of the matrix sample in close proximity to each other increases the probability of observing a constant analyte signal across the sample surface. It also increases the probability that all of the roles of the matrix components are able to come into play at the appropriate time during the experiment. It is the "intimacy" of all of the sample components that is key to creation of a successful MALDI sample.

Certainly the oldest and simplest method of sample deposition is the dried droplet method [48], where a small volume (0.5–2 uL) of sample solution is deposited on the sample plate and the solvent allowed to evaporate at room temperature and atmospheric pressure. Some workers prefer to make a single mixture of all of the sample components; others prefer to mix individual small aliquots of analyte, matrix, and cationization agent directly on the probe surface (note that with fast evaporating solvents the latter method is difficult). It is generally preferred to premix the sample components, as more homogeneous solid samples are generally produced from homogeneous liquid solutions.

6.4.12   Single versus Multilayer Preparation

The choice of single versus multilayer sample preparation is often determined by the relative solubility of the sample components. If all the desired components of a MALDI sample are not soluble in a single solvent, a multilayer preparation will be required. By definition, such a sample will be heterogeneous on the macroscopic scale. However, by creating thin layers (e.g., by using low-concentration solutions or large sample spots), a single laser shot may still penetrate and desorb material from each of the sample layers. Care should, however, be taken in the selection of solvents for the preparation of multilayer samples. We experienced this issue directly when attempting to prepare three-layer samples for use in cationization studies [49]. Although the solubility of a solute in a particular solvent may be vanishingly small, the extreme sensitivity of the MALDI process may result in signal being observed due to remixing in the solution state upon deposition of a new layer using a supposedly "immiscible" solvent. The sandwich [50], two-layer [51] and three-layer [52] methods were all proposed as a means of preparing MALDI samples that yield a more uniform solid sample and most importantly, more constant analyte signal.

In general, smaller droplets of solution, with their correspondingly higher surface area, will dry faster, allowing less time for undesired sample segregation to occur. To enable the creation of smaller droplets, other deposition methods have been developed. Pastor et al. [53] developed what they called an aerospray technique, where the MALDI sample solution is pneumatically nebulized and sprayed onto the surface. The small droplets produced by the aerospray device lead to the production of fairly homogeneous solid samples. The oscillating capillary nebulizer (OCN) device developed by Wang et al. [54] is similar; the MALDI sample solution is passed at low (a few microliter/minute) flow rates through the inside capillary of a concentric fused silica capillary system. High-pressure gas flows through the external capillary. The flowing gas sets up a high-frequency oscillation of the inner capillary, which leads to the ejection of extremely small sample droplets. When used for MALDI sample preparation, these sample droplets are carried to a sample plate by the gas where they deposit to form a very homogeneous sample surface.

In our experience, electrospray deposition [55, 56] is the best method to produce (nearly) homogeneous solid samples exhibiting excellent MALDI signal reproducibility. The extremely small droplets of solution produced by the electrospray process (a few micrometer in diameter) effectively dry in the short distance of travel (approximately 2 cm) between the tip of the electrospray needle and the sample plate. The resulting solid droplets on the surface (which appear as flattened spheres approximately 250 nm in diameter for DHB sprayed from methanol solvent [57]) contain the matrix, analyte, and additives (including the cationization agent) in extremely close proximity. Even if the material is segregated within these solid drops, the distance between the materials, compared to the size of the desorption laser spot, is insignificant. With adequate laser intensity, the entire solid droplets (which are amorphous solids, not small crystallites) are desorbed from the sample surface. If there is a poor match between the properties of the analyte, matrix, and cationization reagent, electrospray deposition decreases the chance of segregation of the MALDI sample upon drying.

Not all samples require use of the electrospray deposition; good quantitative results can be obtained with a good match of matrix, analyte, cationization reagent, and solvent, but a poor match will result in little or no MALDI signal. With a close match of matrix, cationization reagent, solvent, and analyte, a reasonably homogeneous solid solution can be formed *even when using the dried droplet method* of sample preparation. For instance, for a low-molecular-mass PEG such as PEG 1500, using DHB in methanol with an M/A ratio of about 1000 (using adventitious sodium for cationization), a good solid sample is formed, leading to excellent signal reproducibility. However, as the match between matrix, analyte, and cationization reagent becomes less, the advantage of the electrospray deposition technique increases.

6.4.13 Chromatography in Sample Preparation

As will be discussed in greater detail in Chapter 11, the addition of a chromatographic separation prior to MALDI analysis provides a significant advantage. Conventional liquid chromatographic detectors (e.g., UV-visible absorption, refractive index, and light scattering) are generally limited in the sensitivity and molecular information that they can provide as compared to MALDI. More complicated (i.e., real-world industrial) samples often contain multiple sample components with very different chemistries. As discussed in detail above, the optimum MALDI sample preparation for these individual components may be very different. Discrete samples may be created from collected fractions, allowing the MALDI preparation to be custom tailored for each fraction. In addition, continuous deposition of the chromatographic trace may be accomplished with commercially available robotic devices. The specific use of gel permeation chromatography (GPC) to reduce molar mass discrimination that is usually observed with high polydispersity samples will be described in more detail below.

6.5   EVIDENCE OF POOR MALDI SAMPLE PREPARATION

As described above, successful MALDI sample preparation involves creating intimate contact between the matrix (which may consist of a few different molecules) and the analyte. The experiment requires each of the matrix roles to be completed to obtain a usable mass spectrum. To demonstrate the importance of the matrix roles, Figures 6.7 and 6.8 show two examples where a specific role of the matrix was not fulfilled.

Example 1: Figure 6.7 shows a problem with the role of desorber. Here a PEG 1000 sample is prepared for analysis using DHB (panel a) and the potassium salt of DHB (KDHB) in panel b as the matrix. In panel a, strong PEG peaks cationized with Na^+ are observed (by adventitious Na^+ from the matrix). While ample cation is present in the sample shown in panel b, the KDHB is ineffective at desorbing the PEG analyte from the MALDI sample when illuminated with the laser beam.

Example 2: Figure 6.8 shows a problem with the role of ionization agent. Here a PS 5050 sample is prepared for analysis using IAA as a matrix both without (panel a) and with (panel b) the addition of AgTFA as a cationization reagent. Note that as both the proton and the alkali ion affinity of PS is low, only Ag^+ cationized peaks are observed in panel b. When no cationization agent is present in the sample, no sample related peaks are observed. Interestingly, in the case whose data are shown

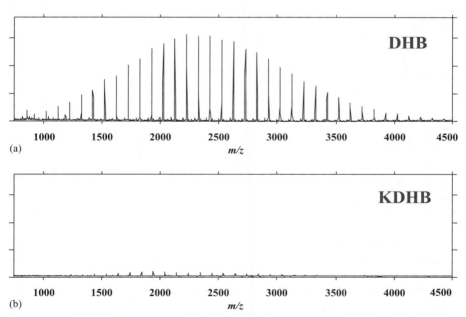

Figure 6.7.   MALDI mass spectra of PMMA 2900 obtained using (a) DHB and (b) KDHB as the matrix. The analyte peaks observed in panel a are cationized with adventitious sodium.

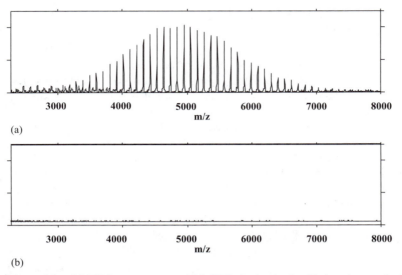

(a)

(b)

Figure 6.8.   MALDI mass spectra of PS 5050 obtained using IAA as the matrix (a) with and (b) without the addition of AgTFA as the cationization reagent.

in panel a, the PS analyte is desorbed from the matrix into the gas-phase; the absence of an appropriate cationization agent allows the preparation of isolated neutral gas-phase molecules of high-molecular-mass analytes for analysis using other techniques.

The heterogeneous incorporation of analyte within the solid matrix preparation, which is characteristic of using the dried droplet method, often results in a wide range of analyte concentrations throughout the MALDI sample. In view of the M/A plots discussed above, it is expected that the analyte signal will therefore also be variable as the laser beam is rastered around the sample. Those positions that yield high analyte signal intensities have been termed "sweet spots." In fact, the simple observation of "sweet spots" on the sample (i.e., implying highly variable ion signal intensities as a function of spatial position) is evidence of a poor sample preparation. Further evidence of a poor preparation is given by the observation of sample discrimination effects. However, in order to uncover these discrimination effects, the true composition of the sample must be known either by preparation or from an alternate analytical analysis. These discrimination effects fall into two categories, molar mass discrimination and structural discrimination.

For polymer samples that cover a broad mass range (i.e., they have a large polydispersity), molar mass discrimination is evidenced by a bias in the observed MALDI spectrum. In most cases, the bias is to low mass; that is, the low-mass oligomers are overrepresented in the observed mass spectrum. This type of discrimination is most likely due to sample preparation issues, in many cases the solubility of the oligomers varies significantly as a function of mass. This was demonstrated for low (<5000 Da) polyethylene terephthalate (PET) samples analyzed using THF (a poor solvent which dissolves only the lower-mass oligomers) and a 70 : 30 (v/v) methylene chloride (CH_2Cl_2) : hexafluorisopropanol (HFIP) azeo-trope [58]. As described above regarding the importance of M/A ratio, it is also possible that different molecular mass species require different M/A ratios for analysis. Molar mass discrimination may also be due to ionization effects; higher molecular mass species are usually ionized more efficiently than lower molecular mass species. Overall, polymer samples with PD > 1.2 are found to exhibit molar mass discrimination. The Montaudo group [59, 60] has demonstrated that this problem may be overcome by preseparation of the polymer sample using gel per-meation chromatography. MALDI analysis is then performed on each of the col-lected fractions, and the data are combined to give a full mass spectrum. It should not be forgotten, however, that TOFMS instruments have a severely restricted dynamic range (the high-speed digitizers used usually have an intensity axis resolu-tion of 256). To accurately reproduce the entire molar mass distribution, the required dynamic range is generally much higher [61].

In addition to molar mass discrimination, structural discrimination may also be observed in MALDI samples. Structural discrimination can result from differ-ences in the solubility of different components in the solid state (e.g., due to the presence of hydrophilic versus hydrophobic end groups), or from differences in ionization of the components of the mixture. This is particularly important in sample mixtures containing different end groups, as an important result of the analysis may be the quantitative determination of the quantity of each end group present in the sample. "Simple" polymer blends can be some of the hardest samples to analyze via MALDI. It may be possible to prepare the sample using one matrix, cationization agent, or set of additives that will produce signal from one component of the mixture, and use a different sample preparation to view the second component (remember the PEG/PTMEG blend example given above). While this may be useful when looking for small changes in low-concentration contaminants that affect the sample properties, it can be an analytical nightmare for determining the composition of even the major components of true unknowns. Without extensive sample knowledge, the intensity of a particular polymer analyte series in the observed mass spectrum cannot be used directly for quantitative analysis.

6.6   A DETAILED EXAMPLE

To illustrate the value of understanding the underlying principles of the matrix roles and the key issues involved in developing new sample preparation methods for polymer MALDI, we provide the following example of a new sample preparation method that was developed in a few minutes for a novel polymer material [18]. In collaboration with Prof. Li Jia of Lehigh University, we developed a MALDI sample preparation method to characterize the chemical structures of novel materials result-ing from the copolymerization of N-allylaziridines and carbon monoxide [62]. One example of this type of material is the copolymerization of N-ethylaziridine with CO. The resulting polymer was soluble in water and water/methanol mixtures, and partially soluble in methanol. This solubility most closely matches that of poly(ethynylformamide) PEF 1500 in Figure 6.3. Using Figure 6.3, PEF 1500 is in

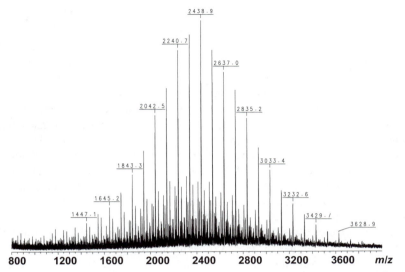

Figure 6.9. MALDI mass spectrum of a novel polymeric material produced from the copolymerization of N-ethylaziridine with CO. The sample was prepared using methanol and DHB. Reproduced from Reference 62 with permission.

the most hydrophilic group, which suggests that either thiourea (TU) or DHB would be the best first choice for a matrix. Using a sample preparation of methanol, DHB, and dried droplet deposition, we acquired the mass spectrum shown in Figure 6.9. The mass spectrum shows the sodium cationized repeat units and end groups expected for this polymer. This is an example of the analysis of a new polymeric material that to our knowledge had never been analyzed previously by MALDI. In a matter of a few minutes, a sample preparation method was developed using the principles described above.

6.7 CONCLUSION

Sample preparation is the key to a successful MALDI experiment. Even before work at the bench begins, a large number of decisions must be made, the goal of which is to produce a homogeneous solid sample solution. One of the key concepts is the incorporation of the analyte into the chosen matrix solid. Heterogeneous incorporation results in the observation of large spot-to-spot variability in the analyte signal, which generally leads experimenters to search for the "sweet spots" on the sample surface. Invoking the concepts of solubility or miscibility of the analyte in the solid matrix enables an organized means of searching for the proper matrix to use for a given analyte. It also is seen to explain many of the practices found in the literature (the use of matrix additives, the application of heat, vacuum, and fast-drying solvents, etc.) suggested as a means of empirically producing better MALDI samples. In fact, anything that affects the rate of precipitation will have an effect on the distribution of analyte, cationization agent, and other impurities in the solid sample.

As described in this chapter, the chosen "matrix" must fulfill a number of distinct roles (dispersant, desorber, ionization agent) in the MALDI analysis. In our experience, dispersion, desorption, and ionization are separate processes that can generally be optimized independently. Most importantly, *each of these roles must be fulfilled* in order to result in a successful MALDI experiment. While many researchers search for a *single molecule* that may be a good matrix for a given class of analytes, it is possible that a *mixture of chemical compounds* may prove more effective if each component of the mixture is selected to optimize its role in the MALDI process.

As development of the MALDI technique continues, we have seen it grow from a qualitative ("what is there?") to a quantitative ("how much is there?") analysis tool. The next development, which is generally occurring now, is to answer the question "how well do we know how much is there?" This can be done by statistically analyzing replicate experiments, the judicious addition of internal standards, and the application of standard additions techniques and factorial design experiments. Careful control of the sample preparation, coupled with the use of techniques other than MALDI to study the prepared MALDI samples, leads to better understanding of the MALDI process, which ultimately leads to greater success in the analysis of real-world samples.

ACKNOWLEDGMENTS

The authors would like to acknowledge the effort of a number of students in the Owens group who have contributed immensely to our understanding of the MALDI sample process over the years, including Richard King, Yansan Xiong (Henry Shion), Robert Goldschmidt, Russell Hensel, Cynthia Chavez-Eng, Andrew Hoteling, and William Erb. We would also like to acknowledge fruitful discussions over the years with William Simonsick, Jr., William Nichols, David Parees, and Richard Knochenmuss. We would also like to thank Air Products & Chemicals, Inc., E.I. DuPont deNemours, Inc., the Eastman Kodak Company, the Xerox Foundation, and the National Science Foundation for research funding and Rohm & Haas, Inc. for the generous donation of a MALDI-TOFMS instrument.

REFERENCES

1. Cotter, R.J., Lasers and mass spectrometry, *Anal. Chem.*, **1984**, 56(3), 485A–504A.
2. Kistemaker, P.G.; Van der Peyl, G.J.Q.; Haverkamp, J., Laser desorption mass spectrometry, *Soft Ioniz. Biol. Mass Spectrom., Proc. Chem. Soc. Symp.*, **1981**, 120–136.
3. Gardella, J.A. Jr.; Graham, S.W.; Hercules, D.M., Structural analysis of polymeric materials by laser desorption mass spectrometry, *Adv. Chem. Series*, **1983**, 203, 635–676.
4. Hillenkamp, F., Laser desorption techniques of nonvolatile organic substances, *Int. J. Mass Spectrom. Ion Phys.*, **1982**, 45, 305–313.
5. Karas, M.; Bachmann, D.; Hillenkamp, F., Influence of the wavelength in high-irradiance ultraviolet laser desorption mass spectrometry of organic molecules, *Anal. Chem.*, **1985**, 57(14), 2935–2939.

6. Cramer, R.; Haglund, R.F., Jr., Hillenkamp, F., Matrix-assisted laser desorption and ionization in the O–H and C=O absorption bands of aliphatic and aromatic matrixes: dependence on laser wavelength and temporal beam profile, *Int. J. Mass Spectrom. Ion Proc.*, **1997**, 169/170, 51–67.

7. Cramer, R.; Burlingame, A.L., IR-MALDI: softer ionization in MALDI-MS for studies of labile macromolecules, *Mass Spectrom. Biol. Med.*, **2000**, 289–307.

8. Zhang, J.; Kinsel, G.R., Quantification of protein-polymer interactions by matrix-assisted laser desorption/ionization mass spectrometry, *Langmuir*, **2002**, 18, 4444–4448.

9. Hillenkamp, F.; Karas, M.; Beavis, R.C.; Chait, B.T., Matrix-assisted laser desorption/ionization mass spectrometry of biopolymers, *Anal. Chem.*, **1991**, 63(24), 1193A–1203A.

10. Juhasz, P.; Costello, C.E.; Biemann, K., Matrix-assisted laser desorption ionization mass spectrometry with 2-(4-hydroxyphenylazo)benzoic acid matrix, *J. Am. Soc. Mass Spectrom.*, **1993**, 4(5), 399–409.

11. Dreisewerd, K., The desorption process in MALDI, *Chem. Rev.*, **2003**, 103(2), 395–425.

12. Beavis, R.C.; Chait, B.T., Velocity distributions of intact high mass polypeptide molecule ions produced by matrix assisted laser desorption, *Chem. Phys. Lett.*, **1991**, 181(5), 479–484.

13. van der Muelen, A.; Rulis, A.M.; deVries, A.E., Molecular beam study of the K+ Br2 reaction in the electronvolt energy region, *Chem. Phys.*, **1975**, 7, 1–16.

14. Litvak, H.E.; Urena, A.G.; Bernstein, R.B., Translational energy dependence of the cross-section for Rb + CH3I → RbI + CH3 from $_0$.12 to 1.6 eV (c.m.), *J. Chem. Phys.*, **1974**, 61, 4091–4100.

15. Ronge, C.; Pesnelle, A.; Perdrix, M.; Watel, G., Ionization of low-Rydberg-state helium atoms by polar molecules: I. velocity dependence of the cross sections, *Phys. Rev. A: Atom. Mol. Opt. Physics*, **1988**, 38, 4552–4559.

16. Hildebrand, J.H.; Prausnitz, J.M.; Scott, R.L., *Regular and Related Solutions: The Solubility of Gases, Liquids, and Solids*. New York: Van Nostrand Reinhold, **1970**.

17. Hansen, C.M., *The Three Dimensional Solubility Parameter and Solvent Diffusion Coefficient: Their Importance In Surface Coating Formulation*, Copenhagen: Danish Technical Press, **1967**.

18. King, R.C. III, Laser Desorption/Laser Ionization Time-of-Flight Mass Spectrometry Instrument Design and Investigation of the Desorption and Ionization Mechanisms of Matrix-Assisted Laser Desorption/Ionization. PhD Thesis, Drexel University, **1994**.

19. Dai, Y.; Whittal, R.M.; Li, L., Confocal fluorescence microscopic imaging for investigating the analyte distribution in MALDI matrixes, *Anal. Chem.*, **1996**, 68(15), 2494–2500.

20. Horneffer, V.; Forsmann, A.; Strupat, K.; Hillenkamp, F.; Kubitscheck, U., Localization of analyte molecules in MALDI preparations by confocal laser scanning microscopy, *Anal. Chem.*, **2001**, 73(5), 1016–1022.

21. Yalcin, T.; Dai, Y.; Li, L., Matrix-assisted laser desorption/ionization time-of-flight mass spectrometry for polymer analysis: solvent effect in sample preparation, *J. Am. Soc. Mass Spectrom.*, **1998**, 9(12), 1303–1310.

22. Hanton, S.D.; Owens, K.G., Using MESIMS to analyze polymer MALDI matrix solubility, *J. Am. Soc. Mass Spectrom.*, **2005**, 16, 1172–1180.

23. Coleman, M.M.; Sermon, C.J.; Bhagwager, D.E.; Painter, P.C., A practical guide to polymer miscibility, *Polymer*, **1990**, 31, 1187–1203.

24. Hoteling, A.J.; Erb, W.J.; Tyson, R.J. III; Owens, K.G., Exploring the importance of the relative solubility of matrix and analyte in MALDI sample preparation using HPLC, *Anal. Chem.*, **2004**, 76(17), 5157–5164.

25. Bauer, B.J.; Byrd, H.C.M.; Guttman, C.M., Small angle neutron scattering measurements of synthetic polymer dispersions in matrix-assisted laser desorption/ionization matrixes, *Rapid Commun. Mass Spectrom.*, **2002**, 16(15), 1494–1500.

26. Sheiko, S.S.; Sun, F.C.; Randall, A.; Shirvanyants, D.; Rubinstein, M.; Lee, H.; Matyjaszewski, K., Adsorption-induced scission of carbon–carbon bonds, *Nature*, **2006**, 440(7081), 191–194.

27. Distler, A.M.; Allison, J., Improved MALDI-MS analysis of oligonucleotides through the use of fucose as a matrix additive, *Anal. Chem.*, **2001**, 73(20), 5000–5003.

28. Kassis, C.M.; DeSimone, J.M.; Linton, R.W.; Lange, G.W.; Friedman, R.M., An investigation into the importance of polymer-matrix miscibility using surfactant modified matrix-assisted laser desorption/ionization mass spectrometry, *Rapid Commun. Mass Spectrom.*, **1997**, 11(13), 1462–1466.

29. Tummala, R.; Limbach, P.A., Effect of sodium dodecyl sulfate micelles on peptide mass fingerprinting by matrix-assisted laser desorption/ionization mass spectrometry, *Rapid Commun. Mass Spectrom.*, **2004**, 18(18), 2031–2035.

30. Zhu, Y.F.; Taranenko, N.I.; Allman, S.L.; Martin, S.A.; Haff, L.; Chen, C.H., The effect of ammonium salt and matrix in the detection of DNA by matrix-assisted laser desorption/ionization time-of-flight mass spectrometry, *Rapid Commun. Mass Spectrom.*, **1996**, 10(13), 1591–1596.

31. Juhasz, P.; Biemann, K., Mass spectrometric molecular-weight determination of highly acidic compounds of biological significance via their complexes with basic polypeptides, *Proc. Natl. Acad. Sci. U S A*, **1994**, 91(10), 4333–4337.

32. Tsarbopoulos, A.; Karas, M.; Strupat, K.; Pramanik, B.N.; Nagabhushan, T.L.; Hillenkamp, F., Comparative mapping of recombinant proteins and glycoproteins by plasma desorption and matrix-assisted laser desorption/ionization mass spectrometry, *Anal. Chem.*, **1994**, 66(13), 2062–2070.

33. Evason, D.J.; Claydon, M.A.; Gordon, D.B., Effects of ion mode and matrix additives in the identification of bacteria by intact cell mass spectrometry, *Rapid Commun. Mass Spectrom.*, **2000**, 14(8), 669–672.

34. Xiong, Y.H., Mechanistic investigation of the cationization process of matrix-assisted laser desorption/ionization time-of-flight mass spectrometry. PhD Thesis, Drexel University, **1997**.

35. Chen, R.; Yalcin, T.; Wallace, W.E.; Guttman, C.M.; Li, L., Laser desorption ionization and MALDI time-of-flight mass spectrometry for low molecular mass polyethylene analysis, *J. Am. Soc. Mass Spectrom.*, **2001**, 12(11), 1186–1192.

36. Lin-Gibson, S.; Brunner, L.; Vanderhart, D.L.; Bauer, B.J.; Fanconi, B.M.; Guttman, C.M.; Wallace, W.E., Optimizing the covalent cationization method for the mass spectrometry of polyolefins, *Macromolecules*, **2002**, 35(18), 7149–7156.

37. Schriemer, D.C.; Li, L., Mass discrimination in the analysis of polydisperse polymers by MALDI time-of-flight mass spectrometry: 1. sample preparation and desorption/ionization issues, *Anal. Chem.*, **1997**, 69(20), 4169–4175.

38. Chavez-Eng, C.M., Quantitative Aspects of Matrix-Assisted Laser Desorption/Ionization Using Electrospray Deposition. PhD Thesis, Drexel University, **2002**.

39. Gantt, S.L.; Valentine, N.B.; Saenz, A.J.; Kingsley, M.T.; Wahl, K.L., Use of an internal control for matrix-assisted laser desorption/ionization time-of-flight mass spectrometry analysis of bacteria, *J. Am. Soc. Mass Spectrom.*, **1999**, 10, 1131–1137.

40. Hoteling, A.J., MALDI TOF PSD and CID: Understanding Precision, Resolution and Mass Accuracy and MALDI TOFMS: Investigation of Discrimination Issues Related to Solubility. PhD Thesis, Drexel University, **2004**.

41. Erb, W.J., Exploration of the Fundamentals of Matrix-Assisted Laser Desoprtion/Ionization Time-of-Flight Mass Spectrometry. PhD Thesis, Drexel University, **2007**.

42. Macha, S.F.; Limbauch, P.A.; Hanton, S.D.; Owens, K.G., Matrix-silver cluster interferences in the matrix-assisted laser desorption/ionization mass spectrometry (MALDI-MS) of non-polar polymers, *J. Am. Soc. Mass Spectrom.*, **2001**, 12(6), 732–743.

43. Keller, B.O.; Li, L., Discerning matrix-cluster peaks in matrix-assisted laser desorption/ionization time-of-flight mass spectra of dilute peptide mixtures, *J. Am. Soc. Mass Spectrom.*, **2000**, 11(1), 88–93.

44. Hoteling, A.J.; Owens, K.G., Improved PSD and CID on a MALDI/TOFMS, *J. Am. Soc. Mass Spectrom.*, **2004**, 15(4), 523–535.

45. McCombie, G.; Knochenmuss, R., Enhanced MALDI ionization efficiency at the metal-matrix interface: practical and mechanistic consequences of sample thickness and preparation method, *J. Am. Soc. Mass Spectrom.*, **2006**, 17(5), 737–745.

46. Fan, X.; Beavis, R.C., A method to increase contaminant tolerance in protein matrix-assisted laser desorption/ionization by the fabrication of thin protein-doped polycrystalline films, *Rapid Commun. Mass Spectrom.*, **1994**, 8(2), 199–204.

47. Vorm, O.; Roepstorff, P.; Mann, M., Improved resolution and very high sensitivity in maldi tof of matrix surfaces made by fast evaporation, *Anal. Chem.*, **1994**, 66, 3281–3287.

48. Cohen, S.L.; Chait, B.T., Influence of matrix solution conditions on the MALDI-MS analysis of peptides and proteins, *Anal. Chem.*, **1996**, 68(1), 31–37.

49. Hanton, S.D.; Owens, K.G.; Chavez-Eng, C.; Hoberg, A.-M.; Derrick, P.J., Updating evidence for cationization of polymers in the gas phase during matrix-assisted laser desorption/ionization, *Euro. Mass Spectrom.*, **2005**, 11(1), 23–29.

50. Xiang, F.; Beavis, R.C., A method to increase contaminant tolerance in protein matrix-assisted laser desorption/ionization by the fabrication of thin protein-doped polycrystalline films, *Rapid Commun. Mass Spectrom.*, **1994**, 8(2), 199–204.

51. Dai, Y.; Whittal, R.M.; Li, L., Two-layer sample preparation: a method for MALDI-MS analysis of complex peptide and protein mixtures, *Anal. Chem.*, **1999**, 71(5), 1087–1091.

52. Keller, B.O.; Li, L., Three-layer matrix/sample preparation method for MALDI MS analysis of low nanomolar protein samples, *J. Am. Soc. Mass Spectrom.*, **2006**, 17(6), 780–785.

53. Pastor, S.J.; Wood, S.H.; Wilkins, C.L., Poly(ethylene glycol) limits of detection using internal matrix-assisted laser desorption/ionization Fourier transform mass spectrometry, *J. Mass Spectrom.*, **1998**, 33(5), 473–479.

54. Wang, L.; May, S.W.; Browner, R.F., Low-flow interface for liquid chromatography-inductively coupled plasma mass spectrometry speciation using an oscillating capillary nebulizer, *J. Anal. At. Spectrom.*, **1996**, 11(12), 1137–1146.

55. Hensel, R.R.; King, R.C.; Owens, K.G., Electrospray sample preparation for improved quantitation in matrix-assisted laser desorption/ionization time-of-flight mass spectrometry, *Rapid Commun. Mass Spectrom.*, **1997**, 11(16), 1785–1793.

56. Axelsson, J.; Hoberg, A.-M.; Waterson, C.; Myatt, P.; Shield, G.L.; Varney, J.; Haddleton, D.M.; Derrick, P.J., Improved reproducibility and increased signal intensity in matrix-assisted laser desorption/ionization as a result of electrospray sample preparation, *Rapid Commun. Mass Spectrom.*, **1997**, 11(2), 209–213.

57. Hanton, S.D.; Owens, K.G.; Blair, W.; Hyder, I.Z.; Stets, J.R.; Guttman, C.M.; Giuseppetti, A., Investigations of electrospray sample deposition for polymer MALDI, *J. Am. Soc. Mass Spectrom.*, **2004**, 15(2), 168–179.

58. Hoteling, A.J.; Mourey, T.H.; Owens, K.G., The importance of solubility in the sample preparation of poly(ethylene terephthalate) for MALD/I TOFMS, *Anal. Chem.*, **2005**, 77(3), 750–756.

59. Montaudo, G.; Garozzo, D.; Montaudo, M.S.; Puglisi, C.; Samperi, F., Molecular and structural characterization of polydisperse polymers and copolymers by combining MALDI-TOF mass spectrometry with GPC fractionation, *Macromolecules*, **1995**, 28(24), 7983–7989.

60. Montaudo, G.; Montaudo, M.S.; Puglisi, C.; Samperi, F., Molecular weight distribution of poly(dimethylsiloxane) by combining matrix-assisted laser desorption/ionization time-of-flight mass spectrometry with gel-permeation chromatography fractionation, *Rapid Commun. Mass Spectrom.*, **1995**, 9(12), 1158–1163.

61. Mourey, T.H.; Hoteling, A.J.; Balke, S.T.; Owens, K.G., Molar mass distributions of polymers from SEC and MALDI-TOF MS: methods for comparison, *J. Appl. Polym. Sci.*, **2005**, 97, 627–639.

62. Jia, L.; Sun, H.; Ding, E.; Allegeier, A.M.; Hanton, S.D., Living alternating copolymerization of N-alkylaziridines and carbon monoxide as a route for synthesis of poly-β-peptoids, *J. Am. Chem. Soc.*, **2002**, 124(25), 7282.

CHAPTER 7

SOLVENT-FREE MALDI SAMPLE PREPARATION

Sarah Trimpin

Department of Chemistry, Wayne State University, Detroit, Michigan

7.1 INTRODUCTION AND SCOPE

Analytical techniques ideally aim for a maximum degree of unbiased analytical information. Mass spectrometry (MS) should be capable of characterizing any synthetic polymer qualitatively and quantitatively. In reality, of course, this is not the case; however, simplification can certainly make techniques more efficient and often more reliable [1]. Such is the case with solvent-free matrix-assisted laser desorption/ ionization mass spectrometry (MALDI) MS which continues to gain in analytical importance [2]. The major breakthrough of the solvent-free MALDI sample preparation method came with the first successful analysis of an insoluble poly(9,9-diphenyl-2,7-fluorene) (polyfluorene) (Figure 7.1) [3]. Recent developmental improvements to the original solvent-free MALDI method were made through industrial and governmental collaborations [4, 5, 6] or directly by industrial laboratories [7], and indicates a general interest in improving the MALDI method for analyzing synthetic polymers. The ability to simultaneously and precisely prepare over 100 synthetic polymer samples by a solvent-free on-target homogenization/ transfer MALDI method [4] offers the powerful opportunity for decongestion of solubility-restricted and even insoluble complex mixtures by coupling to powerful solvent-based separation utilizing liquid chromatography (LC) (see Chapter 11) [6] and/or solvent-free gas phase separation utilizing ion mobility spectrometry (IMS)-MS [8, 9]. These developments provide a likely optimistic and productive future for the field of MALDI-MS.

This chapter will focus on different MALDI concepts, both past and present, all of which are currently valid approaches. It is left to the reader to employ the approach most appropriate to the individual needs and/or applications. The challenges associated with MALDI-MS will be addressed, including segregation due to crystallization effects (e.g., polydimetylsiloxane (PDMS)), unusual solvent-driven problems in highly compatible systems (e.g., polystyrene (PS)/1,8,9-anthracenetriol (dithranol)/tetrahydrofuran (THF)), desorption/ionization preferences in mixture

MALDI Mass Spectrometry for Synthetic Polymer Analysis, Edited by Liang Li
Copyright © 2010 John Wiley & Sons, Inc.

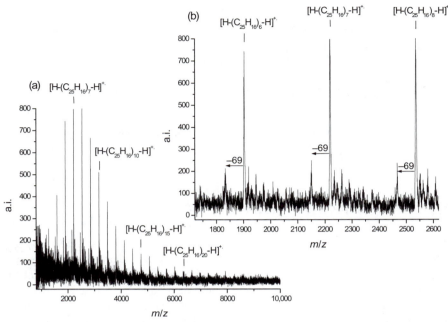

Figure 7.1. MALDI mass spectrum of the insoluble fraction of polyfluorene after Soxhlet extraction in toluene for 5 days; the sample was prepared with dithranol as matrix and BM for homogenization: (a) full scale, (b) inset of the three most intense signals, the signals at −69 Da to each of the expected oligomer ions were characterized by detailed post-source decay (PSD) studies to be fragment ions where one phenyl side chain is cleaved. This fragmentation also occurred already as in-source decay (ISD) when more laser power was employed, which was than additionally to the PSD fragment ion (−69 Da) present in the mass spectra as ISD fragment ion (−77 Da). Mass spectra were recorded using a Bruker Reflex II™ MALDI-TOF mass spectrometer (Bremen, Germany) equipped with a N₂-laser.

analysis (e.g., poly(para-phenylene vinylene) (PPV) versus poly(ethylene glycole) (PEG)), as well as high-throughput issues in an industrial setting. Qualitative and quantitative improvements of analytical results for these examples will be demonstrated from the implementation of the nontraditional solvent-free MALDI method for the analysis of synthetic polymers.

7.2 BACKGROUND AND MOTIVATION

MALDI-MS has been extensively applied to the characterization of soluble macromolecules [1, 10, 11] and has even been used to detect intact soluble molecules with masses higher than 1,000,000 Da [12, 13]. The potential of MALDI for the characterization of synthetic polymers was first demonstrated by Tanaka et al. [10]. The principles of MALDI-MS based on the traditional solvent-based method are described in Chapter 6. One of the central aspects of MALDI analysis is the preparation of a "good" MALDI sample. It is not always as trivial as applying a matrix material and an analyte to the surface of the MALDI plate since it is the special analyte–matrix preorganization which makes the MALDI analysis successful. Many variables influence the integrity of a "good" MALDI sample and may include the concentration of the matrix and analyte, choice of solvent [14], matrix, cationization reagent, sample history, contaminants, and compatible solubilities of matrix and analyte solutions. Hence, in MALDI-MS, sample preparation is one of the key factors that greatly influences the success and the quality of the analytical result. Because there are too many contributing parameters in solvent-based sample preparation for MALDI-MS, there was a need for a more simplistic method that is easily applicable and has fewer critical parameters. One critical parameter is the solvent. There are three crucial aspects to "solvent." First, the chosen solvent must solubilize the entire sample, which may not be an easy task due to varying differences in solubility throughout different parts of the sample. For example, strong adhesion, aggregation, and low solubility of some of the constituents may exist. Subsequently, the analysis of these molecules fails and leads to the erroneous conclusion that they were not present in the sample. The second crucial aspect to the "solvent" is that the solvent system used must initially homogenize a sample with matrix in the liquid MALDI phase. Lastly, the mixture on the MALDI plate must subsequently yield a locally homogenous sample and matrix composition upon evaporation of the solvent, which is ultimately important in achieving reproducibility. Any of these issues can introduce serious limitations to the success and reliability of the analytical results. However, the third aspect is the most complex and often responds poorly in optimization trials. To understand the generality of this problem (often referred to as segregation), one has to keep in mind that (re)-crystallization is commonly used as a purification step in synthetic chemistry. Any crystallization procedure involves separation phenomena of two or more compounds in differing degrees. Intrinsically, any solvent-based MALDI sample preparation suffers a lack of homogeneity caused by solvent evaporation during the crystallization step. For these various reasons, dry sample preparation strategies were introduced [15, 16, 17, 18] and established [19] as nontraditional methods for MALDI-MS for the analysis of synthetic polymers. The strength of the solvent-free approach is its capacity to yield reproducible analytical results for some insoluble and poorly soluble polymers (Figure 7.1) [3].

7.3 EXPERIMENTAL CONCEPTS

7.3.1 Original Concepts

In the year 2000, several laboratories independently published work that focused on the MALDI sample preparation of soluble [16] and solubility-limited synthetic polymers [15], insoluble large polyaromatic hydrocarbons (PAHs) [17], and fragmentation labile peptides [18] prepared in the absence of solvent. Solvent-free methods generally consist of two steps: "dry" homogenization, usually carried out by mechanical mixing of sample, matrix, and salt, and transfer of the resulting powder mixture to the MALDI plate. In solvent-free MALDI-MS, at no point is solvent employed to homogenize analyte, matrix, and salt or in the transfer to the MALDI plate. In the case of a dissolved sample, the sample is dried prior to solvent-free preparation. Hence, homogenization and transfer of the sample in these approaches is not based on the solvent and its properties. The organic or aqueous medium the polymer was originally dissolved in is entirely irrelevant. The MALDI analysis is therefore simplified because fewer combinations and issues of compatibility or solubility have to be considered and explored.

Two general methods were originally employed for homogenization of polymeric material, MALDI matrix, and salt: grinding by mortar and pestle (about 5 min) [15, 19] and shaking by ball-mill (BM) (about 5–10 min) [3, 19, 20], the latter having the advantage in that the process is accomplished automatically. Both methods were applicable over a broad homogenization time period; however, it is not known what is the minimum or maximum period needed to yield satisfactory results. In its infancy, ball-milling was carried out in a metal vessel (tungsten carbide (WC)), into which the matrix and sample is added to be ground, along with a suitably sized ball that is preferably made of the same material as the vessel. Over time, it was shown that the vessel or ball material has no significant impact on performance, thus disposable aluminum vessels were custom-built, avoiding cleaning issues and troublesome carryover. Grinding takes place as a result of the interaction between the ball, sample particles, and grinding vessel wall. The technique works equally well on soft, medium-hard, and even extremely hard, oily, brittle, and fibrous materials. In fact, viscous material (PS, 100 kDa) exerted one of the greatest challenges yet, but was overcome by cooling the vessel during homogenization, which made the polymer more brittle and easier to mix [19]. The grinding ball is free to move around during the shaking process, which gives the technique the ability to produce much finer particles than other methods provide. Generally speaking, smaller balls and longer grinding times yield smaller particles.

Originally, transfer of the dry MALDI sample to the MALDI plate used either a loose powder or a pressed pellet method. In the case of the loose power method, the homogenized MALDI sample is transferred with a spatula [3, 4, 5, 7, 19, 20] or, as later shown, the wooden end of a Q-tip [4], with the latter avoiding undesired scratching of the plate. The prepared sample is gently pressed directly to the MALDI plate [3, 4, 7, 19, 20] or to a conducting carbon disc affixed to the MALDI plate [16] to produce a thin film. In the other approach, a pressed pellet (similar to the potassium bromide (KBr) pellets for infrared analysis) is prepared and attached to the MALDI plate with double-sided adhesive tape [15, 19], which has the intrinsic disadvantage of being more involved and time-consuming in preparation.

The important similarities and differences in the MALDI analytical results are that the different homogenization methods do not produce significantly different analytical results, at least for low-mass polymers; however, the transfer methods of the sample to the plate appear to be important [19]. The loose powder approach, in contrast to the pellet variant, allows the use of a large excess of matrix [3, 19], and permits true matrix-assistance, as evident by its capacity for high-molecular-weight analysis [19], which has not been observed in other solvent-free approaches as reported for example in Reference 21. Further advantages of the loose powder compared to the pellet method include improved sensitivity, higher-mass resolution, and

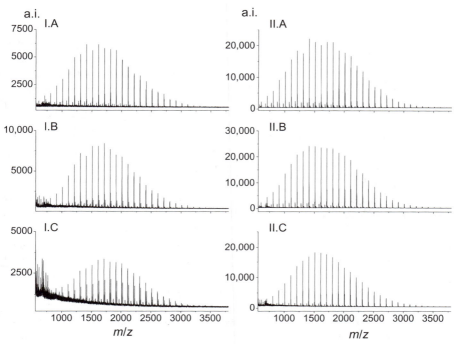

Figure 7.2.   Solvent-free MALDI mass spectrum of PS in comparison with the solvent-based MALDI mass spectrum. With increasing matrix dilution (A 1 : 50, B 1 : 500, and C 1 : 5000) the signal-to-noise-ratio appears to be less affected using (II) the solvent-free method than (I) the solvent-based MALDI method. Mass spectra were recorded using a Bruker Reflex II™ MALDI-TOF mass spectrometer equipped with a N_2-laser.

less restrictive molar analyte : matrix ratios [19]. Hence, only the loose powder method achieves results that are at least comparable to that of the traditional solvent-based MALDI method (Figure 7.2; low-molecular-weight PS). Overall, the original approaches for avoiding solvent in sample preparation seem to vary in methodology only slightly, but the differences in the resulting data are significant. Solvent-free loose powder MALDI [19], irrespective of the employed homogenization procedure, appears to be the most promising sample preparation method [4, 7, 22, 23, 24] to improve the reliability and the potential of MALDI analysis. For this reason, this method was evaluated systematically by comparison with solvent-based MALDI-MS [19]. Numerous experiments established solvent-free analysis as a complementary MALDI method, frequently providing valuable additional information and improving analytical results. Generally, the solvent-free method allows for more homogeneous analyte : matrix mixtures, as well as higher shot-to-shot and sample-to-sample reproducibility [3, 4, 5, 6, 7, 15, 17, 19, 25, 26,27]. As a result, less laser power has to be applied, which yields milder MALDI conditions, as evidenced by reduced analyte fragmentation and lessened background signals, and provides better resolution of the analyte signals [3, 5, 18, 19].

7.3.2 Concept of Vortexing

A wider application of the solvent-free MALDI method for polymer analysis was delayed due to practical aspects of the initial methods. These may include characteristics such as being time-intensive, which is mainly due to intensive cleaning of the equipment to avoid carryover, particularly considering the mortar and pestle application, and availability, specifically in the case of the BM method. An efficient procedure was required to increase the meaningful impact of solvent-free MALDI-MS for the analysis of synthetic polymers. Modifications were made to the original solvent-free loose powder MALDI method which resulted in a simple method that utilizes a common laboratory vortex instrument, a pair of BBs (the type used in air rifles; Zn-plated, 4.5 mm), and common glass vials [7]. This refined solvent-free MALDI method has been found to be efficient, requiring a minimum amount of polymeric sample and preparation time (0.5–1 min). Disposable glass vials and metal balls are used, thereby avoiding carryover and cleaning procedures. Mixing of polymer sample and matrix powder is performed with a vortexer followed by direct measurement of the analyte : matrix : salt powder loosely applied to the MALDI target [7]. In these studies it has been found that this solvent-free method is applicable to liquid, soft, and waxy polymers as demonstrated in the example shown for low-molecular-weight PEG. We are not aware of the vortex method being applied to higher-molecular-weight polymer analysis (>10,000 Da). The BM [19] and mini-ball-mill (MBM) methods [28] were successfully employed for high-molecular-weight analyses. This suggests that the quality of grinding gains in importance with increasing molecular weight of the polymer.

The vortex/BB technique made the workflow of the solvent-free and solvent-based MALDI methods comparable. If a sample is received in a MALDI-appropriate solution (in a vaporizable solvent), solvent-based MALDI can be used, and if the sample comes as a solid, solvent-free MALDI is preferred. However, solvent-free MALDI should be considered at all times since the analytical results of simple synthetic polymer standards generally improves with the implementation of the solvent-free method due to its lower laser threshold which provides less matrix background, better signal-to-noise ratio, less fragmentation, and fewer suppression effects, and overcomes artifacts related to the use of solvents [5, 19, 29]. Most importantly, if one deals with a problematic compound requiring a great deal of sample preparation and optimization, the solvent-free method, due to its simplicity in sample preparation by omission of the intractable solvent and possible oxidation/degradation of the analyte, clearly is the preferred method of choice.

Solvent-based MALDI analysis uses less material than solvent-free approaches (<0.1 mg of analyte). However, when optimization trials have to be performed, which are mainly related to solvent-based MALDI, then solvent-based methods may require more sample and be more time-intensive than the solvent-free MALDI method. In general, the sample amount requirement of ca. 0.1 mg employing solvent-free MALDI analysis is not a significant limitation because most polymeric samples are typically available in sufficient quantities. Biologically relevant amounts of analyte at the femtomole level [26] was recently achieved by utilizing the solvent-free MBM method by making use of smaller vials and beads [28].

Sufficient sample amounts may not be available for all synthetic polymers, as is the case, for example, with fractionated samples in gel-permeation chromatography (GPC) or LC-MALDI analysis. Additionally, high numbers of fractionated sample are generally produced in LC-MALDI which calls for the development of preparing multiple samples solvent-free simultaneously. Solvent-free sample preparation methods can even be simplified and the analytical utility increased beyond the vortex/BB method as is seen with the multi-sample on-target homogenization/transfer method [4] capable of effectively coupling LC to MALDI [6].

7.3.3   Concept of Multi-Sample Solvent-Free On-Target Homogenization/Transfer

Solvent-free MALDI-MS had been shown to simplify the sample preparation by removing the influence of solvent, which provides in cases where the solvent is problematic, considerable time efficiency. For samples in which the solvent is not problematic, the solvent-free method may be less efficient. The two major limitations to any of the previous solvent-free MALDI methods is that only one sample can be prepared at a time and the transfer of the sample from the homogenization vessel to the MALDI-plate is more time-demanding than when solvent-based MALDI methods are employed. In order to perform routine analysis precisely and accurately, and to acquire improved mass spectra faster than with solvent-based methods, it seemed reasonable to propose [5] an automated solvent-free sample preparation platform for high-throughput accurate mass determination (automated workstation and automated mass analysis) designed into a commercial MALDI-TOF mass spectrometer. Solvent-free sample preparation directly on the MALDI plate has also been proposed [7]. These first experiments of on-target solvent-free sample preparation for the MALDI analysis of synthetic polymers were obtained as follows [30]: matrix, salt, and polymer (not weighed) were added to the plate, and the direct pressure from the end of the spatula was used to "grind" the solid materials together. The resulting sample was a thin film of powder, similar to the samples prepared with the vortex method and applied to the target with a spatula. While the resulting mass spectra [7] were not of the same quality as the vortex method, they indicated that further simplifications of these solvent-free methods are feasible and show the potential to speed up the process by eliminating the transfer step.

To further simplify homogenization/transfer and to reduce multi-sample preparation time requirements, different methods were developed for solvent-free MALDI analysis of synthetic polymers that are not only simple but practical [4, 6]. First, the sample volume was reduced, thus increasing the number of samples that can be prepared [4] compared to the original one-sample vortex/BB method [7]. For this, a simple 96-sample holder was used with small metal beads and plastic vials identical to the ones used in the MBM method [25]. This setup homogenized up to 96 analyte and matrix samples simultaneously using simply a vortexer for only 1 min; however, 5 min considerably improved the quality of the mass spectra. The analytical results of the multiple-sample vortex solvent-free MALDI analysis were then essentially identical to the results of the single-sample vortex approach but greatly reduced the time necessary to individually homogenize each sample.

With this approach, each sample needed to be transferred to the MALDI target individually.

A second approach integrated sample transfer into the homogenization step of the previously described multi-sample preparation method [4]. The homogenization of analyte and matrix powder and salt is obtained using a TissueLyser® (QIAGEN) which is conventionally used for disruption of biological samples. A few grains of each synthetic polymer are combined in 2-mL plastic vials followed by the addition of either dithranol or 2,5-dihydroxybenzoeic acid (DHB) matrices, then NaCl, and finally the metal beads (1.1 mm). Weighing of the powders or counting the beads was not employed in these experiments. The approximate amount of powders added were <1 mg (analyte) to ca. 10 mg (matrix) and the volume of the beads is roughly equal to the volume of the MALDI sample. Eight uncapped vials were placed in each adapter set sample holder and each set was capped by a MALDI plate followed by the adapter container lid. Each set was placed in one of the two TissueLyser arms and shaken for 2 min at which time the sample is homogenized and transferred to the MALDI plate simultaneously. The MALDI plates with transferred analyte/matrix/salt were dusted off in a fume hood with difluoroethane gas to remove excess powder. This approach in which analyte/matrix homogenization and transfer occur simultaneously greatly simplifies preparation of multiple MALDI samples and provides considerable time-saving in addition to reducing the amount of analyte consumed for sample preparation. The disadvantage of this approach is that it relies on a cost-intense homogenization devise.

A third approach using a common vortex device was developed which essentially combines the concepts of the two previous methods allowing for on-target homogenization/transfer of multiple samples simultaneously. This approach was first established with 36 samples and then adapted and increased to 100 samples per MALDI plate (Figures 7.3 and 7.4) [4]. Thirty-six wells (organized 6 by 6) in a bacti plate (Nunc, Roskilde, Denmark) exactly fit a PerSeptives Biosystems (Framingham, MA) MALDI plate. Each polymer was transferred either as powder or in solution to each of the wells with addition of 3–10 metal balls (1 mm). If organic solvent was used, it was evaporated at room temperature or at 37 °C in the oven, drying the entire set of samples at the same time. Premixed matrix : salt powder was used according to Reference 5 and was added to the dried analyte. The set was capped by a MALDI-plate, which was tightly affixed with tape and vibrated by the vortexer for ca. 5 min, at which time each sample had been automatically transferred to the MALDI plate. The time to load the wells, homogenize the analyte : matrix : salt with simultaneous transfer to the MALDI target is under 1 min per sample [4]. This method was adapted to 100 samples (organized 10 by 10) [6] and more (unpublished results) [6]; the custom-made adapter set (Figure 7.3a) gives spot sizes of about 1 mm diameter (Figure 7.3b). The mass spectra are of good to excellent quality as exemplified for polymethylmethacrylate (PMMA) 5 kDa using either DHB (Figure 7.4a) or dithranol (Figure 7.4c), each of which were premixed with sodium trifluoroacetate. Carryover was not observed as evaluated by the blank mass spectrum of the target spot with only matrix (DHB) and sodium salt (Figure 7.4b) between the spots containing polymer (Figure 7.4a,c).

All three methods are efficient in that small sample amounts are required; the sample : matrix : salt ratio is not critical, so that there was no need of weighing, and

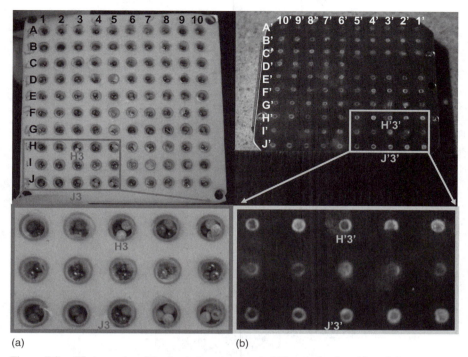

(a)   (b)

Figure 7.3.   Photographs of the loaded custom-made 100-sample plate (a) and the
sample-mirror imaged MALDI plate of the homogenized/transferred MALDI samples
(b) employing the vortex device for 5 min as well as their respective enlargements (bottom).
See color insert.

the time necessary to achieve sufficient homogenization of multi-samples is about
5 min. These methods work well and are reliable for synthetic polymers in a mass
range between ~500 Da (PEG 600) and ~10,000 Da (PS 13,100). The simultaneous
multi-sample approach for direct on-target MALDI sample preparation standardizes
and automates much of the labor-intensive work to prepare a MALDI plate, suggesting that high-throughput solvent-free MALDI-MS analysis of low-molecular-
weight polymers is possible. The method can also be used as a rapid means of
optimizing conditions such as matrix choice, which can be especially important for
difficult samples.

7.4   THEORETICAL CONCEPTS

7.4.1   Derived from the Pressed Pellet MALDI Method

Investigations of the theoretical aspects of the MALDI processes have also been
pursued based on solvent-free MALDI methods. The pressed pellet method usually
employs a low and restricted matrix-to-analyte ratio as recognized for both synthetic

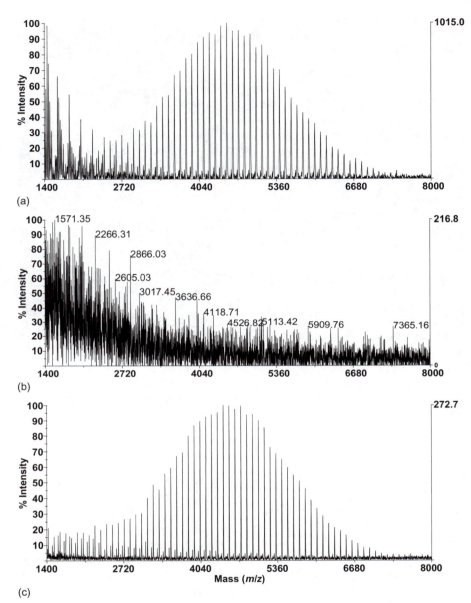

Figure 7.4. Solvent-free MALDI mass spectrum of PMMA 5 kDa. Mass spectra were of good quality and reproducibility using the 100-sample on-target homogenization/transfer solvent-free MALDI method (vortex device) using either DHB (a) or dithranol (c). The mass spectra (b) from the spot (only DHB matrix/sodium trifluor acetate salt employed) in between the two samples reveals that even though this setup is not as a tidily closed system as compared to the TissueLyser® approach, there is no cross-contamination observed. The mass measurements were performed on a PerSeptives Biosystems (Framingham, MA, USA) Voyager-DE STR MALDI mass spectrometer.

polymers and biopolymers. For example, qualitatively best results in pressed pellet MALDI analysis of synthetic polymers were obtained for weight ratios of about 5 : 1 to 1 : 1 of 3-aminoquinoline : low-molecular-weight polyamide as compared to results obtained for the ratios 10 : 1 and 0.2 : 1 [15], which corresponds roughly to an optimal molar analyte : matrix ratio of 1 : 50; in traditional solvent-based and solvent-free loose powder MALDI analysis, optimal molar ratios are, depending on the matrix, approximately 1 : 500 (dithranol) and are nonrestrictive (Figure 7.2; PS : dithranol) [19]. It has been shown that more common molar analyte : matrix ratios of 1 : 500 can be applied using the pressed pellet method accepting reduced qualitative results for simple PS standard, 2 kDa, and dithranol as matrix as evaluated by the achieved resolutions of about 3000 with the pressed pellet, 4000 with the solvent-based, and 5000 with the loose powder MALDI method [19]. As mentioned before, the homogenization time (mortar and pestle, BM, or vortex) is applicable over a broad period.

For peptide and protein analysis utilizing the mortar and pestle/pressed pellet method, molar ratios employed are about 1 : 100 to 1 : 500 [21] and revealed a greatly limited accessible molecular weight range exemplified by the restricted analysis of even small proteins. For example, the signal quality obtained from the pellet was evaluated as weak for bovine insulin (5.7 kDa), and no ion signal was observed for cytochrome c (Cyt c, 12 kDa) either using 2,4-, 2,5-, or 2,6-DHB [21]. Extending the homogenization period improved the spectra quality to some degree [21]. In other studies, the mass measurement of Cyt c using small amounts of 2,5- and 2,6-DHB assessed to provide good mass spectra via the mortar and pestle/pressed pellet method (no further experimental details provided toward the employed ratio) [31]. In solvent-based MALDI analysis, typically used final molar analyte : matrix ratios are about 10^{-5} for these types of compounds [31]. These initial pressed pellet-based studies prompted further investigations toward MALDI mechanistic aspects: One study (mortar and pestle) provided evidence for gas-phase ionization processes of a peptide by employing various different molar matrix : salt : analyte ratios between 500 : 1000 : 1 to 10 : 20 : 1 as well as using different matrices such as sinapinic acid, DHB, 2,4,6-trihydroxyacetophenone, and dithranol [32]. Another study showed that longer BM grinding times (from 5 s to 7 min) extends the accessible mass range up to 12 kDa as evidenced by the mass analysis of Cyt c using either 2,5- or 2,6-DHB as matrix (molar analyte : matrix ratios of 1 : 250); bovine serum albumin (BSA, 66 kDa) was still inaccessible [33].

7.4.2 Derived from the Loose Powder MALDI Method

Theoretical implications using the solvent-free loose powder transfer method, either in combination with mortar and pestle, BM, MBM, or vortex homogenization, have been obtained on peptides [28, 34], small (bovine insulin, Cyt c) [19, 26, 27, 28] and large proteins (BSA, 66 kDa) [28, 34, 35], and small [16, 36] and large molecular weight synthetic polymers (PS 100 kDa, PMMA, 100 kDa) [19] under various different aspects. Studies revealed that the quality of the analytical result strongly depends on the employed grinding times (MBM) as evidenced by mass spectra with increasing grinding times (0.5–10 min) of a peptide employing α-cyano-4-hydroxycinnamic acid (CHCA) [28]. A grinding time of less than 1 min already gives sufficient quality, yet periods between 1 min and 5 min give the best results as shown by the most intense overall signal intensity. Longer grinding times (10 min) seem to assist a higher degree of $[M + Na]^+$ formation as well as a considerably overall signal intensity decrease. A dependence on employed material was also determined, as for example, the least degree of sodiation was observed in the acquired mass spectra using metal beads for homogenization instead of silica or glass beads [25, 28]. In any case, the applicable molar analyte/matrix ratio was not restrictive for biopolymer analysis, specially evidenced by the sensitivity studies [28]; for example, femtomole peptide amounts were accessible [25, 34]. Overall, the general theme is that this nontraditional solvent-free MALDI method allows to employ less laser power and therefore softer MALDI conditions, improving the analytical result, for example, for peptides [18], synthetic polymers [3], and other synthetic macromolecules [5], which triggered more in-depth investigations.

One study employing the loose powder sample transfer method investigated crystallinity dependencies of MALDI samples, and included the analysis of a single crystal of 2,5-DHB incorporating Cyt c, and the powder from ball-milling of the crystal [27]. In addition, comparisons of traditional and nontraditional MALDI sample preparation methods employing other matrices (2,4-DHB, 2,6-DHB, dithranol, and anthracene) were described, all of which gave mass spectra but with different quality [27]. The results of these experiments concluded [26, 27, 34] that the incorporation of analyte in matrix crystals is not helpful for MALDI analysis but obstructive, since it is the crystallinity of the matrix that makes the underlying process energetically less favorable. It was shown that the model of the analyte being embedded in the matrix crystal is not necessary and, in fact, is disadvantageous due to the requirement of increased laser power. The MALDI process is more effective, or "softer," with improved contact between the analyte and the matrix that results from decreased crystallinity. This also gives a theoretical explanation to the poor analytical results obtained when using the solvent-free pellet method [15, 21, 33].

Other studies were directed at investigating the surface morphology of synthetic polymer MALDI samples produced by the solvent-free vortex homogenization/spatula transfer method using microscopy techniques [36]. Low-molecular-weight (~2000 Da) polymers (PS, PMMA, and polyethynyl formamide (PEF)) were prepared with common MALDI matrices (DHB, CHCA, 3-indolylacrylic acid (IAA), and dithranol) and with metal cationization salts (silver or sodium) using molar ratios of about 1/2000/10. The samples were vortexed for 60 s [7] and analyzed using backscattered electrons from scanning electron microscopy (SEM) in a variable pressure instrument and tapping mode atomic force microscopy (AFM). The results of the AFM and SEM images from MALDI samples are shown in Figure 7.5: PS : CHCA : Ag salt and PEF : IAA : Na salt, respectively (Figure 7.5.I.A,B). The SEM images at 5000× magnification (Figure 7.5.I.,II.B) show the samples to be surprisingly homogeneous, considering that only the least powerful of any of the previous electrical homogenization devices was employed and for only 60 s. The samples are composed primarily of particles much smaller than 1 μm, closely packed into the film, with a few much larger (few micrometer in diameter) particles on the surface of the film (data not shown), as seen in SEM images of lower magnification.

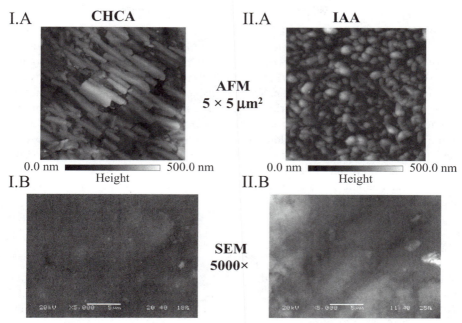

Figure 7.5. Microscopy images of solvent-free MALDI samples prepared by the solvent-free vortex homogenization/spatula transfer method. Images I.A and B are from a sample prepared with PS/CHCA/Ag. The tapping mode AFM images show local, submicron morphology, rods for CHCA (I.A), and spheres for IAA (II.A). The SEM images show surprisingly homogeneous thin films for both samples (I.B and II.B) at high magnification (5000×). See color insert.

A control experiment, imaging the matrix powders only without vortexing, by optical microscopy and SEM, showed that the particle sizes of the matrices before grinding are considerably larger than observed after vortexing and deposition. It can be concluded that a significant grinding effect is obtained by the simple vortex method with 60-s treatments. Additional experiments showed that similarly good homogeneity is observed in as little as 30 s. The reasoning for the unexpected degree in homogeneity and particle size for some of the matrices is speculative. One possibility includes the effectiveness of grinding between two balls in the vial. This is supported by the observation that vortexing with a single ball does not lead to good MALDI spectra as compared to the use of two balls for homogenization [30]. Interestingly, the SEM data showed very distinct charging of the sample, indicating that this film has few defects and behaves as an excellent electrical insulator.

The AFM images of $5 \times 5\,\mu m^2$ regions (Figure 7.5.I.A,II.A) of the samples show that the different matrices have very different local morphology. The CHCA sample (prepared for MALDI with PS polymer and Ag salt) appears as closely packed rods, while the IAA sample (prepared for MALDI with PEF polymer and

Na salt) appears as closely packed spheres. This local morphology does not appear to affect the MALDI experiment.

Another set of experiments, based on the MBM method, focus on ionization processes [25, 35]. In the analysis of hydrophobic peptides, it has been demonstrated that solvent-based analysis fails not only due to possible solubility limitations but also to insufficient ion production. There is also evidence that metal cation attachment to analytes is enhanced during solvent-free sample preparation. This supports the notion that the ionization process in solvent-free MALDI involves desorption of some sort of preformed metal-adducted analyte ions. This helps explain the ease of ionization of synthetic polymer and organic macromolecule in solvent-free MALDI analysis, and the greater difficulty observed for peptide and protein analysis employing the same method. The analyses of biopolymers are optimal when proton attachment occurs.

7.5   MERITS

When the solvent-free MALDI method was first introduced and evaluated, it often yielded equally good or even more descriptive spectra than solvent-based MALDI and typically required less laser power for successful desorption/ionization of the analyte [3, 5, 19, 27]. Systems were examined that were previously shown to be problematic with traditional solvent-based MALDI approaches [3, 5]. Problems frequently encountered with traditional MALDI included solubility, miscibility, and segregation effects during crystallization as a result of unfavorable analyte and matrix polarities. The detection of complex mixtures utilizing solvent-free MALDI-MS was obtained without a priori knowledge of different compound solubilities, without time-consuming optimization procedures and excluded heating or other more severe techniques to increase solubility. The latter is important in special cases (e.g., temperature or oxidatively labile compounds) where heat treatment might lead to decomposition, additional products, or to other misleading MS results (e.g., preferential dissolving of one compound over the other). Solvent-free MALDI-MS simplified the measurement and improved the analysis such that the inaccessibility of insoluble polymers and the analyte segregation phenomena were overcome. The method is facile and reduces the time-consuming optimization of the four-component matrix : salt : solvent : analyte system, fewer matrix : salt : analyte combinations need to be considered, new analyte : matrix combinations are accessible (independent of solubility and compatibility in common solvents), and fewer problems are produced that may arise from trace impurities from binary solvent systems. The rapid and easy measurements of industrial products demonstrate that the solvent-free method is capable of improving throughput of a variety of compounds in routine industrial analyses. The principal improvements for quantitation utilizing solvent-free MALDI-MS sample preparation has been shown [5, 25]. Thus, solvent-free sample preparation appears to be a more comprehensive MALDI method since solubility-limited [5, 15, 17, 20] and insoluble analytes [3] can often be measured, and its reliability is improved because suppression effects due to solubility are eliminated.

7.5.1 Application to Segregation-Intense Poly(dimethylsiloxane)

Solvent-free MALDI analysis produces analyte:matrix mixtures with high homogeneity [19]. For this reason, we applied this method to problems generally occurring in solvent-based MALDI analysis (e.g., segregation of matrix and analyte). Solvent-based MALDI mass measurement of the molecular weight distribution of PDMS appears to furnish errors even for low-molecular-weight samples (<6000 Da) [37]. In our studies, we used a PDMS sample originating directly from industrial production. The sample on the MALDI plate appeared inhomogeneous after evaporation of the solvent in the solvent-based approach. Optimization trials were poor, the visual inhomogeneity was always apparent, and the mass spectra were of low quality. Essentially, the data did not match the values obtained by viscometry. These results agree with Montaudo et al. [37] where the solvent-based MALDI method is problematic for this analyte due to strong segregation phenomena. The same PDMS 450 centistokes (cSt) (correlating to ~20,000 Da) sample was mixed with premixed DHB:silver powder (in 2.5:1 weight ratio.) in a ratio of PDMS:(matrix and salt) of 1:15 via mortar and pestle homogenization. A mass spectrum was readily obtained by solvent-free MALDI-MS (Figure 7.6) from this oily sample. Note that the oily nature of the PDMS sample in this molecular weight range still allowed sufficient homogenization with the matrix and gave high shot-to-shot reproducibility. The molecular weight distribution was broad (mass range from 1000 to 28,000 Da) but showed a good signal-to-noise ratio. The resolved signals verified the repeating units of 74 Da up to n = 450 (Figure 7.6). The measured molecular weight distribution is in good agreement with values from viscometry. These investigations show that errors in the MALDI analysis caused by segregation phenomena, previously attributed to small molecular weight PDMS [37], are overcome even for much higher

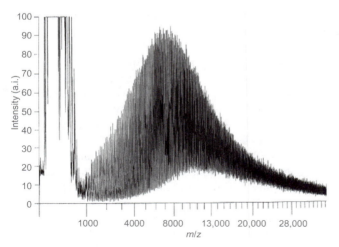

Figure 7.6.   Solvent-free MALDI mass spectrum of industrial PDMS (450 cSt.). The mass spectrum was recorded using a VISION 2000 (Finnigan MAT, Bremen, Germany) instrument equipped with a N_2-laser.

molecular weights by applying the solvent-free method. The advantage of going from a four-component (analyte, matrix, salt, and solvent) to a three-component system (analyte, matrix, and salt) provides fewer combinations that need to be considered and optimized.

The polarity differences between analytes, matrices, and solvents, which cause separation effects during crystallization in solvent-based methods, are problematic as well as a source of trace impurities such as contamination by a binary solvent. Different sources can account for the origin of binary solvents. For example, the dryness and purity of THF used to prepare high-molecular-weight PS samples (>10,000 Da) has been shown to play a critical role in successful mass analysis [12]. Traces of water are also described as impurities in low-molecular-weight PMMA solutions, creating inhomogeneity and significant mass discrimination in solvent-based MALDI-MS [38]. In these cases, precautions such as exclusion of air or redistillation of the solvent are required to avoid segregation. Nevertheless, these precautions are additional steps in sample preparation for successful MALDI analysis and still require optimization, which consumes sample and time, and do not always work, particularly considering azeotropic binary solvents that are less easily removed (e.g. N-methyl pyrrolidone). These kinds of disadvantages were overcome by evaporating the solvent of any problematic analyte to complete dryness and employing solvent-free MALDI analysis.

7.5.2 Application to Suppression Effects in a Polystyrene Polymer Standard

The characterization of solubility-limited compounds and/or incompatible mixtures has the intrinsic advantage that the operator is aware of the challenging situation (e.g., incomplete dissolving of the analyte during sample preparation, inhomogeneities of the sample mixture on the sample holder are visually apparent) and, hence, can take appropriate precautions during sample preparation as well as when acquiring the mass spectrum. Most problematic are effects that are imperceptible to our eye as is the case in highly compatible systems because no inhomogeneities are visible. The following investigation reiterates the reliability of solvent-free MALDI-MS and shows the advantage of employing solvent-free MALDI analysis of PS—a polymer standard normally "easy" in traditional MALDI analysis. The solvent-free MALDI experiments were obtained using the BM device (disposable, custom-made vessels) for homogenization and the loose powder transfer/adhesion of the MALDI sample with a spatula to the MALDI plate.

Routine solvent-based MALDI analysis of the low-molecular-weight PS polymer standard (careful weighing of the sample was omitted) produced noninterpretable results showing two polymer distributions both of which had a repeating unit of 104 Da. The observed mass difference of +60 Da (or conversely −44 Da) from the expected, known polymer distribution (n-Bu- and H-end groups) does not account for any other common end group or analyte-matrix-adduct formation. Repeating the sample preparation and mass measurement only gave the irreproducible result which was that the additional polymer distribution was detected with variable relative signal intensities to the expected polymer distribution, but was always present. A simulation of the isotopic distribution indicated cationization with silver and eliminates the possibility of dealing with simply other common cationizing metals for ionization of the polymer. Complementary methods, including [1]H-nuclear magnetic resonance (NMR) and GPC studies showed no PS oligomer with an additional end group and excluded high-level contaminants (>5%) for this low-molecular-weight PS sample. It was hypothesized that there was a strong suppression effect of PS with the expected end group when applying solvent-based MALDI analysis. Therefore, traditional and nontraditional MALDI were employed as a function of molar analyte:matrix ratios. To ensure uniformity in the employed ratios comparing both methods, a portion of the solvent-free analyte:matrix:salt powder was dissolved in the chosen solvent, THF and dichloromethane respectively, for use in solvent-based MALDI-MS, so that both methods only differed in the presence/absence of solvent.

Solvent-based MALDI-MS, using dithranol as matrix and *THF* as the solvent, and a 1:50:10 molar analyte:matrix:salt (silver trifluoroacetate) ratio showed the expected *n*-Bu- and H-end groups (Figure 7.7.Ia); at a ratio of 1:500:10, both the expected and the additional, unknown polymer distribution were observed simultaneously (Figure 7.7.Ib), whereas for a 1:5000:10 ratio, *only* the unknown polymer distribution was detected (Figure 7.7.Ic). The conventional ratio regularly employed for this analyte is 1:500:10. The additional polymer distribution is observed clearly

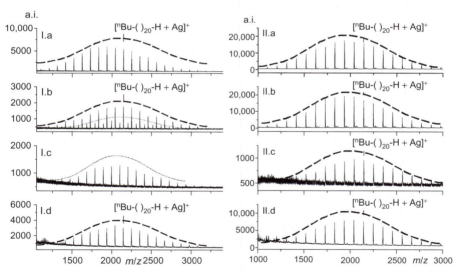

Figure 7.7.   MALDI mass spectra of PS (2000 Da) as a function of molar ratios of I. PS:dithranol:silver salt and solvents applying conventional solvent-based MALDI-MS by dissolving the solvent-free sample (I) in the desired solvent: (a) 1:50:10 dissolved THF; (b) 1:500:10 dissolved THF; (c) 1:5000:10 dissolved in THF; (d) 1:5000:10 dissolved in dichloromethane; II. PS:matrix:silver salt and matrices applying solvent-free MALDI-MS: (a) 1:50:10 and dithranol; (b) 1:500:10 and dithranol; (c) 1:5000:10 and dithranol; (d) 1:5000:10 and IAA. Mass spectra were recorded using a Bruker Reflex II™ MALDI-TOF mass spectrometer equipped with a N_2-laser.

with increasing matrix amounts used. In contrast, the solvent-free MALDI analysis of the $1:50:10$ to $1:5000:10$ samples (Figure 7.7.II) showed only the n-Bu- and H-end groups and even produced signal intensities that were equal to or much higher than with solvent-based MALDI-MS (Figure 7.7.I). Subsequent studies elucidated that the additional distribution only occurs in polymers produced by the same initiator batch. Complementary methods such as NMR spectroscopy failed to detect a polymer with an additional end group. Based on the detection limit of NMR, the quantity can be estimated to be less than 5%.

When analogous solvent-free and subsequent solvent-based MALDI-MS matrix dilution experiments are performed with *dichloromethane* as a solvent, the additional polymer distribution was never detected in any ratio between $1:50:10$ to $1:5000:10$. At this point, a structure assignment to this additional distribution cannot be provided and the origin is also speculative. The fact is that these additional end groups appear in a particular batch and only when using THF as solvent during MALDI sample preparation.

Provided the polymer with the additional end groups is a contamination within the sample and are not MALDI-related artifacts, a rough estimation of the over-representation of the unknown oligomer in the mass spectra (PS oligomer n = 19) obtained by the solvent-based MALDI method leads to a minimum value of +20% for the ratio of $1:500$ and +100% for the molar ratio of $1:5000$. Given the estimated maximum amount of 5% contamination, the solvent-free MALDI method yields an underrepresentation of at most −5% independent of molar analyte:matrix ratios. Although neither method yields quantitative results, application of the solvent-free MALDI method greatly reduced suppression effects.

Other MALDI studies [29, 39] described THF-related artifacts in solvent-based MALDI-MS, which lead to the conclusion that the unknown PS end groups probably relate to MALDI artifacts. For example, it was demonstrated that a PS sample underwent oxidation at the sulfur end group during solvent-based sample preparation employing THF as solvent, whereas this reaction was not observed utilizing solvent-free MALDI-MS [29]. The similarities in terms of analyte, matrix, and solvent of these two studies conclude that the problems occurring during sample preparation in the solvent-based MALDI method may be of general character and are avoidable when employing the solvent-free MALDI method.

7.5.3   Application to High-Throughput Analysis in an Industrial Setting

High-throughput solvent-based MALDI analysis typically involves similar samples—optimal conditions for one sample are assumed to be sufficient for all samples. Studies were first employed to establish whether the solvent-free MALDI method is robust and highly adaptable [5]. Flexibility in sample preparation was explored with studies of previously incompatible analyte:matrix combination [5]. These characteristics triggered high-throughput experiments [5] which essentially tested the potential for the analysis of numerous compounds with various molecular weights with a minimum variation of sample preparation while maintaining sufficient quality in the mass spectra. In these experiments, resolution, signal-to-noise ratio, or accessible molecular weights were not optimized. Instead, the focus was to establish whether an analyte can be assessed with a minimum set of sample preparation conditions. The solvent-free MALDI method was employed to examine the possibility of routinely characterizing poly(alkyleneglycol) (200–10,000 Da), PMMA (500–20,000 Da), PS (500–10,000 Da), PDMS (700–10,000 Da), partially ethoxylated poly(hydroxyester) (up to 3000 Da; resolution is compromised above this molecular weight), and commercially available surfactants (ethoxylated with n = 2 to 100). For cationization of PDMS we employed silver powder and $AgNO_3$ for hydroxyl-functionalized siloxanes which are therefore special cases. All other samples were accessible with the matrix DHB, the cationizing agent sodium chloride (NaCl), and with the same approximate molar ratio of 10 : 1 DHB : NaCl; careful weighing was not required. Premixing matrix and salt improved the sample preparation flow and the analytical result. With the exception of fatty acids which were not readily accessible employing either the solvent-based or the solvent-free MALDI method, these polymers were analyzed via the solvent-free MALDI method, resulting in a much-improved time efficiency, mass spectra with good quality, and a high signal-to-noise ratio (e.g., poly(butylmethacrylate)-1000 diole; BD-1000). Because of the significant range in polarities, thoxylated partial glycerides display highly variable solubility, and thus reveal great difficulties in obtaining results using solvent-based MALDI-MS [5]. However, when a compound mixture containing PEG, ethoxylated glycerine, and up to $C_{16}:C_{18}$ triglycerides was examined utilizing solvent-free MALDI-MS, the mass spectra accurately analyzed the entire compound mixture. Generalizing these results, several advantages for solvent-free MALDI-MS were observed for nonpolar analytes when DHB matrix was used. Contrary to solvent-free MALDI analysis, the separation effect appears to increase when differences in the matrix and analyte polarities increase, resulting in reduced mass spectral quality (e.g., BD-1000) or failure (e.g., PDMS) presumably because of the negative mediation of the solvent. The recently developed solvent-free multi-sample method [4, 6] has the potential to additionally improve automation and standardization of MALDI analysis.

7.5.4 Application to LC-Analysis in an Governmental/Industrial Setting

A solvent-free homogenization/transfer MALDI-MS method has been described for the preparation and precise transfer of up to 100 samples simultaneously onto a MALDI target plate [6]. The method used a mixture of poly-(ethylene oxide) (PEO) consisting of different molecular weights (500–6000) and end groups (PEO, dimethoxy-PEO, monomethoxy monomethacrylate-PEO, and dimethacrylate-PEO) fractionated using liquid adsorption chromatography at critical conditions (LACCC) (see Chapter 11). Offline LACCC fractionation was performed prior to the on-target homogenization/transfer solvent-free sample preparation and MALDI mass analysis. The miniaturization of the solvent-free MALDI method permitted the analysis of less than 2 µg per PEO component per fraction corresponding to about 200 pmol for PEO 6000. This offline method eliminated optimization of spray conditions and spreading of organic solvents on the MALDI plate that occurs with droplet deposition methods. The multi-sample method is well suited for high-throughput analysis and automation as it virtually eliminated the "dead-spot" phenomenon. This hyphenated approach is also ideally suited for solubility-restricted complex mixtures as the removal of the solvent is decoupled from the MALDI sample preparation. Additionally, any precipitate is efficiently recovered from the wall of the well. This now provides the opportunity to explore solvents for optimized separation of extreme complexity that were previously not applicable using solvent-based MALDI (e.g., chlorobenzene).

7.6   INHERENT LIMITATIONS AND PITFALLS OF MALDI ANALYSIS

7.6.1   Quantitation

Problems that arise by the use of "solvent" in the sample preparation, such as segregation and/or suppression, are circumvented by the solvent-free method and can improve the MALDI analysis and reliability. Of course, desorption/ionization differences of polymers or polymer end groups that arise during MALDI analysis are not affected. Model experiments on the effects of desorption/ionization differences in polymer mixtures are described in the following section and demonstrate intrinsic limitations of MALDI-MS analysis that do not depend on sample preparation methods but instead on polymer properties which can be manipulated only to a limited extent, if at all, by proper sample preparation (matrix, salt) and MS conditions (laser power). This is exemplified with the following model experiments of two defined polymer mixtures, one focusing on the end groups, the other one on the backbone property, both of which influence the ionization of polymers. The provided response factors are rough estimations based on signal intensities.

The influence of the polymer end group is demonstrated through defined polymer mixtures employing PS with two different end groups. The silver cation is known to sandwich with phenyl ring side chains which are present in both polymers. The sodium cation prefers to attach to heteroatoms which therefore target the end group of $PSCO_2H$ preferentially. $PSCO_2H:PS$ is prepared in a molar ratio of $53:47$ and homogenized (BM device) with dithranol. A relatively high laser power was necessary to acquire mass spectra which only showed $[PSCO_2H + 2Na-H]^+$, indicating that sodium is already present in the sample. This MALDI sample was split into two, sodium trifluoroacetate was added to one and silver trifluoroacetate to the other part and rehomogenized. The solvent-free mass spectra (Figure 7.8) obtained with increasing laser power (A to E) showed in both of the samples the best response factor employing the lowest laser power. In case of silver salt, this led to a response factor of $46:54$ of $PSCO_2H:PS$ (Figure 7.8.I.A), and for sodium salt to $72:28$ (Figure 7.8.II.A). The $PSCO_2H$ component of the polymer mixture was dependent on the salt used and was either underrepresented as seen employing the silver salt, or overrepresented using the sodium salt. The false-negative result was greater using sodium salt for ionization of the polymers even though that sodium was already present in the sample. These differences in results were not unexpected. The ionization is more

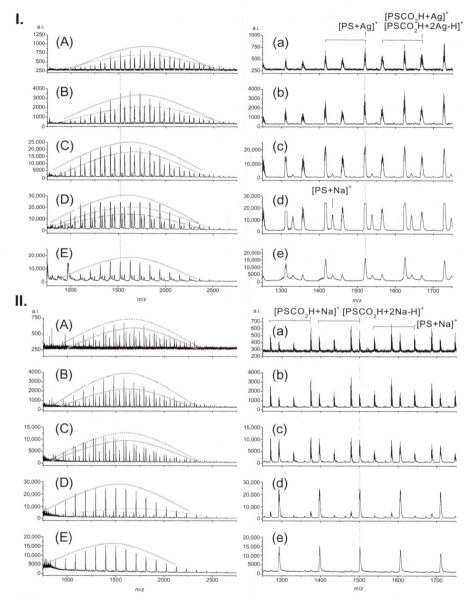

Figure 7.8.   Solvent-free MALDI mass spectra of a polymer model mixture with different end groups, $PSCO_2H/PS$ (53/47%), prepared using dithranol matrix and (I) silver trifluoroacetate salt and (II) sodium trifluoroacetate salt; each sample was measured in dependence of the employed laser power: (A) low, (B) medium low; (C) medium; (D) medium high; (E) high. (a) to (e) denotes the respective inset with repeat units and ionization. Mass spectra were recorded using a Bruker Reflex II™ MALDI-TOF mass spectrometer equipped with a N_2-laser.

equally induced for both polymers, which only differ in the end groups, with silver cationization. With increasing laser power, the error in response becomes greater in both samples. With the highest laser power employed, the addition of silver salt gives a response factor of $16:84$ of $PSCO_2H:PS$ (Figure 7.8.I.E) and sodium salt of $96:4$ (Figure 7.8.II.A). The analytical results greatly depend on both the salt used to provide ionization of the polymers and the laser power applied. These are only two of many variables that influence the MALDI results in synthetic polymer mixture analysis; others are similarly important and are illustrated in the following section for the matrix influence on the ionization of the backbone of the polymer.

Studies were employed to demonstrate suppression effects in a defined polymer mixture caused by matrix properties and desorption/ionization differences of polymers during the MS analysis. Two polymers with extremely different ionization preferences were analyzed by solvent-free MALDI-MS as follows: PPV, assisted by the 7,7,8,8-tetracyanoquinodimethane (TCNQ) matrix, undergoes radical cation formation (Figure 7.9.I.a); PEG, assisted by the IAA matrix and sodium trifluoroacetate, shows Na-pseudo-molecular-ions (Figure 7.9.I.b). Both produce, when measured individually, mass spectra with good signal intensities (Figure 7.9.I). The analytes were mixed in a $1:1$ molar ratio to simulate mixture analysis and were divided into two parts. First, IAA matrix along with sodium trifluoroacetate was mixed with the PPV:PEG analyte mixture. Only PEG was detected, and PPV was entirely suppressed (Figure 7.9.II.b). This results from the IAA matrix, which cannot assist radical cation formation, and therefore leaves PPV unionized. Second, TCNQ was used as matrix, and homogenized with the other half of the polymer mixture, resulting in both polymers being detected correctly (Figure 7.9.II.a). The observed signal intensities were about $4:96$ PPV:PEG, which is not representative of the actual molar ratios employed. Clearly, PPV and its ionization mechanism are less favorable in this competing ionization process than the Na-pseudo-molecular-ionization of PEG.

Overall, the qualitative aspect of MALDI analysis, including reliability and reproducibility, is greatly improved by the solvent-free MALDI method, which therefore improves quantitative analysis. However, it must be stressed that the solvent-free MALDI does not have a direct influence on the individual greatly varying polymer properties. The only aspect that is changed with solvent-free versus solvent-based MALDI is that solvent is omitted, hence, matrix and analyte properties are unchanged, which is crucial to, for example, ionization efficiencies, each contributing to the outcome of the quantitative MALDI result. Based on the compiled data of the noted improvements in various studies (e.g., better shot-to-shot and sample-to-sample reproducibility, higher sensitivity and resolution, less laser power required, insoluble compounds detectable, and mixture analysis improved), the hypothesis can be drawn that when solvent-free MALDI is used, sample preparation and MS conditions are more conducive for easier and better optimized quantitative analyses. Initial work on solubility-restricted biopolymers made use of these improvements [25]. Other studies demonstrated [5] that previously incompatible analyte:matrix combinations are applicable. Thus, we hypothesize that essentially any analyte:matrix:salt combination can be employed, thus the most appropriate sample preparation conditions for MALDI quantitation can be used.

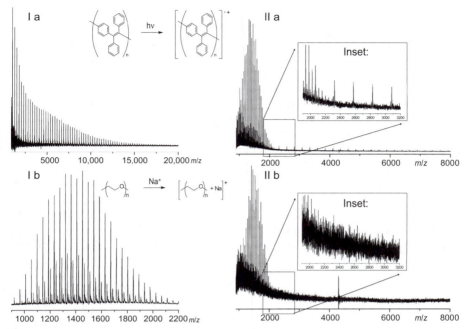

Figure 7.9. Solvent-free MALDI mass spectra of a polymer model mixture, PPV/PEG, with different ionization preferences as a function of the sample preparation conditions: I. pure, individual polymer samples: (a) $1:500$ PPV:TCNQ; (b) $1:500:10$ PEG:IAA:sodium salt. The employed conditions give good quality mass spectra for both analytes, PPV undergoes radical cation formation and ionization which is assisted by the TCNQ matrix, and PPV undergoes sodium adduction which is assisted through the addition of the sodium salt. II. Model polymer mixture PPV:PEG (molar ratio $1:1$) was prepared: (a) $1:500$ (PPV:PEG $1:1$):TCNQ; (b) $1:500:10$ (PPV:PEG $1:1$):IAA:sodium salt. In any of the measurements of the defined polymer mixture, PEG was highly overrepresented in the acquired mass spectra and demonstrates the limitation of accurate mass measurements of polymer mixtures if different ionizations occur. It also reflects the little influence sample preparation can have as seen in II (a) where at least both polymers are detected because sample preparation conditions were applied that assisted the ionization of both polymers. Mass spectra were recorded using a Bruker Reflex II™ MALDI-TOF mass spectrometer equipped with a N_2-laser.

7.6.2 Insoluble Polymers

Solubility problem can hamper polymer analysis by GPC or NMR spectroscopy. A previous study [3] has shown that the solvent-free MALDI method measures insoluble polymers that were obtained after polyfluorene was Soxhlet extracted for 5 days in toluene until weight consistency was observed in the soluble, extracted fraction. In these studies, other inherent problems became evident.

Compared to the oligomers ($n = 12$) in the soluble fraction (<10% of the total weight), the ones present in the insoluble fraction ($n = 20$) differed greatly and also included some of the lower-molecular-weight oligomers observed in the soluble fraction. This indicates that the solubility of these oligomers is drastically reduced with only a few additional repeating units. However, these results also indicate that it is tremendously difficult even for low-molecular-weight polymers to separate soluble and insoluble constituents of a sample, which essentially corresponds to reducing the polydispersity of a sample. Even the rigorous Soxhlet extraction only enriched the soluble constituents from the insoluble fraction. Astonishingly, solvent-free MALDI sample preparation (ball-milling for 10 min) "freed" soluble from the insoluble constituents, which was not possible soxhletting the polymer for 5 days in hot toluene. Thus, ball-milling was more efficient in much less time than Soxhlet extraction.

In those studies [3], it was emphasized that the MALDI analysis does not likely show the real upper molecular weight of the insoluble polymer fraction. The higher-molecular-weight fractions are most likely suppressed as is commonly observed in the MALDI analysis of high-molecular-weight polymers [40, 41]. This reveals the intrinsic complications of the MALDI analysis. If the extraction could be improved beyond enriching soluble and insoluble constituents of a sample, it may enhance sensitivity of the high-molecular-weight components by removing low-molecular-weight soluble oligomers that suppress the detection of the higher-molecular-weight constituent. This approach in essence corresponds to simple fractionation which is commonly used for soluble, polydisperse polymers using GPC. This can effectively narrow the polydispersity of the polymer, which is essential to MALDI analyses (PD > 1.2 give errors for simple soluble synthetic polymers [42]). Combining the advantages from both worlds, solvent-based extraction assisted by ball-milling may show great improvements in the efficiency of separating soluble and insoluble constituents of a polymer.

Phenyl side-chain fragmentation of the polyfluorene was detected in the mass spectra (Figure 7.1) [3]. After employing post-source decay (PSD) analysis, it was apparent that the side-chain fragmentation of the phenyl rings increased with increasing molecular weight of the oligomer. Additional linear mode mass measurements further identified that the phenyl side-chain cleavage occurs already in the source (ISD fragmentation). These results identified the abundant signals observed in the mass spectra as being ISD and PSD fragment ions of phenyl side-chain cleavage. Since the phenyl rings contribute to the insolubility of this polymer, it can be hypothesized that the properties providing the insolubility of a synthetic polymer may also lead to great difficulties in the MALDI mass measurement of insoluble polymers with increasing molecular weight. Optimizing the sample preparation conditions by employing TCNQ matrix allowed the employment of less laser power that prevented the fragmentation [3]. It is likely that this will not always be possible.

A few cases are discussed exemplifying where the use of solvent in the MALDI sample preparation of solubility-limited polymers is constructive and may be ultimately required for successful MALDI mass measurements. It is unlikely that solvent-free MALDI analysis will be able to directly acquire mass spectra of highly entangled polymers such as commercial polyester or nylon. Solvent-based MALDI analysis using hexafluoroisopropanol as solvent has been reported to measure nylon polymers beyond 25,000 Da and, in an experiment where only high-mass ions were able to reach the detector, signal was observed up to about 70,000 Da [43]. The use of solvent can untangle these polymer chains. It is unlikely that an entirely solvent-free MALDI sample preparation can achieve the same results. Further adjustments to the solvent-free sample preparation would be required to access these types of highly intractable synthetic polymers in the future. An evaporation and grinding MALDI sample preparation approach has been presented to produce a MALDI spectrum of low-molecular-weight Kevlar® (DuPont, Wilmington, DE) fibers [44]. It has not been demonstrated how representative the results are toward the real sample composition. Clearly, the sample preparation for the MALDI analysis of fibers is more involved than for other synthetic polymers and warrants further research. Of course, these approaches are only applicable to polymers that can be (partially) solubilized in a solvent, which is not the case for some polymers.

One ultimate challenge for MALDI mass measurements is solubility-limited or insoluble polymers, especially commercial polymers such as Teflon® (DuPont) that exhibit very high polydispersities and often high molecular weights. The only known approach to access broad polydisperse synthetic polymers is by sample pretreatment, which includes GPC to collect and measure polymer fractions. It has been shown for a commercial highly polydisperse and water-soluble synthetic polymer (polyvinylpyrrilidone (PVP)) that the polydispersity can be finally narrowed by GPC refractionations so that MALDI mass measurements of the multiple fractionated sample can be obtained [45]. This approach was applicable to low-molecular-weight PVPs up to 14 kDa (PD 10.87) but was already limited for the analysis of PVP 22.1 kDa (PD 7.37). This approach will not be applicable to insoluble, or likely even solubility-limited, highly polydisperse synthetic polymers of even low molecular weights. However, the incorporation of IMS along with MS allows, with the use of solvent-free sample preparation, a total solvent-free MS methodology that embraces solvent-free ionization and solvent-free gas phase separation. Most recent work shows the powerful ability of IMS-MS for decongestion of sample complexity of even isomeric blends [8] while providing detailed insights to conformational aspects of polymers using cross-section analysis [9]. The concept of total solvent-free MS analysis will enable addressing chemically challenging questions ranging from insolubility, high mass components, and large polydispersity of polymeric samples to structure–function relationships.

The final frontier is perhaps polydisperse, high-molecular-weight polyethylene. These samples are barely soluble in chlorobenzene, suffer the detection problems of polydisperse and high-molecular-weight components, and are tremendously difficult to ionize even for low-molecular-weight compounds [46]. Synthetic polymers with these extreme characteristics may never be measurable by MALDI analysis.

7.7 CONCLUSION AND FUTURE DIRECTIONS

Solvent-free MALDI is being used in >50% of the routine MALDI analyses performed in some laboratories [5, 47, 48]. Most likely, the solvent-free MALDI method will be important for accurate mass measurements of solubility-limited and laser power-sensitive synthetic polymers in a broad manner as long as common MALDI limiting factors (large polydispersity, inability to ionize) are met. Solvent-free MALDI-MS may be most appropriate for polymers <15 kDa, especially if the simple vortex method is used for homogenization of the MALDI sample. Future applications of solvent-free MALDI-MS may include compounds such as solid-state synthesized products (e.g., to exclude unknown solvent effects such as oxidation or degradation) [5, 29, 39], which are sometimes only accessible by time-consuming sample pretreatments or by LD MS, if at all. On their own, these capabilities have tremendous significance. Other aspects improved by the solvent-free MALDI method include reliability in the results and potential quantitative information on mixtures. All in all, solvent-free MALDI analysis accomplishes a broadened repertoire in synthetic polymer analysis where MALDI-MS approaches still dominate [1]. This technique allows hitherto unknown analyte : matrix combinations. The solvent parameter is eliminated from the complex number of variables, leaving only the interaction of the most important components, the analyte and the matrix, which directly affects the MALDI process and therefore has the greatest impact on the quality of the mass spectrum. Hence, a matrix displaying the best performance for a particular analyte can be chosen regardless of solubility or compatibility, so that optimization attempts are much simpler and more focused, which may prove invaluable for quantitative analysis. For this, focused quantitative MALDI-MS strategies must be further developed and will certainly be a difficult task, as shown previously [49]. Problems in quantitative investigations such as polydisperse samples, over-representation of lower-molecular-weight compounds, saturation of signals and detection inefficiency [40, 41], and suppression effects due to preferential desorption/ionization still exist, making accurate quantitation of mixtures difficult. However, intrinsic overestimation of parts of a mixture, for example, due to preferential desorption/ionization, is more reproducible in solvent-free experiments and, therefore, can be recalculated mathematically for semi-quantitative or possibly even quantitative analysis.

The improvements in analysis gained by utilizing solvent-free MALDI-MS are striking. The uniqueness and widespread accessibility to MALDI mass spectrometers makes the solvent-free method particularly important as it expands the capabilities for mass spectrometric analysis to synthetic polymers that previously may not have been possible. The on-target sample preparation of multiple samples simultaneously combines all the benefits important to the MALDI analysis of polymers, including simplifying sample preparation, avoiding carryover, and cleaning of equipment, is cost efficient, minimizes required sample amounts, standardizes and automates much of the work-flow, and reduces the time required per sample, thus simplifying optimization experiments. Further simplifications of the solvent-free MALDI technique along with implementation of hyphenated approaches are likely the key to wider applications. The groundwork for decongestion of solubility-restricted complex polymeric blends has been accomplished utilizing multi-sample solvent-free MALDI coupled to solvent-based separation techniques such as LACCC. The prospect of decongestion of insoluble complex polymeric blends is the next logical step, utilizing total solvent-free analysis consisting of solvent-free MALDI coupled to solvent-free gas phase separation employing IMS-MS.

REFERENCES

1. Weidner, S.M.; Trimpin, S., *Anal. Chem.*, **2008**, 80, 4349–4361.
2. Trimpin, S., *Encyclopedia of Mass Spectrometry: Molecular Ionization*, Vol. 6 (Gross, M.L.; Caprioli, R.M., editors). New York: Elsevier, **2007**, pp. 683–689.
3. Trimpin, S.; Grimsdale, A.C.; Räder, H.J.; Müllen, K., *Anal. Chem.*, **2002**, 74, 3777–3782.
4. Trimpin, S.; McEwen, C.N., *J. Am. Soc. Mass Spectrom.*, **2007**, 18, 377–381.
5. Trimpin, S.; Keune, S.; Räder, H.J.; Müllen, K., *J. Amer. Mass Spectrom.*, **2006**, 17, 661–671.
6. Trimpin, S.; Weidner, S.M.; Falkenhagen, J.; McEwen, C.N., *Anal. Chem.*, **2007**, 79, 7565–7570.
7. Hanton, S.D.; Parees, D.M., *J. Amer. Mass Spectrom.*, **2005**, 16, 90–93.
8. Trimpin, S.; Clemmer, D.E., *Anal. Chem.*, **2008**, 80, 9073–9083.
9. Trimpin, S.; Plasencia, M.D.; Isailovic, D.; Clemmer, D.E., *Anal. Chem.*, **2007**, 79, 7965–7974.
10. Tanaka, K.; Waki, H.; Ido, Y.; Akita, S.; Yoshida, Y.; Yoshida, T., *Rapid Commun. Mass Spectrom.*, **1988**, 2, 151–153.
11. Karas, M.; Hillenkamp, F., *Anal. Chem.*, **1988**, 60, 2299–2301.
12. Schriemer, D.C.; Li, L., *Anal. Chem.*, **1996**, 68, 2721–2725.
13. Wenzel, R.J.; Matter, U.; Schultheis, L.; Zenobi, R., *Anal. Chem.*, **2005**, 77, 4329–4337.
14. Yalcina, T; Daia, Y.; Li, L., *J. Amer. Mass Spectrom.*, **1998**, 9, 1303–1310.
15. Skelton, R.; Dubois, F.; Zenobi, R., *Anal. Chem.*, **2000**, 72, 1707–1710.
16. Marie, A.; Fournier, F.; Tabet, J.C., *Anal. Chem.*, **2000**, 72, 5106–5114.
17. Przybilla, L.M.; Brand, J.D.; Yoshimura, K.; Räder, H.J.; Müllen, K., *Anal. Chem.*, **2000**, 72, 4591–4597.
18. Trimpin, S.; Klok, H.-A.; Mayer-Posner, F.-J.; Räder, H.J.; Müllen, K., *Proceedings of the 48th ASMS Conference on Mass Spectrometry and Allied Topics, Long Beach*, California, **2000**.
19. Trimpin, S.; Rouhanipour, A.; Az, R.; Räder, H.J.; Müllen, K., *Rapid Commun. Mass Spectrom.*, **2001**, 15, 1364–1373.
20. Leuninger, J.; Trimpin, S.; Räder, H.J.; Müllen, K., *Macromol. Chem. Phys.*, **2001**, 202, 2832–2842.
21. Glückmann, M.; Pfenninger, A.; Krüger, R.; Thierolf, M.; Karas, M.; Horneffer, V.; Hillenkamp, F.; Strupat K., *Int. J. Mass Spectrom.*, **2001**, 210/211, 121–132.
22. Dolan, A.R.; Wood, T.D., *J. Am. Soc. Mass Spectrom.*, **2004**, 15, 893–899.
23. Räder, H.J.; Rouhanipour, A.; Talarico, A.M.; Palermo, V.; Samori, P.; Müllen, K., *Nat. Mater.*, **2006**, 5, 276–280.
24. Eelman, M.D.; Blacquiere, J.M.; Moriarty, M.M.; Fogg, D.E., *Angew. Chem. Int. Ed.*, **2008**, 47, 303–306.
25. Trimpin, S.; Deinzer, M.L., *J. Am. Soc. Mass Spectrom.*, **2007**, 18, 1533–1543.
26. Trimpin, S.; Räder, H.J.; Müllen, K., *Proceedings of the 50th ASMS Conference on Mass Spectrometry and Allied Topics*, Orlando, Florida, **2002**.
27. Trimpin, S.; Räder, H.J.; Müllen, K., *Int. J. Mass Spectrom.*, **2006**, 253, 13–21.
28. Trimpin, S.; Deinzer, M.L., *J. Amer. Mass Spectrom.*, **2005**, 16, 542–547.
29. Polce, M.J.; Wesdemiotis, C.; Kim, H.; Quirk, R.P., *Proceedings of the 53rd ASMS Conference on Mass Spectrometry and Allied Topics*, San Antonio, Texas, **2005**.
30. Private communication from Dr. Scott Hanton (Air Products).
31. Horneffer, V.; Forstmann, A.; Strupat, K.; Hillenkamp, F; Kubitscheck, U., *Anal. Chem.*, **2001**, 73, 1016–1022.
32. Zhang. J.; Zenobi, R., *J. Mass Spectrom.*, **2004**, 39, 808–816.
33. Horneffer, V.; Glückmann, M.; Krüger, R.; Karas, M.; Strupat, K.; Hillenkamp, F., *Int. J. Mass Spectrom.*, **2006**, 249/250, 426–432.
34. Trimpin, S.; Deinzer, M.L., *Biotechniques*, **2005**, 39, 799–805.
35. Trimpin, S.; McEwen, C.N.; Ji, H.; Deinzer, M.L., *Molecular & Cellular Proteomics*, **2006**, 5.10 (Suppl.), S300.
36. Hanton, S.; McEvoy, T.M.; Stets, J.R., *J. Amer. Mass Spectrom.*, **2008**, 19, 874–881.
37. Montaudo, G.; Montaudo, M.S.; Puglisi, C.; Samperi, F., *Rapid Commun. Mass Spectrom.*, **1995**, 9, 1158–1163.
38. Chen, H.R.; Guo, B.C., *Anal. Chem.*, **1997**, 69, 4399–4404.

39. Hait, S.B.; Sivaram, S., *Macromol. Chem. Phys.*, **1998**, 199, 2689–2697.

40. McEwen, C.N.; Jackson, C.; Larsen, B.S., *Int. J. Mass Spectrom. Ion Proc.*, **1997**, 160, 387–394.

41. Byrd, H.C.M.; McEwen, C.N., *Anal. Chem.*, **2000**, 72, 4568–4576.

42. Martin, K.; Spickermann, J.; Räder, H.J.; Müllen, K., *Rapid Commun. Mass Spectrom.*, **1996**, 10, 1471–1474.

43. Private communication with Dr. Charles McEwen (DuPont).

44. Gies, A.P.; Hercules, D.M.; Ellison, S.T.; Nonidez, W.K., *Macromolecules*, **2006**, 39, 941–947.

45. Trimpin, S.; Eichhorn, P.; Räder, H.J.; Müllen, K.; Knepper, T.P., *J. Chromatogr. A*, **2001**, 938, 67–77.

46. Chen, R.; Yalcin, T.; Wallace, W.E.; Guttman, C.M.; Li, L., *J. Amer. Mass Spectrom.*, **2001**, 12, 1186–1192.

47. Hanton, S.D.; Parees, D.M., *Proceedings of the 52nd ASMS Conference on Mass Spectrometry and Allied Topics*, Nashville, Tennessee, **2004**.

48. Private communication with Patricia Peacock (DuPont).

49. Wallace, W.E., *Proceedings of the 52nd ASMS Conference on Mass Spectrometry and Allied Topics*, Nashville, Tennessee, **2004**.

CHAPTER 8

MALDI MASS SPECTROMETRY FOR THE QUANTITATIVE DETERMINATION OF POLYMER MOLECULAR MASS DISTRIBUTION[1]

Charles M. Guttman and William E. Wallace

Polymers Division, National Institute of Standards and Technology, Gaithersburg, MD

8.1 INTRODUCTION

As an unavoidable consequence of their synthesis, synthetic polymers are never obtained as a single molecular mass but rather as a distribution of molecular masses. The molecular mass distribution (MMD), and the mass moments (MM) derived from it, is used in areas of polymer science as diverse as fundamental physical studies, polymer processing, and consumer product design. For example, the observed MMD is often compared to predictions from kinetic or mechanistic models of the polymerization reaction itself [1, 2]. Similarly, the tensile strength, melt flow rate, and many other physical properties are dependent on its MMD. Determination of the MMD is used for quality control in polymer synthesis and as specification in international commerce.

Various averages, or molecular moments, can be defined as useful summaries of the MMD. Measuring and computing these summary statistics has comprised a major part of traditional polymer analysis. The two most common measures of the MMD are the number-average relative molecular mass, M_n, and the mass-average relative molecular mass, M_w:

$$M_n = \sum_j m_j n_j \Big/ \sum_j n_j \qquad \text{(Eq. 8.1)}$$

[1] Official contribution of the National Institute of Standards and Technology; not subject to copyright in the United States of America

MALDI Mass Spectrometry for Synthetic Polymer Analysis, Edited by Liang Li
Copyright © 2010 John Wiley & Sons, Inc.

$$M_w = \sum_j m_j^2 n_j \bigg/ \sum_j m_j n_j \qquad\qquad \text{(Eq. 8.2)}$$

$$PD = M_w / M_n \qquad\qquad \text{(Eq. 8.3)}$$

where m_j = mass of a discrete oligomer, n_j = number of molecules at the given mass m_j, and PD defines the polydispersity index which is a measure of the breadth of the polymer distribution. When the polydispersity index is equal to one (i.e., in statistical terms the variance of MMD is zero), all of the polymer molecules in a sample are of the same molecular mass and the polymer is referred to as monodisperse. Natural polymers such as proteins are often monodisperse. This is seldom the case for synthetic polymers, which have a non-point-mass MMD, whose width depends in a complex fashion on the polymerization chemistry, specifically its mechanism, kinetics, and reaction conditions.

From classical polymer methods of analysis of the MMD, one obtains one or two moments of the MMD. M_w can be obtained traditionally by the classical measurements of light scattering or ultracentrifugation, and M_n can be obtained by the classical measurement of colligative properties like osmotic pressure, or by end-group analysis like nuclear magnetic resonance (NMR) or titration techniques. Generally, these moments often give an incomplete description of the overall MMD. Properties like melt viscosity, tensile strength, and impact strength often depend on the tails of the MMD rather than the central portion as defined by the two central moments. Thus, it is critically important that we make an effort to obtain the entire MMD. Furthermore, it is not uncommon, due to purposeful blending or to the polymer chemistry (resulting from two different mechanisms going on during the preparation), that the polymer MMD will be bimodal; thus, the central moments, M_n and M_w, are exceedingly poor representations of the MMD. Gel permeation chromatography (GPC) and matrix-assisted laser desorption/ionization time-of-flight mass spectrometry (MALDI-TOF MS) offer the prospect of obtaining the entire MMD. Particularly for MALDI-TOF MS at molecular masses below about 25,000 Da for polystyrene (PS), one can obtain a molecular mass spectrum (MMS) with distinct oligomer peaks. Such equivalent averages as described above are then applied to MMS of polymers where it is assumed that the area under a peak of an oligomer is proportional to the number of molecules at a given mass m_j. From the MMS one can, under a proper understanding of the limitations and corrections of the spectrum, derive a good approximation to the true MMD of the polymer. It is the purpose of this chapter to consider how one can use the MMS to approximate the MMD. For this we shall consider the repeatability of the MMS; the accuracy or precision of MALDI MS for polymer analysis; the use of MALDI-TOF MS for development of polymer standards; and international efforts on standardization of experimental and instrumental protocols for MALDI-TOF MS polymer characterization. Furthermore, a new Taylor's expansion method for signal-axis intensity calibration of polydisperse materials is described in detail. This method is applied to earlier data to show that it gives predictions consistent with the idea that MALDI-TOF obtains the correct MMD for low polydispersed materials.

8.2   WHAT CAN MALDI DO FOR QUANTITATION?

When one examines the usefulness of an instrumental method for development as a protocol for general use, there are a variety of issues one must consider. First, one can ask how repeatable is the method in a single laboratory by one, or by several, operators? How well does it give results that are the same from experiment to experiment? This is considered to be the within-laboratory, or intralaboratory, repeatability. Second, we can ask how good is the agreement when the same sample is used in different laboratories? How well do the measured spectra agree with one another among the laboratories? This is usually discussed in terms of the "robustness" of the method. One can obtain estimates of this by an interlaboratory comparison. The former two types of uncertainty are called in the International Standards Organization (ISO) "Guide to the Expression of Uncertainty in Measurement" [3] type A uncertainties—that is, statistical uncertainties. (However, laboratory-to-laboratory variation may sometimes be viewed as type B uncertainties, a descriptor of systematic uncertainties.) A useful summary version of the Guide can be obtained as NIST Technical Note 1297 [4].

Finally, one can ask the precision or accuracy of the method. How well does the method measure the quantities the method is supposed to measure (in this case the MMD or its moments)? From one point of view, this can be considered as how well it agrees with other methods measuring the same quantity. These other values can be obtained from an independent measurement on the sample using a separate method, or by using the same method and comparing against a certified value. We later call these methods of quantitation "soft quantitation."

From a second point of view, we can ask how well does the method measure the "true" value of the quantity we are hoping to measure? How we know what is the true value is often unclear, so we ask another question. What are the uncertainties in the measurement, other than statistical, which affect the measurement or the quantity we are trying to measure? We would like to be able to say something about systematic uncertainty in the MMS measured by MALDI-TOF MS arising from such instrument attributes as laser energy, detector voltage, and sample preparation parameters (such as choice of matrix, salt concentration, and matrix : polymer ratio). These uncertainties are often called type B or systematic uncertainties. In order to estimate these uncertainties, one needs a model or theoretical construct of the experiment, that is, how the experiment in some way relates to the measurand. This is called "hard quantitation."

8.3   REPEATABILITY AND ROBUSTNESS
OF MALDI-TYPE A UNCERTAINTY

It is not easy to obtain within-laboratory repeatability from a technical paper because such numbers are typically not reported. Furthermore, within-laboratory repeatability depends strongly on the ease of which spectra are obtained (or conversely the determination of the experimentalist when it is difficult to obtain good spectra). The key factor here is the signal-to-noise ratio. The first NIST interlaboratory comparison on a low-mass PS allowed us to estimate the within-laboratory repeatability, as well as between-laboratory repeatability, of an easily done polymer using a well-defined method by having each participating laboratory perform three repeats of the same experiment. First, we will describe the interlaboratory comparison itself and then we will compare intra- to interlaboratory results.

The interlaboratory comparison was conducted by NIST among 23 laboratories using MALDI-TOF MS on well-characterized PS [5]. An example of a MALDI-TOF MS spectrum from this polymer is given in Figure 8.1. The PS was synthesized to have a tertiary butyl end group at one end and a proton at the other. NMR characterization of M_n was found to give (7050 ± 400) Da. The M_w was found to be (7190 ± 600) Da by light scattering. The expanded uncertainty values in the ( ) represent both type A uncertainties (statistical uncertainty) and type B uncertainties (systematic uncertainties) [4]. Fourier-transform infrared spectroscopy (FTIR) confirmed the presence of a single pair of end groups and no other end groups in measurable amounts. Each participating laboratory was asked to perform MALDI mass spectrometry using two distinct protocols. Each laboratory was asked to do three repeats of each protocol to check for intralaboratory variability. The M_w and M_n by MALDI-TOF MS averaged among all laboratories were found to be 6740 Da and 6610 Da, respectively.

The polymer from this interlaboratory comparison was later released as NIST Standard Reference Material (SRM) 2888 certified for M_w by light scattering but with MALDI-TOF and NMR offering supplemental numbers.

The overall intralaboratory repeatability standard deviation is defined as the square root of the sum of the squares of the standard deviation of each laboratory divided by the number of laboratories following ASTM E691 "Standard Practice for Conducting an Interlaboratory Study to Determine the Precision of a Test Method" [6]. The overall intralaboratory repeatability standard deviation for the M_n is 43 Da using retinoic acid as the matrix and 40 Da using dithranol as the matrix. The overall

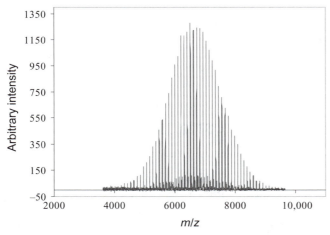

Figure 8.1.   Typical spectrum of a polystyrene (SRM 2888) with $M_n = 6800$ Da.

intralaboratory repeatability standard deviation for the M_w is 38 Da using retinoic acid as the matrix and 34 Da using dithranol as the matrix. This shows that each institution had very good control of its MALDI-TOF MS measurement within its laboratory. We observed for the between-laboratory standard deviation of the M_n following the protocol using retinoic acid, 132 Da and for that using dithranol, 116 Da. For M_w we obtain 118 Da using retinoic acid and 101 Da using dithranol. Notice the between-laboratory variation is greater than the within-laboratory variation, as expected. Also notice that the values of the standard deviations are small compared to the M_n and M_w, indicating that MALDI-TOF MS has the ability to give credible MMDs of low-mass polymers. Furthermore, the uncertainties are small compared to the overall uncertainty from the equivalent numbers from light scattering and NMR. This is mainly due to the inclusion of the type B uncertainties (systematic uncertainties) in the light scattering and NMR estimates of uncertainty. Such values are not yet available for MALDI-TOF MS.

8.4   MALDI SYSTEMATIC UNCERTAINTIES-TYPE B UNCERTAINTIES

Many of the quantities measured on which uncertainties are determined, for example, mass of an object, wavelength of light, or viscosity of a solution, are single-point measurands so the calibration need only be on a single axis. The MMS is a two-dimensional representation and so both the mass axis and number of ions at each mass (which we call the signal axis) have to be calibrated separately. We are then confronted with a much more complicated calibration problem since we must find a calibration method for both axes (the mass axis and signal axis). We discuss each axis separately.

8.4.1   Mass Axis Quantitation Uncertainty

Mass axis quantification is the most easily performed of the two. Calibration of most TOF instruments is usually done with biopolymers of known molecular masses. These biopolymers are selected because they typically provide a single major peak whose mass is known accurately; thus, mass axis quantification is quite straightforward. Calibration usually can be done using three of these biopolymers. Collecting data with 2 ns time intervals, one can get better than single mass unit resolution on an instrument with a 1.5 m flight tube in reflectron mode at about 7000 Da.

Calibration of the mass axis can also be done by combining a single biopolymer with a homopolymer calibrant. The oligomeric masses, m_j, with n repeat units of mass r and masses of the end group, m_{end}, of the polymer calibrant are given by

$$m_j = nr + m_{end} + m_{salt} \qquad \text{(Eq. 8.4)}$$

where n is the number of repeat units in the n-mer of the polymer and the m_{salt} refers to the mass of the metal cation adducted to the polymer n-mer.

Thus, calibration of the mass axis using a PS calibrant reduces to determining n for one of the peaks; this is accomplished through the use of the biopolymer mass as follows. The main peak from the biopolymer is assigned to its mass. The biopolymer peak will either lie between the masses of two n-mers of the PS calibrant, or exactly correspond to the mass of an n-mer. If it is at exactly the same mass as one of the n-mers of the PS calibrant, use Equation 8.4 to find the degree of polymerization, n, for the n-mer. If the peak of the biopolymer lies between the masses of two n-mers of the PS calibrant, use Equation 8.4 to find n_1, the mass of the n-mer whose mass is less than that of the repeat unit lower than the mass of biopolymer. Find additional calibration points by selecting PS peaks at intervals between 5 to 10 repeat units less than and greater than n_1 and compute masses from Equation 8.1. Generally, a total of four or five calibration points are selected.

For obtaining an MMD of a homopolymer, mass accuracy of better than a few mass units may not be necessary since polymer MMD are often not critically dependent on such accurate masses. Good mass resolution and calibration is required for mixtures of homopolymers with different end groups, for copolymer composition determination, and for polymer architecture studies. Since many of the above problems have even more signal axis uncertainties than we can currently deal with, no further definition of the mass axis is necessary.

8.4.2 Signal Axis Quantitation Uncertainty

Comparison with Other Methods—Soft Quantitation Generally, most laboratories obtained reasonable agreement for narrow MMD polymers between moments of MMD obtained from MALDI and those obtained by other methods, for example, light scattering to obtain M_w; end-group analysis or osmometry to get M_n; or GPC to get both. This kind of quantitation we call soft quantitation—by this we mean the agreement or near agreement of the MMD or some of its moments by two methods. Montaudo et al. have shown good agreement between classical methods and MALDI for polymers with $M_w/M_n < 1.3$ [7]. In particular, they studied PS, poly(methylmethacrylate) (PMMA), and poly(ethylene glycol) (PEG) up to molecular masses of 100,000 Da and showed as long as the PD was less then 1.3, agreement between MALDI and GPC for the peak of each (close to the M_n of a Gaussian distribution) were excellent. Danis et al. [8] too have shown good agreement between M_n and M_w from GPC and MALDI for narrow polydispersity polymers of PS, polybutadiene (PBD), and polyisoprene. Lloyd et al. [9] in their study of PMMA by GPC and MALDI support this view. While most of the previous work was done on molecular masses of less than 100,000 Da, Schriemer and Li [10] showed that they obtained good agreement with PS for narrow MMD from 100,000 Da to 1,300,000 Da between classical methods and MALDI in the measurement of M_n and M_w. Schnoll-Bitai et al. [11], using values of moments of polymers and assuming theoretical distributions of polymers, showed the effect of various matrices and machine conditions on the moments and distributions obtained for these polymers. They also looked at the effect of mixing on the mass discrimination effects on these properties.

Signal Axis Quantitation—Hard Quantitation Many systematic uncertainties can arise in the signal axis quantitation. Here we review some of the literature on this issue.

Uncertainties Arising from Sample Preparation   In order to quantitate a polymer MMD, the sample preparation method must allow each oligomer in the MMD to be laid down on the sample target in a consistent manner. Sample preparation for MALDI is typically performed by dissolving the matrix, salt, and polymer in a common solvent. The solution is then hand-spotted or sprayed onto the MALDI target. There have been discussions of problems arising from sample preparation, specifically the use of solvents in which the polymer or part of the MMD are not soluble to the concentrations expected or the use of mixed solvents in which the polymer or part of the polymer MMD phase separated out of the solution [12, 13].

For many matrices, hand spotting leads to the growth of large crystals. Although it is often thought that biopolymers are incorporated into the matrix crystals, it is unlikely that synthetic polymers, such as atactic PS which themselves do not crystallize, can incorporate themselves into a matrix crystal. The work of Bauer et al. [14] suggests that the polymers are not in the matrix but rather form in clusters outside the matrix. Thus, the larger the crystals, the less uniformly dispersed are the polymers and perhaps the more likely the polymers of differing MM behave differently in this environment. Spraying (either by electrospraying or nebulizing) was initiated to better disperse the polymer. It allows one to deposit small almost dry matrix crystals and polymer on the target surface. This leads to more consistent overall signal for the polymer [15] and a better representation of the whole polymer MMD.

Uncertainties Arising from Ablation, Ionization, and Drift Regions   Little is known about the desorption/ionization process in MALDI. Work on the same type of polymer of different molecular masses suggests that higher ablation energies are required for polymers of higher molecular masses. Thus, since the normal procedure is to adjust the laser ablation energy to just above the threshold needed to obtain the polymer spectra in MALDI, the desorption process of MALDI could also cause the moments determined by mass spectrometry to be lower than the true moment. It is not known whether low-mass molecules desorb better than higher-mass molecules, but, if preferential desorption of low-mass molecules occurs, it could account for the low MALDI values often reported when the moments from MALDI are compared to those obtained by classical methods [16–19].

For homopolymers, we commonly see the dominant polymer repeat units with a high signal intensity. We may also see less intense intermediate peaks adjacent to the major high intensity repeat units. These intermediate peaks more often than not have the same repeat unit as the major peaks. These intermediate peaks may arise from a variety of sources. It is most obvious that end groups or cations may differ. One can have two different sets of end groups on linear chains or many different types of end groups on branched chains. This gives rise to a main group series of peaks arising from the major pair of end groups and a second series of peaks from the second set of end groups.

One may see additional peaks from different salts attaching to the polymer. As noted before for polar polymers like PEG and PMMA, one does not need to add salts to the matrix. Impurities in the matrix in the form of Na or K salts seem to be in sufficient concentration to bring about the addition of Na or K to the polymers

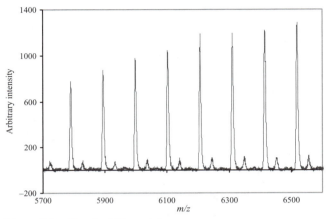

Figure 8.2.   Detail of Figure 8.1.

without the use of additional salts. Rashidezadeh and Guo [20] have shown PS responds differently to differing metals. Ag seems to give the best response of all the metals tried.

Some intermediate peaks will often appear differently in linear and reflectron modes for reasons other than sensitivity differences between the two modes. If the reflectron spectra are different from the linear spectra, this may be due to fragmentation during the flight down the linear flight tube. Of course, fragmentation occurring in the ablation region (the source region) cannot be detected this way. Ablation region fragmentation may be detected by the appearance of more n-mers in the low-mass regions of the MMD as the laser energy is increased. This has been looked at by Wetzel et al. [21, 22] who showed for PEG and other backbone ether linkage polymers that the fragmentation increases as the laser intensity increases.

Finally, the intermediate peaks may arise from matrix adducts to the polymer. Goldschmidt et al. [23] have found matrix adducts of PS polymers of 3900 Da and 7000 Da of the form Matrix + Ag + Polymer and Matrix + Ag + Ag + Polymer. Figure 8.1 shows the entire spectrum and Figure 8.2 shows these adduct peaks in a PS of about 7000 Da used in the NIST Interlaboratory Comparison [5].

Uncertainties Arising from the Detector   The detector is another influence on the MMD that may impact the moments calculated by MALDI-TOF MS. The detector in most of the instruments is a microchannel plate (MCP) detector. The smaller molecular mass molecules hit the detector first, and the larger molecules follow later. If there were saturation occurring in the detector, then there would be a discrimination against the high-mass molecules. The issue of detector saturation has been discussed by many authors [24, 25]. Since the matrix molecules or low-molecular-mass PS molecules could saturate the detector before the higher-molecular-mass molecules arrive at the detector, this would cause the computation of a lower M_n and M_w than the true values.

The detector may also be less sensitive to larger mass polymer molecules due to the way the ions are detected. MCP detectors count the number of ions by an ion-to-electron conversion when the ionized polymer collides with the detector plate. A bias occurs against the high-mass species when the ion-to-electron conversion is diminished due to a decrease in the impact velocity of the larger ionic species [25, 26, 27]. If a discrimination against the high-mass polymer molecules exists, this would cause the estimated moments of the distribution to be less than the true value of the moment.

Finally, there is a problem with the detection for higher n-mer masses in MALDI. Since the detector is an event detector, a counter of the number of n-mers, peaks disappear into the noise before peaks would disappear into the noise on a normal GPC detector, which detects the total mass associated with each n-mer. The consequences of this signal-to-noise problem on the polydispersity (M_w/M_n) studied by MALDI-TOF MS has been discussed by Saucy and Zhu [28].

Uncertainties from Data Analysis Obtaining peak areas depends on drawing of baselines, smoothing of data, and often assuming shapes of peaks. Such assumptions are similar to those made in other forms of spectroscopic or chromatographic system. Such assumptions may be a cause of the disagreement between the moments determined by classical methods and those determined by MALDI. Furthermore, there is much more noise in the low-mass end of the spectra obtained by MALDI-TOF MS. Some of this noise may arise from fragments of the polymer, clusters of matrix and salt, or metal clusters. If the software does not account for these differences in baseline noise, then the data may be overweighted in the lower part of the MMD, and thus the calculated M_n and M_w would be lower than the true moments. These issues and a description of an effort to create an unbiased peak analysis program are described in another chapter of this book.

Assuming correct peak areas are obtained, the question arises as to how to compare the MMD estimated from a time of flight mass spectrometer to that obtained from GPC using a UV or differential refractive index (DRI) detector. GPC is the instrument most commonly used to obtain the MMD for synthetic polymers. The most obvious difference between the two techniques is that the TOF detector counts numbers of ions of n-mers while the DRI or UV detectors determine mass concentration of the polymer. The mass concentration of the polymer is proportional to the product of the molecular mass of the n-mer and the number of molecules of that n-mer per unit volume. Thus, in the mass spectrometry we obtain the number MMD while in the GPC we normally obtain the mass MMD. Furthermore, the raw signal of each must be corrected for the transformation from time to mass, which is different for each instrument. Guttman [29] has discussed this transformation in detail.

With all these complications, it seems impossible that we can find a simple way to look at quantitation on the signal axis. But taking the clue from the work of Montaudo et al. [30] and others that the polydispersity of a polymer must be fairly low to do quantitation by MALDI alone, a number of authors have looked at this issue.

Gravimetric Blending for Hard Quantitation The works discussed below all use blending of polymers of different molecular mass to obtain some estimate of how gravimetric blending affects the predicted versus found properties of the MMD or moments thereof.

Perhaps the most unique of these is the work of Shimada et al. [31, 32]. Here the authors fractionate lower relative molecular mass PS's and polyethylene glycols into a single oligomer using preparative supercritical chromatography. Oligomers with degree of polymerization up to 25 were fractionated. Then with five or six of these from a given polymer MMD, they obtain the calibration curve as a function of machine parameters and laser energy and molecular mass. They then use this calibration curve to obtain the MMD of a whole polymer from the spectra obtained at a given set of machine setting and laser energies. This is certainly unique but depends on being able to fractionate the polymers into oligomers at any molecular mass, which seems very unlikely. Further as we shall note later, it seems important that we are linear or near linearity in signal intensity versus analyte concentration to assume that we can make quantitative the entire MMD. They are able to do this for molecular masses up to 1800 Da for the PEG and up to 3000 Da for the PS.

Yan et al. [33] have looked at mass ratios of two polydimethylsiloxane polymers by MALDI, 6140 Da and 2200 Da. After initially optimizing the laser power to give the strongest signal intensity for both the high and low-molecular-mass polymers, they found that the mass ratios estimated from MALDI were very close to those predicted from gravimetric ratios. Chen and He [34, 35] performed a similar experiment using polyethylene glycols with a variety of end groups. They found that for molecular masses from 800 Da to 3300 Da, the gravimetric amount of polymer seemed to map well into the MALDI predicted values.

Li and coworkers [16, 25, 36] started by looking at equimolar blends of polymers, PS of molecular masses below 20,000 Da, to optimize the instrument and sample preparation parameters. They did this by varying machine and matrix and salt parameters finding where the M_n best matched the M_n expected of the equimolar mixtures. They made a careful study of PS of molecular mass 5050 Da, 7000 Da, and 11,600 Da and blends of the three. Fitting their polymer MMD to Gaussian distributions and using the PS 5050 Da and 11,700 Da, they studied changes in the MMD of the polymer of 7000 Da with blends of the 5050 Da polymer. They found they could detect no systematic uncertainties within 0.5% in the MMD or the moments of the MMD assuming MMD were Gaussian distributions.

Model for Gravimetric Blending As we stated in Section 8.3, to estimate the level of uncertainty in an instrumental method, one needs a theoretical construct to allow one to consider how uncertainties affect the final measurand. A detailed model of the MALDI-TOF MS instrument and process for signal intensity seems unlikely at this time. However, the work of Yan et al. [37], Chen and He [35], and Zhu et al. [38] described previously each implicitly assume that there is a point in the parameter space of the instrument and the sample preparation where we can assume that the signal intensity, S_i, for an oligomer of mass m_i is linearly proportional to n_i, the number of polymer molecules at that oligomer mass. Mathematically this is given by

$$S_i = kn_i \quad \text{for} \quad n_i < n_0 \qquad \text{(Eq. 8.5)}$$

where for a narrow enough range of m_i we presume k is a constant independent of m_i and the range of linearity, $n_i < n_0$, is about the same for all polymers in the range. From the work of Goldschmidt et al. [39], we find, for example, for PS in retinoic acid or dithranol, there are large regions of the linearity in the overall signal to the ratio of polymer to matrix where there is a large molar excess of Ag salt for PS of molecular mass around 8000 Da. From that work we also know that if we get to a high polymer-to-matrix ratio, we see the curve become nonlinear and approach saturation. This cutoff too seems to be a weak function of molecular mass.

If we assume we are in the linear region for all the oligomers of a polymer, the overall signal from a mass of polymer is given by

$$\sum S_i m_i = k \sum n_i m_i \qquad \text{(Eq. 8.6)}$$

with $n_i m_i$ summed over all i. This is the total mass of the polymer as long as the polymer is evenly distributed throughout the matrix on the target spot. Then we can obtain:

$$\sum S_i m_i / \sum S_i = k \sum n_i m_i / k \sum n_i \qquad \text{(Eq. 8.7)}$$

The right-hand side of the equation is the exact M_n of the polymer (see Eq. 8.1 independent of k since the k in numerator and denominator cancel out. The same holds for equations for M_w and all higher moments. This is generally true if we are in the linear range.

We know from the work of Owens [40], Montaudo et al. [7], McEwen et al. [41], and Li and coworkers [11, 38, 42] that if we get the m_i too far apart, the values for k or the n_0 cutoffs must change dramatically otherwise the MALDI would be able to obtain MMD correctly for very broad distributions.

In general, if we are in the linear range for each oligomer i of a polymer

$$S_i = k_i n_i \qquad \text{(Eq. 8.8)}$$

where k_i is a slowly varying function of i or more simply a slow varying function of m_i, the mass of the oligomer for a fixed set of instrument parameters and sample preparations parameters (e.g., matrix, polymer matrix concentration, and salt and salt concentration). An equation similar to Equation 8.7 has been assumed by Sato et al. [43] for the relationship between fractions obtained from an analytical GPC and signals measured from each fraction by MALDI-TOF MS. This is simply a course graining assumption on ours.

If we assume in the above equation that k_i is a slowly varying function of i or of m_i, as we stated before, then we may make a Taylor's expansion around a mass peak toward the center of the MMD. This is done so that the function is changing as little as possible over the entire MMD. For now we call the mass we are expanding around M_0 we obtain

$$S_i = k_o n_i + B(m_i - M_o) n_i \qquad \text{(Eq. 8.9)}$$

We shall now explore this function. Functions for the total signal (S_T) and total mass (M_T) multiplied by signal are

$$S_T = \sum_i S_i = k_o \sum n_i + B(M_n - M_o) \sum_i n_i \qquad \text{(Eq. 8.10)}$$

$$M_T = \sum_i m_i S_i = k_o M_n \sum_i n_i + B M_n (M_w - M_o) \sum_i n_i \qquad \text{(Eq. 8.11)}$$

where M_n and M_w are defined in Equations 8.1 and 8.2 and are the true number-average and mass-average relative molecular masses and we notice

$$M_w M_n = \sum_i m_i^2 n_i \Big/ \sum_i n_i \qquad \text{(Eq. 8.12)}$$

Then taking the ratio of these two equations and assuming $M_n = M_0$ (because we are doing the Taylor's expansion around M_n), we obtain

$$M_n^{\text{exp}} = M_n (1 + M_n B / k_o (PD - 1)) \qquad \text{(Eq. 8.13)}$$

where

$$M_n^{\text{exp}} = \sum m_i S_i \Big/ \sum S_i \qquad \text{(Eq. 8.14)}$$

is the experimentally measured M_n from MALDI.

One step further and we obtain

$$(M_n^{\text{exp}} - M_n)/M_n = M_n (B/k_o)(PD - 1) \qquad \text{(Eq. 8.15)}$$

Equivalent results can be obtained for higher moments. The result on M_n can be compared with the data from Montaudo et al. [44], who found that for a polymer of low enough polydispersity, the sums over the moments of the signal give a good representation of the moments of the MMD from GPC. In Table 8.1 we use the results of Montaudo et al. [7] to estimate B/k_o. We estimate the error in M_n from their estimated error in the peak of the GPC versus that of the MALDI. Following Montaudo et al. [7], we defined Δ by

$$\Delta = ((M_n^{\text{exp}} - M_n)/M_n) \qquad \text{(Eq. 8.16)}$$

where we take M_n to be the M_n obtained from some other measure like GPC or osmometry (i.e., by soft quantitation). We can use their data to estimate the Δ in Equation 8.16 if we assume the polymers are narrow PD and Gaussian (which the polymers they used more likely are because they are made by an anionic polymerization). But even for the Gaussian distribution, this approximation has a number of errors in it. First, using a refractive index detector, GPC gives the fraction of total mass of the polymer at each oligomer mass (the classical mass distribution) so the peak is more closely related to M_w while MALDI-TOF MS gives the fraction of the total number of oligomers at each oligomer mass (the classical number distribution) so the peak is more closely related to M_n. Thus, the fraction discrepancy found by Montaudo et al. is off by a factor of PD $-$ 1, which we correct for in Table 8.1. Second, the peak in the chromatogram or the spectra is not the peak of the actual distribution since the Jacobean of transform from time to mass (see Guttman [29]) in each case can shift the position of the peak of the chromatogram or of the spectrum. These corrections depend on the ratio of the mass to time calibration constants in each of the instruments used to take the data. Not having access to the raw data in time, no correction can be made by us. In the final column of Table 8.1 we

TABLE 8.1.   Data of Montaudo et al. [7] to Estimate B/k_0 and the Correction to 1 in the Taylor's Expansion of k_j in Equation 8.9 for Masses 30% Above or Below Maximum Mass in Distribution for Poly(Methyl Methacrylate) (PMMA), Polystyrene (PS), and Poly(Ethylene Glycol) (PEG)[a]

Polymer	M_p (GPC) Da	M_p (MALDI) Da	$\Delta - (PD - 1)$	B/k_0 Da^{-1}	$m_j = 0.70 M_n$ or $1.30 M_n$ Correction on 1 to S_j Da
PMMA	2,400	2,100	0.045	0.00023	0.16
	3,100	2,700	0.039	0.00014	0.13
	4,700	4,200	0.0063	1.3E-05	0.02
	6,540	5,200	0.114	0.00019	0.38
	9,400	7,500	0.102	0.00010	0.30
	12,700	10,400	0.101	9.9E-05	0.37
	17,000	15,000	0.057	5.6E05	0.28
	29,400	27,000	0.021	1.2E-05	0.11
	48,000	47,000	−0.029	−1.2E-05	−0.17
PS	5,050	5,100	−0.060	−0.00023	−0.35
	7,000	7,020	−0.042	−0.00015	−0.32
	9,680	9,600	−0.012	−6.0E-05	−0.17
	11,600	11,300	−0.004	−1.2E-05	−0.04
	22,000	20,800	0.024	3.7E-05	0.24
	30,300	28,000	0.046	5.0E-05	0.45
	52,000	46,000	0.085	5.4E-05	0.85
PEG	4,100	3,900	−0.001	−5.9E-06	−0.01
	7,100	7,400	−0.072	−0.00034	−0.72
	8,650	8,610	−0.025	−9.8E-05	−0.25
	12,600	12,790	−0.055	−0.00011	−0.41
	23,600	23,700	−0.064	−4.5E-05	−0.32

[a] M_p, M_n, and m_j in Da, B/k_0 in Da^{-1}, Δ, PD, and S unitless.

estimate the correction to the 1 in the equation for the signal rewritten from Equation 8.9 as

$$S_j = n_j k_o (1 + B/k_o (m_j - M_n))$$ (Eq. 8.17)

where the oligomer m_j has molecular mass 0.7 times M_n and 1.3 times M_n (e.g., from Table 8.1 for a polymer with $M_n = 10,000$ Da the oligomer of MM 13,000 Da or 7000 Da) yield terms contributing to the value in the column compared to 1. They all have values less than 1, indicating the expansion is expected to be good for this range.

Schriemer and Li [45] compared M_n from GPC and MALDI for PS from about 100,000 Da to 1,300,000 Da. They acquired MALDI on these polymers and obtained estimates of the M_n and PD from GPC from the suppliers. We can use that data directly in Table 8.2. We find essentially the same result from their data as we find in Table 8.1 from the Montaudo et al. data. The sign and the exact values depend on the instrument parameters and the sample preparation the authors used, which are unlikely the same for any of the polymers. What we can say is that Taylor's expansion seems to be okay since the correction to 1 is $M_n B/k_o$, which for low PD is small compared to 1.

TABLE 8.2. Data of Schriemer and Li [45] to Estimate of B/k_0 and the Correction to 1 in the Taylor's Expansion of k_j in Equation 8.9 for Masses 30% Above or Below Maximum Mass in Distribution for Polystyrene (PS)[a]

Polymer	M_n (GPC) Da	M_n (MALDI) Da	Δ	B/k_0 Da^{-1}	$m_j = 0.70 M_n$ or $1.30 M_n$ Correction on 1 to S_j Da
PS	152,000	153,700	−0.011	−2.7E-06	−0.12
	198,000	202,500	−0.022	−2.2E-06	−0.13
	316,000	308,900	0.022	2.0E-06	0.19
	340,000	402,400	−0.183	−3.4E-06	−0.35
	481,000	469,800	0.023	9.7E-07	0.14
	618,000	589,100	0.046	1.51E-06	0.27
	728,000	738,400	−0.014	−5.1E-07	−0.11
	892,000	916,200	−0.027	−5.4E-07	−0.14
	1,378,000	1,391,000	−0.009	−1.1E-07	−0.047

[a] M_n, and m_j in Da, B/k_0 in Da^{-1}, Δ, PD, and S unitless.

We close this section saying that the gravimetric calibration of the signal axis can avoid some of the issues discussed above in the section on the uncertainties arising from ablation, ionization, and drift regions. However, many of the issues arising from uncertainties in sample preparation and data analysis still plague the gravimetric calibration techniques. In sample preparation, the problem of phase separation as a function of molecular mass (each component of the blend is usually of a different molecular mass) appears. In solution, the solubility of some polymers is very molecular-mass dependent (e.g., polyethylene glycol's solubility in tetrahydrofuran). As the solvent evaporates in the evaporation droplet technique or the electrospray method, the oligomers of different molecular masses come out of solution at different rates and thus may find themselves at different matrix:polymer ratios. In data analysis, the issues of determining the baseline consistently and drawing areas consistently is still an issue and may incorrectly represent the amount of mass in one or the other polymer in the blend. Further, if one polymer is a small fraction of the other, the noise will cause major integration problems.

8.5   INTERNATIONAL EFFORTS ON STANDARDIZATION OF EXPERIMENTAL PROTOCOL POLYMER CHARACTERIZATION BY MALDI-TOF MS

With the results of Montaudo et al. [7] and Li and coworkers [11, 36, 42] showing that for a narrow MMD one can obtain good agreement with classical methods, one might expect for such narrow MMD polymer to be able to obtain quantitative agreement among various labs. As described above, an interlaboratory comparison was done to look at the robustness of the method. With the successful completion of the interlaboratory comparison showing the method was robust enough to give consistent results among various labs, it was decided to develop a standard method using the protocol designed for the interlaboratory comparison.

This was done under the auspices of The Versailles Project on Advanced Materials and Standards (VAMAS). The main purpose of VAMAS is to encourage trade in high technology products by providing the technical basis for methods of practice and specifications for advanced materials. (VAMAS is technically a "Memorandum of Understanding" between Canada, France, Germany, Italy, Japan, the United Kingdom, the United States, and the European Union, although laboratories from other countries can participate in the research activities.)

A new technical working activity (TWA-28) was established in VAMAS to develop mass spectrometry as a quantitative method to determine the MMD of synthetic polymers. The principal goals were to develop a method which would provide (1) sample preparation techniques that minimize sampling bias in terms of molecular mass, (2) a standard method for mass calibration of mass spectrometers useful for polymer MS, (3) data analysis procedures, and (4) a detailed protocol anyone could follow. Research in the TWA was built on knowledge and results gained in the interlaboratory comparisons of MALDI-TOF MS sponsored by NIST. The protocols from these studies served as the starting point for extending the method to PS possessing different MMDs.

After a number of years of activity, a method was developed by VAMAS TWA-28. A modification of this method was adopted as an ASTM International Method ASTM D7034 "Standard Test Method for Molecular Mass Averages and Molecular Mass Distribution of Atactic Polystyrene by Matrix Assisted Laser Desorption/Ionization (MALDI)-Time of Flight (TOF) Mass Spectrometry (MS)" by the Analytical Methods subcommittee of the polymer committee, D20.70. A second modification of the VAMAS method was released from the "Deutsches Institut für Normung" as DIN 55674 "Synthetic Polymers—Determination of Molecular Mass and Molecular Mass Distribution of Polymers by Matrix Assisted Laser Desorption/ Ionization-Time-Of-Flight-Mass Spectrometry."

REFERENCES

1. Lee, W.; Lee, H.; Cha, J.; Chang, T.; Hanley, K.J.; Lodge, T.P., Molecular weight distribution of polystyrene made by anionic polymerization, *Macromolecules*, **2000**, 33, 5111–5115.
2. Clay, P.A.; Gilbert, R.G., Molecular weight distribution in free-radical polymerization: 1. model development and implications for data interpretation, *Macromolecules*, **1995**, 28, 552–556.
3. BIPM, IEC, IFCC, ISO, IUPAC, IUPAP, OIML, *Guide to the Expression of Uncertainty in Measurement*, International Organization for Standardization, Geneve, **1995**, pp. 1–100.
4. Taylor, B.N.; Kuyatt, C.E., Guidelines for evaluating and expressing the uncertainty of NIST measurement results. NIST Technical Note 1297, 1994 Edition, **1994**, NIST, U.S. Government.
5. Guttman, C.M.; Wetzel, S.J.; Blair, W.R.; Fanconi, B.M.; Girard, J.E.; Goldschmidt, R.J.; Wallace, W.E.; VanderHart, D.L., NIST-sponsored interlaboratory comparison of polystyrene molecular mass distribution obtained by matrix-assisted laser desorption/ionization time-of-flight mass spectrometry: statistical analysis, *Analytical Chemistry*, **2001**, 73, 1252–1262.

6. ASTM, E691-05 Standard Practice for Conducting an Interlaboratory Study to Determine the Precision of a Test Method, **2005**, ASTM International, West Conshohocken, PA, www.astm.org.

7. Montaudo, G.; Montaudo, M.S.; Puglisi, C.; Samperi, F., Characterization of polymers by matrix-assisted laser-desorption ionization time-of-flight mass-spectrometry—molecular-weight estimates in samples of varying polydispersity, *Rapid Communications in Mass Spectrometry*, **1995**, 9, 453–460.

8. Danis, P.O.; Karr, D.E.; Xiong, Y.S.; Owens, K.G., Methods for the analysis of hydrocarbon polymers by matrix assisted laser desorption/ionization time of flight mass spectrometry, *Rapid Communications in Mass Spectrometry*, **1996**, 10, 862–868.

9. Lloyd, P.M.; Suddaby, K.G.; Varney, J.E.; Scrivener, E.; Derrick, P.J.; Haddleton, D.M.A., Comparison between matrix-assisted laser-desorption ionization time-of-flight mass-spectrometry and size-exclusion chromatography in the mass characterization of synthetic-polymers with narrow molecular-mass distributions—poly(methyl methacrylate) and poly(styrene), *European Mass Spectrometry*, **1995**, 1, 293–300.

10. Schriemer, D.C.; Li, L., Detection of high molecular weight narrow polydisperse polymers up to 1.5 million daltons by MALDI mass spectrometry, *Analytical Chemistry*, **1996**, 68, 2721–2725.

11. Schnoll-Bitai, I.; Hrebicek, T.; Rizzi, A., Towards a quantitative interpretation of polymer distributions from MALDI-TOF spectra, *Macromolecular* Chemistry and Physics, **2007**, 208, 485–495.

12. Schriemer, D.C.; Li, L.A., Mass discrimination in the analysis of polydisperse polymers by MALDI time-of-flight mass spectrometry: 1. sample preparation and desorption/ionization issues, *Analytical Chemistry*, **1997**, 69, 4169–4175.

13. Kassis, C.M.; DeSimone, J.M.; Linton, E.W.; Lange, G.W.; Friedman, R.M., An investigation into the importance of polymer-matrix miscibility using surfactant modified matrix-assisted laser desorption/ionization mass spectrometry, *Rapid Communications in Mass Spectrometry*, **1997**, 11, 1462–1466.

14. Bauer, B.J.; Byrd, H.C.M.; Guttman, C.M., Small angle neutron scattering measurements of synthetic polymer dispersions in matrix-assisted laser desorption/ionization matrixes, *Rapid Communications in Mass Spectrometry*, **2002**, 16, 1494–1500.

15. Hanton, S.D.; Hyder, I.Z.; Stets, J.R.; Owens, K.G.; Blair, W.R.; Guttman, C.M.; Giuseppetti, A.A., Investigations of electrospray sample deposition for polymer MALDI mass spectrometry, *Journal of the American Society for Mass Spectrometry*, **2004**, 15, 168–179.

16. Schriemer, D.C.; Li, L.A., Mass discrimination in the analysis of polydisperse polymers by MALDI time-of-flight mass spectrometry: 1. sample preparation and desorption/ionization issues, *Analytical Chemistry*, **1997**, 69, 4169–4175.

17. Lehrle, R.S.; Sarson, D.S., Polymer molecular-weight distribution—results from matrix-assisted laser-desorption ionization compared with those from gel-permeation chromatography, *Rapid Communications in Mass Spectrometry*, **1995**, 9, 91–92.

18. Larsen, B.S.; Simonsick, W.J.; Mcewen, C.N., Fundamentals of the application of matrix assisted laser desorption ionization mass spectrometry to low mass poly(methylmethacrylate) polymers, *Journal of the American Society for Mass Spectrometry*, **1996**, 7, 287–292.

19. Montaudo, G.; Garozzo, D.; Montaudo, M.S.; Puglisi, C.; Samperi, F., Molecular and structural characterization of polydisperse polymers and copolymers by combining MALDI-TOF mass-spectrometry with GPC fractionation, *Macromolecules*, **1995**, 28, 7983–7989.

20. Rashidezadeh, H.; Guo, B.C., Investigation of metal attachment to polystyrenes in matrix-assisted laser desorption ionization, *Journal of the American Society for Mass Spectrometry*, **1998**, 9, 724–730.

21. Wetzel, S.J.; Guttman, C.M.; Girard, J.E., The influence of matrix and laser energy on the molecular mass distribution of synthetic polymers obtained by MALDI-TOF-MS, *International Journal of Mass Spectrometry*, **2004**, 238, 215–225.

22. Wetzel, S.J.; Guttman, C.M.; Girard, J.E., Influence of laser energy and matrix of MALDI on the molecular mass distribution of poly(ethylene glycol), *Abstracts of Papers of the American Chemical Society*, **2003**, 225, U666.

23. Goldschmidt, R.J.; Wetzel, S.J.; Blair, W.R.; Guttman, C.M., Post-source decay in the analysis of polystyrene by matrix-assisted laser desorption/ionization time-of-flight mass spectrometry, *Journal of the American Society for Mass Spectrometry*, **2000**, 11, 1095–1106.

24. Mcewen, C.N.; Jackson, C.; Larsen, B.S., Instrumental effects in the analysis of polymers of wide polydispersity by MALDI mass spectrometry, *International Journal of Mass Spectrometry and Ion Processes*, **1997**, 160, 387–394.

25. Schriemer, D.C.; Li, L.A., Mass discrimination in the analysis of polydisperse polymers by MALDI time-of-flight mass spectrometry: 2. instrumental issues, *Analytical Chemistry*, **1997**, 69, 4176–4183.

26. Geno, P.W.; Macfarlane, R.D., Secondary-electron emission induced by impact of low-velocity molecular-ions on a microchannel plate, *International Journal of Mass Spectrometry and Ion Processes*, **1989**, 92, 195–210.

27. Gilmore, I.S.; Seath, M.P., Ion detection efficiency in sims: dependencies on energy, mass and composition for microchannel plates used in mass spectrometry, *International Journal of Mass Spectrometry*, **2000**, 202, 217–229.

28. Saucy, D.A.; Zhu, L., Comparison of polymer molecular weight polydispersity measured by MALDI-TOF/MS and conventional GPC. *Proceedings of the 47th ASMS Conference on Mass Spectrometry*, Dallas, Texas, June 13–17, **1999**.

29. Guttman, C.M., The relationship between the signals from size exclusion chromatography and time of flight mass spectrometry to a polymer molecular weight distribution, *Polymer Preprints*, **1996**, 37, 837–838.

30. Montaudo, G.; Montaudo, M.S.; Puglisi, C.; Samperi, F., Characterization of polymers by matrix-assisted laser-desorption ionization time-of-flight mass-spectrometry—molecular-weight estimates in samples of varying polydispersity, *Rapid Communications in Mass Spectrometry*, **1995**, 9, 453–460.

31. Shimada, K.; Nagahata, R.; Kawabata, S.; Matsuyama, S.; Saito, T.; Kinugasa, S., Evaluation of the quantitativeness of matrix-assisted laser desorption/ionization time-of-flight mass spectrometry using an equimolar mixture of uniform poly(ethylene glycol) oligomers, *Journal of Mass Spectrometry*, **2003**, 38, 948–954.

32. Shimada, K.; Lusenkova, M.A.; Sato, K.; Saito, T.; Matsuyama, S.; Nakahara, H.; Kinugasa, S., Evaluation of mass discrimination effects in the quantitative analysis of polydisperse polymers by matrix-assisted laser desorption/ionization time-of-flight mass spectrometry using uniform oligostyrenes, *Rapid Communications in Mass Spectrometry*, **2001**, 15, 277–282.

33. Yan, W.Y.; Gardella, J.A.; Wood, T.D., Quantitative analysis of technical polymer mixtures by matrix assisted laser desorption/ionization time of flight mass spectrometry, *Journal of the American Society for Mass Spectrometry*, **2002**, 13, 914–920.

34. Chen, H.; He, M. Y.; Pei, J.; He, H. F., Quantitative Analysis of Synthetic Polymers Using Matrix-Assisted Laser Desorption/Ionization Time-of-Flight Mass Spectrometry. *Analytical Chemistry* **2003**, 75, 6531–6535.

35. Chen, H.; He, M., Quantitation of synthetic polymers using an internal standard by matrix-assisted laser desoprtion/ionization time of flight mass spectrometry, *American Society for Mass Spectrometry*, **2005**, 16, 100–106.

36. Zhu, H.H.; Yalcin, T.; Li, L., Analysis of the accuracy of determining average molecular weights of narrow polydispersity polymers by matrix-assisted laser desorption ionization time-of-flight mass spectrometry, *Journal of the American Society for Mass Spectrometry*, **1998**, 9, 275–281.

37. Yan, W.Y.; Gardella, J.A.; Wood, T.D., Quantitative analysis of technical polymer mixtures by matrix assisted laser desorption/ionization time of flight mass spectrometry, *Journal of the American Society for Mass Spectrometry*, **2002**, 13, 914–920.

38. Zhu, H.H.; Yalcin, T.; Li, L., Analysis of the accuracy of determining average molecular weights of narrow polydispersity polymers by matrix-assisted laser desorption ionization time-of-flight mass spectrometry, *Journal of the American Society for Mass Spectrometry*, **1998**, 9, 275–281.

39. Goldschmidt, R.; Guttman, C., Response saturation of polystyrene in MALDI-TOF-MS. *Proceedings of the 47th ASMS Conference on Mass Spectrometry and Allied Topics*, Dallas, Texas, June 13–17, **1999**.

40. Owens, K.G., Sample preparation issues in the analysis of synthetic polymers by MALDI-TOF MS, *Abstracts of Papers of the American Chemical Society*, **2000**, 219, U363–U364.

41. McEwen, C.N.; Jackson, C.; Larsen, B.S., Instrumental effects in the analysis of polymers of wide polydispersity by MALDI mass spectrometry, *International Journal of Mass Spectrometry and Ion Processes*, **1997**, 160, 387–394.

42. Schriemer, D.C.; Li, L.A., Mass discrimination in the analysis of polydisperse polymers by MALDI time-of-flight mass spectrometry: 2. instrumental issues, *Analytical Chemistry*, **1997**, 69, 4176–4183.

43. Sato, H.; Ichieda, N.; Tao, H.; Ohtani, H., Data processing method for the determination of accurate molecular weight distribution of polymers by SEC/MALDI-MS, *Analytical Sciences*, **2004**, 20, 1289–1294.

44. Montaudo, G.; Montaudo, M.S.; Puglisi, C.; Samperi, F., Characterization of polymers by matrix-assisted laser-desorption ionization time-of-flight mass-spectrometry—molecular-weight estimates in samples of varying polydispersity, *Rapid Communications in Mass Spectrometry*, **1995**, 9, 453–460.
45. Schriemer, D.C.; Li, L., Detection of high molecular weight narrow polydisperse polymers up to 1.5 million daltons by MALDI mass spectrometry, *Analytical Chemistry*, **1996**, 68, 2721–2725.

APPENDIX A.   CURRENT POLYMER MOLECULAR MASS STANDARDS AVAILABLE FROM NIST

Currently all Polymer Molecular Mass Standard Reference Materials (SRM) are certified using traditional polymer molecular mass methods like light scattering, osmometry, and ultracentrifugation. They are, of course, certified for M_w and M_n and are generally relatively narrow MMD polymers (low PD). One polyethylene, SRM 1475, is a broad PD material and has its entire MMD certified. This was done by taking fractions of the whole polymer and obtaining the M_w by light scattering on each fraction and from which the MMD was reconstructed. For one standard, SRM 2888, light scattering certifies M_w and supplemental values of M_n and are given by the MALDI-TOF MS.

Most of these MM standards are used as calibrants or checks on calibration when performing GPC.

TABLE 8.A1.   Polymer Molecular Mass Standard Reference Materials Available from NIST

SRM	Polymer Type	Molecular Mass (Da)	Polydispersity
705a	Polystyrene	$M_w = 179,300$	$M_w/M_n = 1.07$
706a	Polystyrene	$M_w = 285,000$	broad MMD
1475a	Polyethylene, linear	$M_w = 52,000$	$M_z : M_w : M_n = 7.54 : 2.90 : 1$
1478	Polystyrene	$M_w = 37,400$	$M_w/M_n = 1.04$
1479	Polystyrene	$M_w = 1,050,000$	narrow MMD
1480	Polyurethane	$M_w = 47,300$	
1482a	Polyethylene, linear	$M_w = 13,600$	$M_w/M_n = 1.19$
1483a	Polyethylene, linear	$M_w = 32,100$	$M_w/M_n = 1.11$
1484a	Polyethylene, linear	$M_w = 119,600$	$M_w/M_n = 1.19$
1487	Poly(methylmethacrylate)	$M_w = 6000$	narrow MMD
1488	Poly(methylmethacrylate)	$M_n = 29,000$	narrow MMD
1923	Poly(ethylene oxide)	$M_w = 26,900$	$M_w/M_n = 1.04$
1924	Poly(ethylene oxide)	$M_w = 120,900$	$M_w/M_n = 1.06$
2885	Polyethylene, linear	$M_w = 6280$	
2886	Polyethylene	$M_w = 87,000$	
2887	Polyethylene	$M_w = 196,400$	
2888	Polystyrene	$M_w = 7190$	M_n and M_w by MALDI as supplemental values

CHAPTER 9

NEW APPROACHES TO DATA REDUCTION IN MASS SPECTROMETRY[1]

William E. Wallace,[1] *Anthony J. Kearsley,*[2] *and Charles M. Guttman*[1]

[1]Polymers Division and [2]Mathematical and Computational Sciences Division, National Institute of Standards and Technology, Gaithersburg, MD

9.1 INTRODUCTION

After extensive sample preparation, careful instrument tuning, and thorough data collection, the mass spectrometrist is presented with a spectrum (or series of spectra) that he hopes will contain the answer to the original analytical question. At this point, a critical series of decisions must be made as to how to process the spectrum to reduce it from a large array of mass-versus-intensity data pairs (perhaps 50,000 and often more) to a restricted number of spectrum metrics which could be as few as one (e.g., the number average relative molecular mass) or as many as several hundreds (e.g., a collection of peak locations and peak integrals). A distillation of the information present in the original data must be performed even in cases where only a qualitative result is desired. For example, some type of spectrum matching must be performed when determining if the current spectrum is like a previous spectrum or appears similar to spectrum contained in a reference library of spectra [1]. Often, a binary "yes or no" question must be answered (e.g., "Is the sample what I anticipate it should be?"). Sometimes, probabilistic measure or confidence level determination is the desired outcome. For quantitative analysis, the result will be in the form of a numerical value with a certain precision and accuracy determined by the character of the measurement uncertainty (systematic and random). The cautious mass spectrometrist quickly learns that data analysis is fraught with ambiguity. One numerical analysis method gives one result, while a second gives a contradic-

---

[1] Official contribution of the National Institute of Standards and Technology; not subject to copyright in the United States of America.

MALDI Mass Spectrometry for Synthetic Polymer Analysis, Edited by Liang Li

tory result. Which one is to be believed? Slightly smoothing a spectrum may cause a peak-picking algorithm to appear to work better because peaks and troughs become more clearly defined and separated. But smoothing too much will preclude peak-picking algorithms from identifying minor but possibly significant peaks. With substantial smoothing, all peak structure could become opaque to peak-picking algorithms. Clearly this begs the question, how does one determine when "a little" is enough and "a lot" is too much? Operator-independent methods based on well-understood numerical techniques that can serve as touchstones are needed to reduce the ambiguity and, therefore, the uncertainty, in the analytical process. Reference methods that can be used in most situations will give confidence to the analyst and credibility to the analytical process.

In general, the mass spectrometrist is concerned with "peaks," defined commonly as "significant" excursions in the spectrum intensity from its baseline value in the positive direction. Genuine peaks are assumed to be the result of ions of a given mass-to-charge ratio (i.e., *m/z*) being detected by the instrument. Spurious peaks may arise from purely random events (i.e., "noise"). Peaks may also arise from systematic instrumental artifacts. From a statistical point of view, these may be impossible to distinguish from genuine peaks. The analyst needs answers to the following questions:

(a) When is a given excursion correctly classified as a genuine peak?

(b) At what mass-to-charge ratio is the peak most likely located?

(c) Does it overlap with other nearby peaks?

(d) Where does a peak begin and end?

(e) What is the area of the spectrum underneath the peak?

An answer to the first question is used to separate true peaks from noise spikes. An answer to the second question is required for species identification and is used predominantly in qualitative analysis. The third question must be answered to determine if two or more peaks overlap as a result of insufficient mass-to-charge resolution. Knowledge of the location of the peak beginning and end, the fourth question, is required to determine peak area. Peak area, in turn, is typically required for quantitative analytical results [2]. Succinct answers to these questions will result in a reliable translation between the spectrum and the metrics the analyst wishes to determine.

9.2 PREVIOUS APPROACHES

Standard approaches to the reduction of mass spectral data have focused on calculating local derivatives and/or intensity thresholds of the data. A few of the many reviews in the literature can be found in References 3–6. Typically, excursions from the baseline are found at increases in the first derivative. As the algorithm proceeds sequentially through the data (typically but not necessarily from low to high mass-to-charge ratio), an initial excursion of the derivative, or an increase in intensity above a preset threshold, indicates a peak beginning. Peak maximum is found when the derivative after an initial increase flattens out to zero. As the algorithm proceeds sequentially, the derivative will change sign and then flatten out to zero again, or the intensity will drop below the preset threshold value, as the baseline is restored.

Many variations of this basic method exist. For example, second derivatives may be used to find peak maxima. In some cases, third derivatives may also be employed. There are two major problems that one encounters when using these derivative-based approaches. First, the function whose derivative must be approximated is only available at discrete prescribed points; that is, one has access only to (x, y) pairs of data, not to a continuous function. Second, random noise results in inaccurate derivative estimates. It is well known that the availability and accuracy of derivative approximations decrease as noise in a function increases. The result is that noisy data analyzed with algorithms that employ derivative approximations may fail to realize genuine peaks and may identify peaks caused purely by artifacts. Furthermore, the higher the derivative, the greater its sensitivity to random noise [7]. In this case, smoothing or filtering of the data is one way to insure existence and computability of needed derivative estimates. Running or windowed averages, Savitsky-Golay smoothing [8–10], or Fourier filtering [11] are the most common of the many methods possible and have been extensively discussed in the literature. However, the success of these methods relies on a circular logic in which the type and degree of smoothing determine the effectiveness of the peak finding algorithm, and the effectiveness of the peak finding algorithm determines the amount of smoothing required. The problem is compounded when the noise is variable across the mass-to-charge ratio range, or when noise is not constant between spectra, but the analyst wishes to apply the same data analysis methods to all spectra. Different kinds of, or degrees of, smoothing may be required in different parts of the spectrum. Likewise, derivative computation (or other estimates) may be more feasible in one part of the spectrum than another. The analyst is left with no clear guidance on how to proceed, potentially leading him back to performing the experiment anew with a view to producing spectra that are more amenable to the data analysis tools available.

9.3 TWO NEW CONCEPTS

Here we present two new approaches to data reduction in mass spectrometry: one for qualitative peak identification, and one for quantitative peak identification and integration. Each method is operator independent. This has the dual virtues of eliminating operator bias (intentional or unintentional) and of allowing for full automation of the data analysis. Elimination of operator bias is paramount in work on standard reference materials, as well as in other areas such as forensic science. Full automation is highly useful in high-throughput situations such as found in manufacturing quality-control environments where the cost of analyzing data by hand becomes prohibitive. The first method presented focuses on using autocorrelation to identify small periodic peaks buried in random noise. The second method uses concepts rooted in time-series segmentation to find the beginning and end of nonoverlapping peaks. We will discuss each in turn and provide illustrative examples.

9.4 MASS AUTOCORRELATION

Signal autocorrelation has an extensive history in the communications field. The mass autocorrelation function is defined as

$$G(L) = \Sigma_i S(m_i) S(m_{i+L}) / \Sigma_i S(m_i) S(m_i) \qquad \text{(Eq. 9.1)}$$

where $S(m_i)$ is signal at mass m_i taken on equal intervals of mass, Δm, and L is the lag which is equal to a given number of mass units. Equal intervals of mass are used because most correlation algorithms require the signal to be evenly spaced points on the scale of interest. Time-of-flight (TOF) mass separation is the technique most often applied to synthetic polymers due to their high-molecular masses, typically in excess of 1000 Da and often much greater (into the 100,000 Da range and beyond). No other mass separation technique can reach such high masses. The TOF signal, $S(t_i)$, is collected on equal intervals of time. The transformation from this time-base signal $S(t_i)$ to a mass-base signal $S(m_i)$ involves both an interpolation and a change of the signal itself by a Jacobean transform. The mathematical methods to effect this transformation are discussed in References 12 and 13.

9.5 MASS AUTOCORRELATION EXAMPLE

In our work to create intact gas-phase macromolecular ions of polyethylene [14, 15], we found that there was an abrupt upper mass limit beyond which we were unable to produce intact ions. Questions as to whether this upper limit was dictated by polymer crystallization effects (specifically, chain bridging between crystallites [16]) in the MALDI sample preparation lead us to inquire what was the absolute highest mass intact ion we were able to produce. At these high-mass levels, the signal descends down to the noise level, forcing us to process noisy data to find our answer. Various common smoothing techniques were applied, but the results left considerable ambiguity in the fact that peaks could be "lost" or "gained" by smoothing when examined by eye.

The inset to Figure 9.1 shows the middle- and high-mass range of the mass spectrum of Standard Reference Material® 1482 (certified mass-average molecular mass (M_w): 13,600 g/mol ± 1500 g/mol; certified number-average molecular mass (M_n): 11,400 g/mol ± 300 g/mol), the highest relative-molecular-mass polyethylene on which we could produce intact oligomers. The arrow in the inset between 14,500 Da and 17,500 Da shows the region where the autocorrelation analysis was performed. In the main panel of Figure 9.1 is a detail of the spectrum in the high-mass region. A separate autocorrelation was performed for each of the six windows shown by white arrows. In this region the signal to noise ratio drops below unity. Determining the presence of peaks in this region by traditional means is quite difficult. Smoothing tended to replace any noisy peaks with a wandering baseline.

Figure 9.2 shows the autocorrelation functions for the six regions in Figure 9.1. In the lowest mass, highest signal intensity region (14,500–15,000 Da), there is a clear mass correlation at 28 Da, which is the mass of the polyethylene repeat unit. There are also correlations at integer multiples of 28 Da. As the windows move

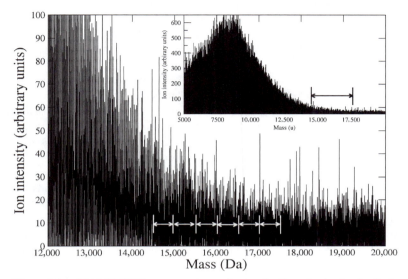

Figure 9.1. MALDI-TOF mass spectrum of a covalently cationized polyethylene [14, 15]. The figure inset shows the high-mass region of the spectrum with the black arrow indicating the area of autocorrelation. The full figure shows where the polymer signal devolves into noise. The white arrows give the autocorrelation windows corresponding to Figure 9.2. The MALDI target was prepared by the solventless method by grinding at room temperature in a 10 : 1 ratio by mass the *all-trans* retinoic acid matrix with the polyethylene analyte.

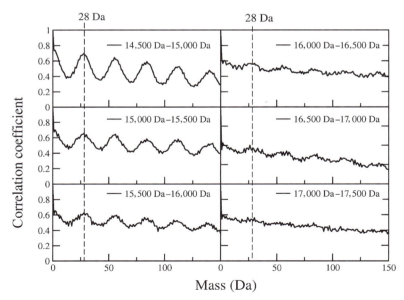

Figure 9.2. Autocorrelation of the data in Figure 9.1 in the mass windows indicated showing the gradual loss of the autocorrelation peak at 28 Da.

higher in mass in 500 Da increments, the correlation peak at 28 Da slowly decreases in intensity. In the 17,000–17,500 Da window, the peak is just barely visible. From this we conclude that intact polyethylene ion peaks are observable up to a few repeat-unit masses above 17,000 Da.

It was observed by Voigt-Martin and Mandelkern [16] that there is a quantitative change in linear polyethylene crystallization behavior as a function of molecular mass. In the low- to intermediate-mass range (defined as 10^3–10^4 Da), crystallites tend to be thick and long. In the intermediate- to high-mass range (defined as 10^4–10^6 Da), the crystallites tend to be thinner and shorter. This was attributed to the increased number of entanglements per chain being an impediment to crystallization. Higher-molecular-mass polyethylenes will have more material in the amorphous interstitial regions between crystallites which, in turn, leads to increased bridging between crystallites. This may explain our difficulty in observing intact polyethylene ions above 1.7×10^4 Da: bridging of chains between crystallites may interfere with the disentanglement of chains to form single-chain intact ions in the gas phase during the MALDI process. Increasing laser energy to enable the chains to disentangle (by adding more thermal energy) only serves to fragment the chains, leaving no intact molecules in the gas phase. However, recent work [17] has shown that heating the MALDI target to melt the crystalline polymer greatly enhances ion yield.

9.6 TIME-SERIES SEGMENTATION

An alternative to calculating local derivatives is to consider the spectrum as a whole and to reduce it to a set of concatenated line segments based on its features. As shown in Figure 9.3, by connecting the first (x, y) pair to the last (x, y) pair in the spectrum, we create a crude baseline for the entire spectrum. From this line, the (x, y) pair that is the greatest normal distance from the line is determined. This yields two line segments spanning the spectrum. This procedure is continued until the spectrum is replicated by a series of line segments with each peak determined (at the minimum) by two line segments and the intervening baseline determined (also at a minimum) by a single line segment. After the spectrum has been segmented, least squares and orthogonal distance regression may be used to adjust the line segments to best fit the data; however, caution must be exercised because if the random noise level varies across the spectrum, the quality of the fit will also vary across the spectrum. For this reason, our method uses a background spectrum taken at the same instrumental conditions as the spectrum to be analyzed with a sample that is free of analyte (but, e.g., in the case of MALDI, contains the matrix and the cationizing salt). A background spectrum requires additional experimental effort but yields significant dividends when analyzing the data to determine quantitative measures.

A nonlinear programming algorithm using an L2 (least squares) approximation to an L1 (least absolute-value) fit was employed [18–22]. L1 fits are superior to L2 fits due to their increased tolerance for outliers; that is, outlying points do not exert as much control over the final fit. Given a data set of N points we find a collection of strategic points and the unique optimal piece-wise linear function passing through

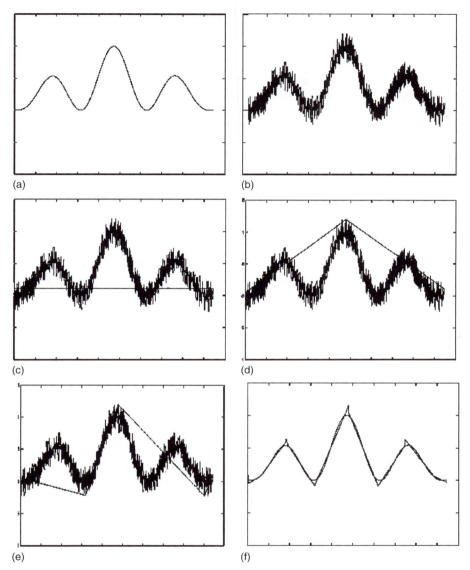

Figure 9.3.   A schematic example of the time-series segmentation method as described in
the text. Panel a is the noise-free data; panel b is with added random noise; panel c shows
the line segment connecting the first and last data points of the spectrum; panel d shows the
two new line segments created by connecting the endpoints of the original segment to the
data point with the greatest normal distance from the original line segment; panel e shows
the second step of the segmentation routine; and panel f shows the final segmentation of the
three-peak spectrum overlaid on the original noise-free data.

the x coordinate of each strategic point. This defines a set of function maxima and minima corresponding to the peak maxima and the peak limits. The original data are then integrated by finding the area of the polygon determined by the strategic points.

Our segmentation method is a two-step algorithm. The first portion requires the selection of strategic points and is derived from the earlier work of Douglas and Peucker [23]. Strategic points are selected based on an iterative procedure that identifies points whose orthogonal distance from the endpoint connecting line segment is greatest. Once a point with greatest orthogonal distance from the mean has been identified, it joins the collection of strategic points and, in turn, becomes an endpoint for two new line segments from which a point with the greatest orthogonal distance is found. This numerical scheme is performed until the greatest orthogonal distance to any endpoint connecting line segment drops beneath a prescribed threshold value. This threshold value is the only algorithmic parameter and is based on a statistical analysis of the data and its corresponding analyte-free spectrum. Clearly, the selection of these points does not require equally spaced data; therefore, the method is equally well suited for TOF data expressed in either time or mass space. We generally choose to work in time space with the data in its most basic state and to eliminate for doing a point-by-point correction of intensity using partial integrals [13, 24]. The second phase of the algorithm, developed specifically for this work, requires the solution of an optimization problem, specifically, locating strategic point heights (i.e., adjusting strategic point y-axis values at their associated strategic x-axis value) that minimize the sum of orthogonal distance from raw data. This problem is a nonlinear (and nonquadratic) optimization problem that can be accomplished quickly using a recently developed nonlinear programming algorithm [25].

The algorithm works as follows [21, 26]. The first two strategic points chosen are always the first and last points of the data set. A line is drawn connecting these two points and the data point the greatest orthogonal distance from this line is selected as a new strategic point. This process is iterated over all line segments until the orthogonal distance falls below a threshold parameter calculated from the statistical analysis of the data set and its congruent analyte-free data set as described in the next section. Finally, the strategic point heights are adjusted to minimize the distance from the original (full) data set. Clearly, this method requires no knowledge peak shape and no preprocessing of the data (e.g., smoothing), nor does it require equal spacing of data points.

Once the data set is fully segmented, strategic points are discarded in accordance with the statistical analysis of the original data set and its corresponding analyte-free data set. This "deflation" of strategic points using statistically derived thresholds is performed by first analyzing the analyte-free spectrum for peaks and peak areas. Once a collection of peaks and peak areas has been accumulated, the spectrum with sample is then analyzed. Each peak identified from the spectrum with analyte is compared to peaks found in close relative proximity from the analyte-free spectrum algorithm output (i.e., peaks that appear with similar time or mass coordinates). If any peak in the spectrum with analyte has a smaller peak height or smaller peak area than most (about 95%) of the background-spectrum peaks in close proximity, then that peak is ignored. Likewise, any peak that falls outside the statistically significant measure for area and height is also discarded. Thus, no peak is identified from the sample spectrum that could have been identified by height or area from the background spectrum. This discarding of strategic points also serves to prevent the inadvertent subdivision of larger peaks into a set of smaller peaks. This can sometimes occur if the noise in the analyte spectrum is much greater than the noise in the corresponding background spectrum.

Once the final set of strategic points has been found, the area of the polygon defined by these points is calculated. (The polygon is often, but not always, a triangle. The algorithm will work on polygons of any number of vertices connected by line segments.) The line connecting the first and last strategic points for a given peak determines a "local baseline." The mathematical basis for the polygonal area calculation algorithm is Green's theorem in the plane and can be interpreted as repeated application of the trapezoidal rule for integration [27]. The method returns the exact area of the polygon.

9.7   TIME-SERIES SEGMENTATION EXAMPLES

Figure 9.4 shows the analysis by our computer code *MassSpectator* using time-series segmentation to find peaks in a typical polystyrene MALDI-TOF mass spectrum. Notice that the ion intensity is plotted on a logarithmic scale. This is necessary because the peak heights span three orders of magnitude. The background spectrum sits at the bottom of the plot. Recall that the background spectrum allows *MassSpectator* to build a model of the noise. The calculated local noise is used to discard peaks that arise from statistical fluctuations in the spectrum, that is, fluctuations that are not statistically significant. The circles in the plot indicate where a peak begins,

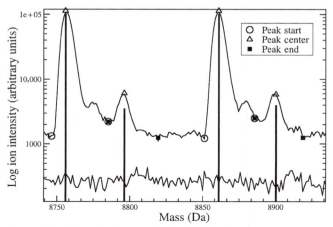

Figure 9.4.   Result of an application of *MassSpectator* to a polymer mass spectrum. Note the logarithmic scale on the ion intensity axis. See text for full description.

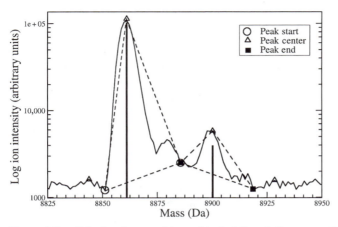

Figure 9.5.   Calculated trapezoids used to find the relative areas of the peak in Figure 9.4.

the triangle where a peak is centered, and the solid squares where a peak ends. Notice that in several instances, the point where one peak ends is where an adjacent peak begins. Also notice that the shoulders on the high-mass side of the main series peaks are not identified as separate peaks. This is because *MassSpectator* uses a 90% peak-to-valley rule to determine whether there are two peaks or one. Lowering this value to, say, 50% would result in small noisy peaks being divided into several smaller peaks. Furthermore, for the case of shoulders, without choosing some heuristic rule or mathematical model of the peak shape [28], peak integration is not possible. Since we wished to avoid such choices in making the software operator independent, we chose the conservative 90% rule.

Figure 9.5 shows a detail of the spectrum shown in Figure 9.4. The dotted lines that form the two triangles are the boundaries *MassSpectator* uses to find the area of each of the two peaks visible. The bottom side of each triangle forms a de facto local baseline for each peak. In this the program does not try to fit a global baseline to the entire spectrum. Instead, it produces a discontinuous function of straight lines that approximates a global baseline for the spectrum in a piece-wise fashion. In Figure 9.5 it appears as if the triangles may produce an overestimate of the peak area. Two points need to be emphasized in this regard. First, if there is an error, it is consistent across peaks and proportional to the total peak area. Most important to standards work (and to unbiased data analysis in general) is the notion that the program will work consistently for all situations. The choice of a simple trapezoidal rule for finding peak area fits this criterion. We have considered other options for finding peak area (such as Romberg integration) but felt that adding complexity would decrease consistency. The second point regarding bias in estimation of area is the possible underestimation of the area from the shape of the baseline. Most global baseline algorithms would considerably increase the peak area by attempting to draw a smooth line across the bottom of the spectrum. That is, the baseline would be drawn connecting the first circle on the far left of the spectrum to the filled square on the far right.

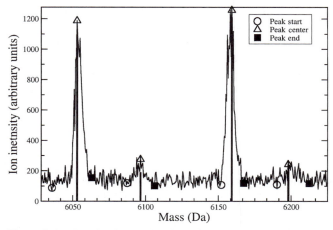

Figure 9.6.   Result of an application of *MassSpectator* to a noisy polymer mass spectrum. In this case, the ion intensity axis scale is linear. Note that for clarity, the background spectrum is not shown.

Figure 9.6 shows the application of *MassSpectator* to a spectrum with a poor signal to noise ratio. (Notice that in this case the ion intensity is plotted on a linear scale.) The program is still able to pull out peaks that are dominated by noise. The smaller peaks have a signal to noise ratio of approximately 1 : 1; that is, the signal height is approximately the same height as the baseline noise. This is the benefit of having a background spectrum to work with. Making a model of the noise of the measurement is a powerful tool in peak picking, much more powerful than any smoothing algorithm. It does require more work on the part of the analyst but rewards him with a more robust tool to work with less than ideal data.

Since *MassSpectator* uses as few assumptions as possible, and is conceived as a general method for peak picking and integration, we find it is a useful touchstone in our research. We have used it in comparison to commercial software where the peak picking and integration algorithms are proprietary, and therefore unknown to us. We have also used *MassSpectator* to systematically study the effect of instrument parameters on mass spectra. We have found the algorithms provide a robust, consistent method able to handle a wide range of spectra from high to low signal to noise ratios.

9.8   FUTURE DIRECTIONS

There are several issues the authors wish to address in the near future. A more sophisticated integration method will be implemented in *MassSpectator*; however, the new method must be as robust as trapezoidal integration to be able to handle peaks with areas spanning several orders of magnitude within one spectrum (and sometime immediately adjacent to one another). Second, a robust, consistent *global* baseline correction algorithm is desired. Baseline correcting a spectrum, especially one on a steeply sloping background, would need to be implemented if the peak integration routine to be implemented does not provide for a local baseline as the trapezoidal rule does. Finally, the authors are working on a protocol and associated mathematical theorems to create a method of data binning (where adjacent points are summed to create a single point out of several). Such a method may be able to provide a consistent method of "smoothing" rooted in a statistical analysis of the spectrum and its accompanying background spectrum.

ACKNOWLEDGMENT

The authors would like to thank Stefan Leigh (Statistical Engineering Division, NIST) for critical reading of the manuscript and many helpful recommendations.

REFERENCES

1. Ausloos, P.; Clifton, C.L.; Lias, S.G.; Mikaya, A.I.; Stein, S.E.; Tchekhovskoi, D.V.; Sparkman, O.D.; Zaikin, V.; Zhu, D., The critical evaluation of a comprehensive mass spectral library, *J. Am. Soc. Mass Spectrom.*, **1999**, 10, 287–299.
2. Guttman, C.M.; Flynn, K.M.; Wallace, W.E.; Kearsley, A.J., Quantitative mass spectrometry and polydisperse materials: creation of an absolute molecular mass distribution polymer standard, *Macromolecules*, **2009**, 42, 1695–1702.
3. Smit, H.C.; van den Heuvel, E.J., Signal and data analysis in chromatography. In *Chemometrics and Species Identification*, Topics in Current Chemistry, Vol. 141 (Dewar, M.J.S., ed.) Berlin: Springer-Verlag, **1987**.
4. Zupan, J., *Algorithms for Chemists*. New York: Wiley, **1989**.
5. Papas, A.N., Chromatographic data systems: a critical review, *CRC Critical Reviews in Analytical Chemistry*, **1989**, 20, 359–404.
6. Kateman, G.; Buydens, L., *Quality Control in Analytical Chemistry*. New York: Wiley, **1993**, chapter 4, Data processing.
7. Hippe, Z.; Bierowska, A.; Pietryga, T., Algorithms for high level data processing in gas chromatography, *Anal. Chim. Acta*, **1980**, 122, 279–290.
8. Savitsky, A.; Golay, M.J.E., Smooting and differentiation of data by simplified least squares procedures, *Anal. Chem.*, **1964**, 36, 1627–1639.
9. Eilers, P.H.C., A perfect smoother, *Anal. Chem.*, **2003**, 75, 3299–3304.
10. Vivó-Truyols, G.; Schoenmakers, P.J., Automatic selection of optimal Savitsky-Golay smoothing, *Anal. Chem.*, **2006**, 78, 4598–4608.
11. Kast, J.; Gentzel, M.; Wilm, M.; Richarson, K., Noise filtering techniques for electrospray quadrupole time of flight mass spectra, *J. Am. Soc. Mass Spectrom.*, **2003**, 14, 766–776.
12. Owens, K.G., Application of correlation analysis techniques to mass spectral data, *Appl. Spectrosc. Rev.*, **1992**, 27, 1–49.
13. Wallace, W.E.; Guttman, C.M., Data analysis methods for synthetic polymer mass spectrometry: autocorrelation, *J. Res. Natl. Inst. Stand. Technol.*, **2002**, 107, 1–17.
14. Bauer, B.J.; Wallace, W.E.; Fanconi, B.M.; Guttman, C.M., Covalent cationization method for analysis of polyethylene by mass spectrometry, *Polymer*, **2001**, 42, 9949–9953.
15. Lin-Gibson, S.; Brunner, L.; Vanderhart, D.L.; Bauer, B.J.; Fanconi, B.M.; Guttman, C.M.; Wallace, W.E., Optimizing the covalent cationization method for the mass spectrometry of polyolefins, *Macromolecules*, **2002**, 35, 7149–7156.
16. Voigt-Martin, I.G.; Mandelkern, L., A quantitative electron microscopy study of the crystalline structure of molecular weight fractions of polyethylene, *J. Polym. Sci.: Polym. Phys. Ed.*, **1984**, 22, 1901–1917.

17. Wallace, W.E.; Blair, W.R., Matrix-assisted laser desorption/ionization mass spectrometry of covalently cationized polyethylene as a function of sample temperature, *Int. J. Mass Spectrom.*, **2007**, 263, 82–87.

18. Barrondale, I., L1 approximation and analysis of data, *Appl. Stat.*, **1968**, 17, 51–57.

19. Barrondale, I.; Roberts, F.D.K., An improved alogorithm for discrete L1 linear approximation, *SIAM J. Numer. Anal.*, **1973**, 10, 839–848.

20. Duda, R.O.; Hart, P.E., *Pattern Classification and Scene Analysis*. New York: John Wiley and Sons, **1973**.

21. Kearsley, A.J.; Wallace, W.E.; Guttman, C.M., A numerical method for mass spectral data analysis, *Appl. Math. Lett.*, **2005**, 18, 1412–1417.

22. Kearsley, A.J., Projections onto order simplexes and isotonic regression, *J. Res. Natl. Inst. Stand. Technol.*, **2006**, 111, 121–125.

23. Douglas, D.H.; Peucker, T.K., Algorithms for the reduction of the number of points required to represent a digitized line of its character, *Canadian Cartographer*, **1973**, 10, 112–122.

24. Guttman, C.M., Polymer preprint, *ACS Polym. Preprints*, **1996**, 37, 837.

25. Boggs, P.T.; Kearsley, A.J.; Tolle, J.W., A practical algorithm for general large-scale nonlinear optimization problems, *SIAM J. Opt.*, **1999**, 9, 755–778.

26. Wallace, W.E.; Kearsley, A.J.; Guttman, C.M., An operator-independent approach to mass spectral peak identification and integration, *Anal. Chem.*, **2004**, 76, 2446–2452.

27. Beyer, W.H., *CRC Standard Mathematical Tables*. Boca Raton, FL: CRC Press, **1981**.

28. Kempka, M.; Sjödahl, J.; Björk, A.; Roeraade, J., Improved method for peak picking in matrix-assisted laser desorption/ionization time-of-flight mass spectrometry, *Rapid Commun. Mass Spectrom.*, **2004**, 18, 1208–1212.

CHAPTER 10

MALDI-MS/MS FOR POLYMER STRUCTURE AND COMPOSITION ANALYSIS

Anthony T. Jackson

AkzoNobel RD&I, Deventer, The Netherlands

MATRIX-ASSISTED LASER desorption/ionization-tandem mass spectrometry (MALDI-MS/MS) has played an important role in polymer characterization. In this chapter, a brief introduction to MALDI-MS/MS is followed by details of experimental approaches and a number of examples of applications of the technique. This chapter focuses on a description of the information that can be gleaned from MALDI-MS/MS data, especially for the determination of end-group structures. Examples are given for a number of polymer systems, including methacrylates, styrenes, polyethers, and polyesters. A detailed example for a block copolymer shows how MALDI-MS/MS may be used to sequence copolymers (as well as homopolymers). All data are presented using recently introduced annotation for polymer fragmentation.

10.1 INTRODUCTION TO MALDI-MS/MS

MALDI-MS/MS is typically used as a secondary tool, alongside MALDI-time-of-flight (MALDI-TOF) mass spectrometry and nuclear magnetic resonance (NMR) spectroscopy, for the characterization of backbone and end-group functionality of synthetic polymers. Masses of the combined end groups of synthetic polymers can be inferred from MALDI-TOF data. Information on individual end-group functionality can often be gleaned from MALDI-MS/MS data. Furthermore, sequence data for polymers and copolymers can often be generated using MALDI-MS/MS. The high specificity of the data obtained, especially when high mass accuracy experiments are performed, can provide information (e.g., about end-group functionality and detailed copolymer sequencing) that is not available by any other technique.

MALDI Mass Spectrometry for Synthetic Polymer Analysis, Edited by Liang Li
Copyright © 2010 John Wiley & Sons, Inc.

One of the limitations compared to MALDI-TOF is the lower molecular weight cutoff for analysis by means of MALDI-MS/MS. The latter technique has been shown to be generally applicable for oligomers with molecular weights of up to approximately 5000 Da. Selection of lower molecular weight oligomers from a polymer with a high polydispersity can enable data to be generated, possibly after extraction of a low molecular fraction (such as by solvent extraction or using gel-permeation chromatography [GPC]).

A range of systems have been characterized by means of MALDI-MS/MS, including polyethers [1–10], polystyrenes [11–17], polymethacrylates [1, 7, 13, 18–21], polyacrylates [22], polyesters [7, 23], and polyesteramides [24, 25]. A number of these and other polymers have also been analyzed by means of MS/MS allied with other techniques [6, 26–36], such as electrospray ionization (ESI) [6, 26, 27, 31]. Initial MS/MS experiments on synthetic polymer were mostly obtained from ionization techniques such as liquid secondary ion mass spectrometry (LSIMS) [33–37] or field desorption (FD) [28–30, 38], prior to developments in MALDI and ESI. The greater precursor ion abundances generated by the latter two ionization methods, along with developments in TOF-based instrumentation, has led to the domination of use of these techniques for MS/MS characterization of synthetic polymers.

10.2 TECHNIQUES OF MALDI-MS/MS

The common techniques for generating MALDI-MS/MS data for polymers can be separated into two classes: (1) post-source decay (PSD) using standard MALDI-TOF instruments fitted with a reflectron [2–4, 7, 13] and (2) use of tandem mass spectrometers [5, 7–14, 18–21, 23, 24, 39], such as quadrupole-orthogonal TOF mass spectrometers or TOF-TOF instrumentation. More details about instrumentation for MS/MS are given in Chapter 6. The use of Fourier transform mass spectrometry (FTMS) for MALDI-MS/MS for the structural characterization of polymer systems is discussed in Chapter 4.

Many MALDI-MS/MS studies are performed using PSD as a consequence of the greater proliferation of the reflectron MALDI-TOF instrumentation that is employed in these studies, when compared to "real" tandem mass spectrometers. The latter instruments, though, often enable spectra of greater signal-to-noise ratios and a much higher selectivity in the precursor ion [40]. The ion gates that are typically used for PSD experiments only allow a broad mass range of precursor ions to be selected, whereas most tandem mass spectrometers enable the selection of a much narrower window of ions (which can just be the mono-isotopic peak of the analyte of interest).

Most of the MALDI-MS/MS data described in this chapter were obtained using a hybrid magnetic sector-orthogonal TOF instrument [41, 42]. A schematic of this tandem mass spectrometer is displayed in Figure 10.1. The instrument allows selection of precursor ions with high resolving power and high sensitivity detection of fragment ions in the TOF analyzer. Collision-induced dissociation (CID) of

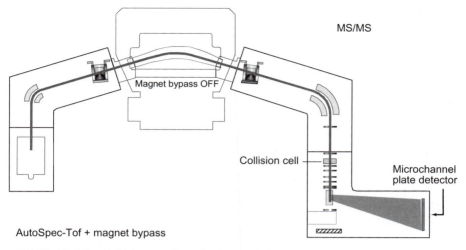

MALDI: MS/MS with high parent ion selection resolution

Figure 10.1.   Schematic of MALDI-MS/MS instrument (Autospec-oa-TOF) used to generate most of the data shown in this chapter. See color insert.

selected precursor ions occurs in the collision cell, with xenon as the collision gas and collision energy of 800 eV. More details of this instrument have been described elsewhere [41, 42].

An example of spectra obtained from both a four-sector (LSI-MS/MS) and the hybrid magnetic sector-TOF (MALDI-MS/MS) instrument are shown in Figure 10.2. A much higher signal-to-noise ratio was noted in the MALDI-MS/MS fragment ion spectra, mainly as a consequence of using both MALDI for ionization, plus TOF analysis of fragment ions [21]. The precursor ion in both cases was the sodiated 25-mer of poly(methyl methacrylate) (PMMA).

The higher selectivity of tandem mass spectrometers can be a real advantage when characterizing "real-world" polymers, which are often complex (often due to complexity of feedstocks used in manufacture or side reactions that can occur during polymerization). Separation of precursor ions with saturated and unsaturated end-group functionality (differing by a mass-to-charge ratio of 2, such as for free radical polymers that are terminated by disproportionation), for example, would require use of a tandem mass spectrometer with good resolving power.

The capability for analysis of precursor ions with relatively high mass-to-charge ratios, but still retaining very good signal-to-noise ratios in the fragment ion spectrum, is shown in Figure 10.3 [21]. The maximum mass-to-charge ratio is approximately 5000 for precursor ions in the hybrid magnetic sector-TOF instrument [41, 42]. Precursor ions of much higher mass-to-charge ratio have been characterized using PSD in a MALDI-TOF instrument [43], but the selection capabilities of this experiment are very poor (such that a wide range of precursor ions were selected for analysis with no specificity).

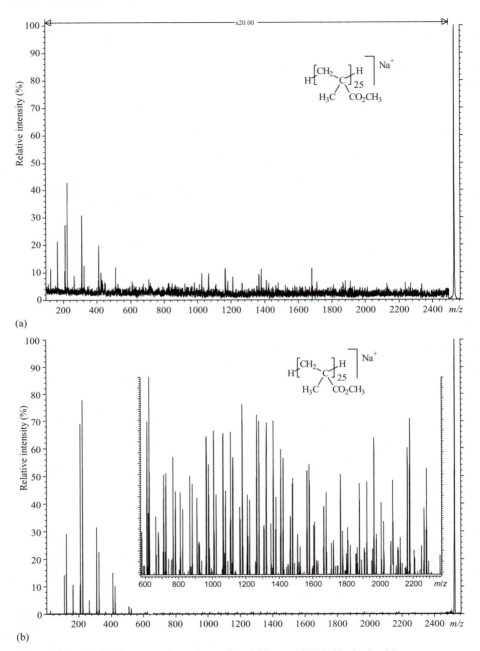

Figure 10.2. MS/MS spectra from the sodiated 25-mer of PMMA obtained by: (a) LSI-MS/MS using a four-sector instrument and (b) MALDI-MS/MS in a magnetic sector-orthogonal acceleration TOF. Note the high signal-to-noise ratios even for fragment ions between *m/z* 600–2300 in the MALDI-MS/MS spectrum. Reproduced from Reference 7 with permission.

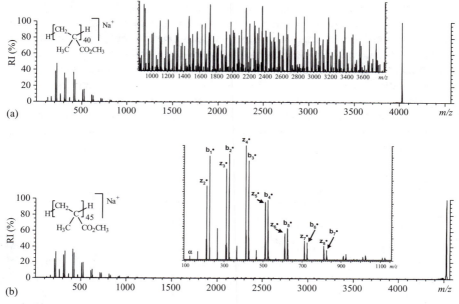

Figure 10.3.   MALDI-MS/MS spectra from CID in a magnetic sector-orthogonal acceleration TOF of (a) the sodiated 40-mer of PMMA and (b) the sodiated 45-mer of PMMA. Note the high signal-to-noise ratios even for fragment ions between m/z 900–3800 shown in the expansion of the spectrum from the 40-mer. Modified from Reference 7.

10.3   EXPERIMENTAL DETAILS

Selection of matrix is important in order to get significant enough signals to generate MS/MS spectra with good signal-to-noise ratios. The same matrices that are used to generate MALDI-TOF mass spectra with good signal-to-noise levels can also be employed for MALDI-MS/MS experiments. We have typically used matrices such as dithranol and retinoic acid in our laboratory, as materials can be used to generate very abundant ion signals for a range of polymer classes, including polyethers, polystyrenes, polymethacrylates, and polyesters.

The selection of the ionizing cation is very important in order to generate the optimal fragment ion spectra. A general recommendation for cation selection is as follows: (1) Ag^+ for styrenics such as poly(styrene) and poly(α-methyl styrene); (2) Cu^+ or Ag^+ for dienes such as poly(butadiene) and poly(isoprene); (3) Li^+ for polyethers such as poly(ethylene glycol) (PEG), poly(propylene glycol) (PPG), and poly(tetramethylene glycol); and (4) Li^+ or Na^+ for acrylic polymers, including poly(methyl methacrylate) (PMMA). Selection of cations for other polymer systems is recommended to be made based on the functional groups present. Polymers containing, for example, ester groups (such as poly(ethylene terephthalate) [PET]) would normally be analyzed by addition of alkali metal salts during sample preparation, to promote the formation of lithiated or sodiated precursor ions.

It should be noted that the choice of cation is not only performed based on the ion signals generated for the precursor ion. The cation often plays an important role in directing the fragmentation of polymer chains, so careful thought about selection of the most appropriate metal ion (or proton in some cases) is required. A good example of this is for polyethers such as PEG, where a range of alkali metal cations can be used to generate abundant precursor ions, but only the cations with lower radii (typically only Li⁺) may be employed to obtain structurally useful fragment ion spectra. Combining results obtained by MALDI-MS/MS with ion mobility techniques has enabled an understanding of the role of the metal cation in directing fragmentation of some polymer systems [1, 14, 39].

10.4 USING MS/MS FOR POLYMER STRUCTURE AND COMPOSITION ANALYSIS

This section describes how MALDI-MS/MS data can be used to generate information on both the end groups and sequence from synthetic polymers. A number of specific polymer types are used as examples, where these systems have been well characterized by MALDI-MS/MS. This information may well be more generally applicable to many other polymers that are not described below. More detailed descriptions of the fragmentation mechanisms of synthetic polymers are provided in Chapter 6 (Section 4).

10.4.1 Methacrylate Polymers

Very intense peaks at low mass-to-charge ratios in the MALDI-MS/MS spectra of acrylic polymers, such as PMMA, can be used to generate information on the end groups [7, 13, 20, 21]. Two series of peaks, one originating from fragments containing the initiating (α) end group and the other from ions with the terminating (ω)-end functionality intact, are typically very intense in the fragment ion spectra. Furthermore, ions of higher mass-to-charge ratios (but typically much lower abundance) can be very useful in providing sequence information, such as for copolymers [20, 21].

The recently introduced annotation for fragmentation of polymers (see more detailed description in Chapter 6, Section 6.3) can be summarized for methacrylates as shown below in Scheme 10.1. Fragment ions containing the initiating end group are typically observed as the $\mathbf{b_n}^{\bullet}$ series and the terminating functionality as the $\mathbf{z_n}^{\bullet}$ series in the spectra of methacrylates, as can be noted from the example for PMMA in Figure 10.4.

The expected mass-to-charge ratios of fragment ions that enable end-group information to be inferred from methacrylates can be summarized as described below for the $\mathbf{b_n}^{\bullet}$ series (α end group) and the $\mathbf{z_n}^{\bullet}$ series (ω end group):

$$M(\mathbf{b_n}^{\bullet}) = M(R') + nM(P) + M(Cat) \qquad \text{(Eq. 10.1)}$$

$$M(\mathbf{z_n}^{\bullet}) = M(R'') + nM(P) + M(Cat) - 14 \qquad \text{(Eq. 10.2)}$$

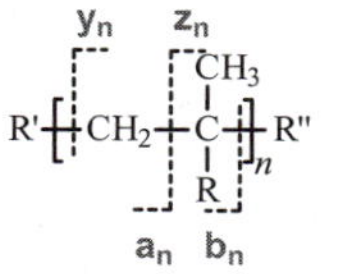

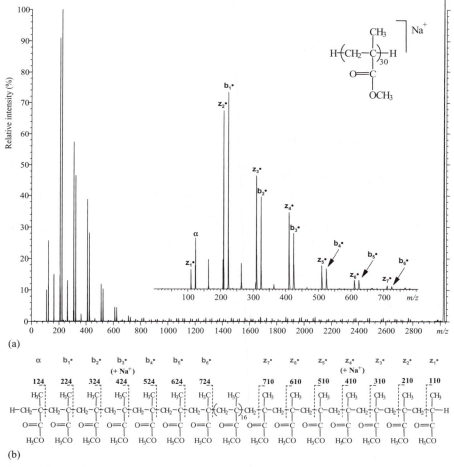

R is $-CO_2X$, e.g., where X is $-CH_3$ or $-CH_2CH_2CH_2CH_3$
R' is α end group and R" is ω end group

Scheme 10.1. Proposed annotation for fragmentation of methacrylate polymers such as PMMA and PBMA.

Figure 10.4. (a) MALDI-MS/MS spectrum from 30-mer of PMMA, annotated (in expansion of m/z 0–800 as inset) with new fragmentation notation for synthetic polymers and (b) proposed fragmentation scheme indicating how the masses and structures of the end groups can be inferred from the data. Note the intensity of peaks from the $b_n^\bullet$ and $z_n^\bullet$ series in the spectrum.

where $M(b_n^{\bullet})$ and $M(z_n^{\bullet})$ are the mass-to-charge ratios of the fragment ions of the $b_n^{\bullet}$ and $z_n^{\bullet}$ series respectively, $M(R')$ and $M(R'')$ are the masses of the α and ω end group, respectively, $M(P)$ is the mass of polymer repeat unit (i.e., 100.1 Da for PMMA), and $M(Cat)$ is the mass of the cation. Furthermore, the presence of peaks from both the $b_n^{\bullet}$ and $z_n^{\bullet}$ series can be verified by comparing their experimental mass-to-charge ratios with that of the precursor ion using the relationship described by the following equation:

$$M(precursor) = M(b_n^{\bullet}) + M(z_n^{\bullet}) + nM(P) - M(Cat) + 14 \quad \text{(Eq. 10.3)}$$

where $M(precursor)$ is the mass-to-charge ratio of the precursor ion. Sequence-specific ions of lower abundance may be found at higher mass-to-charge ratios in the spectra of methacrylates, as may be seen in the MALDI-MS/MS spectrum from poly(butyl methacrylate) (PBMA) [21] shown in Figure 10.5. Four series of peaks that contain either the initiating (a_{nb} and b_n,) or terminating (y_{nb} and z_n) end groups are noted between m/z 300–1380, with another series (K_n) resulting from loss of both end groups. It is the former four series of peaks that may be used for sequence information, such as for the differentiation between block and random copolymers [20, 21].

10.4.2 Styrene Polymers

Styrenic polymers such as poly(styrene) have also been shown to produce fragment ion spectra that are dominated by peaks in the low mass-to-charge ratio region [11–13]. Some of the peaks in this region of the spectrum may be used to infer both the masses and the structures of the end groups of the polymer, with peaks observed at higher mass-to-charge ratios being useful for sequencing of polymers and copolymers. The new annotation for fragmentation of poly(styrene) may be represented as shown in Scheme 10.2, with similar fragmentation noted for other styrene based-polymer systems such as poly(α-methyl styrene) (see Chapter 6).

An example MALDI-MS/MS spectrum from poly(styrene) is shown in Figure 10.6, with silver used to cationize the oligomers [11]. This polymer was generated by anionic polymerization using ethylene oxide as the end-capping agent. Note the intensity of peaks between m/z 50–725. The end-group functionality can be inferred from the mass-to-charge ratios of the peaks from the $b_n^{\bullet}$ and $z_n^{\bullet}$ series, in an analogous way to that from methacrylates.

In a similar fashion to that described (*vide infra*) for methacrylate polymers, the masses of the end groups of styrene-based polymers may be calculated from the fragment ion spectra using the mass-to-charge ratios of peaks from the $b_n^{\bullet}$ and $z_n^{\bullet}$ series and the equations below:

$$M(b_n^{\bullet}) = M(R') + nM(P) + M(Cat) \quad \text{(Eq. 10.4)}$$

$$M(z_n^{\bullet}) = M(R'') + nM(P) + M(Cat) - 14 \quad \text{(Eq. 10.5)}$$

where $M(b_n^{\bullet})$ and $M(z_n^{\bullet})$ are the mass-to-charge ratios of the fragment ions of the $b_n^{\bullet}$ and $z_n^{\bullet}$ series, respectively, $M(R')$ and $M(R'')$ are the masses of the α and ω end group, respectively, $M(P)$ is the mass of polymer repeat unit (i.e., 104.1 Da for

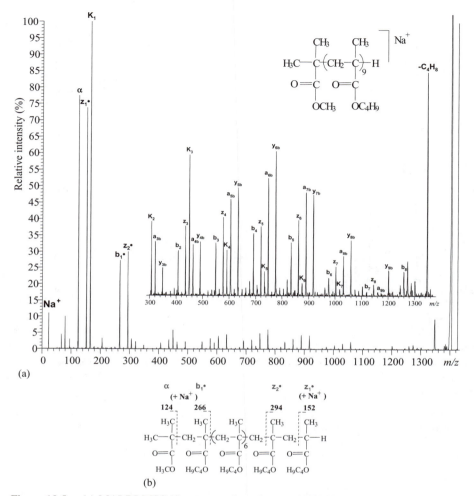

Figure 10.5. (a) MALDI-MS/MS spectrum from 9-mer of PBMA (expansion of m/z 300–1380 as inset), annotated with new fragmentation notation for synthetic polymers and (b) proposed fragmentation scheme indicating how the masses and structures of the end groups can be inferred from the data. Note the peaks from the a_{nb}, b_n, y_{nb}, and z_n series in expansion of the spectrum. Modified from Reference 21.

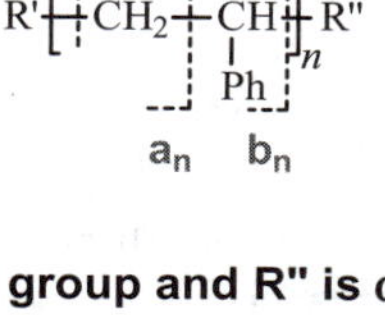

R' is α end group and R'' is ω end group

Scheme 10.2. Proposed annotation for fragmentation of poly(styrene).

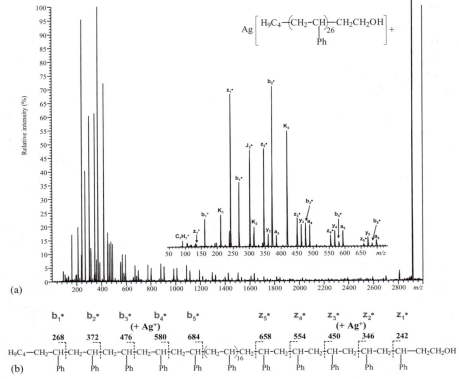

(a)

Figure 10.6. (a) MALDI-MS/MS spectrum from 26-mer of poly(styrene) end-capped with ethylene oxide, annotated (in expansion of m/z 50–725 as inset) with new fragmentation notation for synthetic polymers and (b) proposed fragmentation scheme indicating how the masses and structures of the end groups can be inferred from the data. Note the intensity of peaks from the $b_n^{\bullet}$ and $z_n^{\bullet}$ series in the spectrum. Modified from Reference 11.

poly(styrene)), and M(Cat) is the mass of the cation. The relationship between the mass-to-charge ratio of precursor ion and the fragment ions of the $b_n^{\bullet}$ and $z_n^{\bullet}$ series may be represented as for methacrylates by the following equation:

$$M(\text{precursor}) = M(b_n^{\bullet}) + M(z_n^{\bullet}) + nM(P) - M(\text{Cat}) + 14 \qquad (\text{Eq. } 10.6)$$

where M(precursor) is the mass-to-charge ratio of the precursor ion. This equation may be used to verify that observed peaks in the MALDI-MS/MS spectra from styrenes are indeed from the $b_n^{\bullet}$ and $z_n^{\bullet}$ series and may, therefore, be employed to infer end-group information.

Information about the sequence of styrene polymers may be derived, from MALDI-MS/MS spectra, using series of peaks (that have retained either the initiating or terminating end group) observed at lower intensity (and typically higher mass-to-charge ratios) than those from the $b_n^{\bullet}$ and $z_n^{\bullet}$ series. An expansion (m/z 750–1150) of the spectrum from the 26-mer of poly(styrene) is shown in Figure 10.7 (full spectrum is Figure 10.6a), with a number of series of peaks annotated.

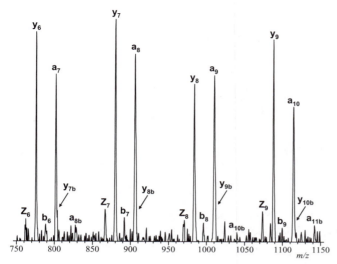

Figure 10.7.   Expansion of m/z 750–1150 from MALDI-MS/MS spectrum from 26-mer of poly(styrene) end-capped with ethylene oxide (see full spectrum in Figure 10.6a), annotated with new fragmentation notation for synthetic polymers. Modified from Reference 11.

These six series, with three from ions retaining the initiating end group (a_n, a_{nb}, and b_n) and the other three with retention of the terminating functionality (y_n, y_{nb}, and z_n), may be used to infer sequence information in a similar fashion to that for methacrylate polymers (e.g., to differentiate between block and random copolymers). A more detailed description of the origin of these fragment ions is described in Chapter 6.

10.4.3   Polyethers

Polyethers include 1,2-epoxide-derived polymers such as poly(ethylene glycol) (PEG), poly(propylene glycol) (PPG), and poly(butylene glycol) (PBG). These polymers (especially PEG) have been relatively well studied by MS/MS [2–10, 26, 27, 32–36, 44], after initial pioneering studies by Lattimer et al. [33–36]. The spectra obtained when lithium is employed for cationization are typically rich in structurally significant fragment ion peaks. The nomenclature for the fragmentation of polyethers, generated from 1,2-epoxide-based monomers, can be described as shown below in Scheme 10.3.

The main series of fragment ions that are typically detected in fragmentation studies of polyethers derived from 1,2-epoxide monomers are also shown in Scheme 10.3. Three series of peaks are normally detected if the initiating and terminating end groups are the same, with six series often noted to be present if the end-group functionalities are different. It has been noted that further series of fragment ions can be generated if either or both of the end groups of the polyethers are esters, such as those derived from reaction of polyethers with acids. The mass-to-charge ratios

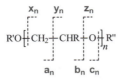

R is –H, –CH₃ or -CH₂CH₃
R' is α end group and R" is ω end group

R'O-(CH₂CHRO)$_n$-H c_n'' H-(OCH₂CHR)$_{n-1}$-OR" z_n''

R'O-(CH₂CHRO)$_{n-1}$-CH=CHR b_n H₂C=CR-(OCH₂CHR)$_{n-1}$-OR" x_n

R'O-(CH₂CHRO)$_{n-1}$-CH₂CR=O c_n O=CHCHR-(OCH₂CHR)$_{n-1}$-OR" z_n

Scheme 10.3.   Proposed annotation for fragmentation of polyether polymers from 1,2-epoxide monomers such as PEG, PPG, and PBG. Structures of typically observed fragment ions are also shown.

of the main six series of fragment ions that are typically observed may be described by the equations shown below:

$$M(\mathbf{b_n}) = M(R') + nM(P) + M(Cat) - 1 \qquad \text{(Eq. 10.7)}$$

$$M(\mathbf{c_n}) = M(R') + nM(P) + M(Cat) + 15 \qquad \text{(Eq. 10.8)}$$

$$M(\mathbf{c_n''}) = M(R') + nM(P) + M(Cat) + 17 \qquad \text{(Eq. 10.9)}$$

$$M(\mathbf{x_n}) = M(R'') + nM(P) + M(Cat) - 1 \qquad \text{(Eq. 10.10)}$$

$$M(\mathbf{z_n}) = M(R'') + nM(P) + M(Cat) + 15 \qquad \text{(Eq. 10.11)}$$

$$M(\mathbf{z_n''}) = M(R'') + nM(P) + M(Cat) + 17 \qquad \text{(Eq. 10.12)}$$

where $M(\mathbf{b_n})$, $M(\mathbf{c_n})$, $M(\mathbf{c_n''})$, $M(\mathbf{x_n})$, $M(\mathbf{z_n})$, and $M(\mathbf{z_n''})$ are the mass-to-charge ratios of the fragment ions of the $\mathbf{b_n}$, $\mathbf{c_n}$, $\mathbf{c_n''}$, $\mathbf{x_n}$, $\mathbf{z_n}$, and $\mathbf{z_n''}$ series, respectively, $M(R')$ and $M(R'')$ are the masses of the α and ω end group respectively, $M(P)$ is the mass of polymer repeat unit (i.e., 44.0 Da for PEG), and $M(Cat)$ is the mass of the cation.

Analogous fragmentation was noted from another similar polymer system, namely poly(tetramethylene glycol) (PTMEG) [5]. The proposed fragmentation for this polymer is shown in Scheme 10.4, with new notation for annotation of peaks based on that described for other polyethers above (Scheme 10.3) and polyoxetanes in Chapter 5 (Figure 5.13).

Similar equations, to those for polyethers from 1,2-epoxide-based monomers (*vide infra*), can be used to calculate the expected mass-to-charge ratios of fragment ions from PTMEG:

$$M(\mathbf{d_n}) = M(R') + nM(P) + M(Cat) - 1 \qquad \text{(Eq. 10.13)}$$

$$M(\mathbf{e_n}) = M(R') + nM(P) + M(Cat) + 15 \qquad \text{(Eq. 10.14)}$$

$$M(\mathbf{e_n''}) = M(R') + nM(P) + M(Cat) + 17 \qquad \text{(Eq. 10.15)}$$

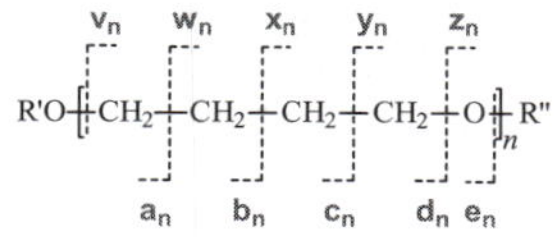

$$v_n \quad w_n \quad x_n \quad y_n \quad z_n$$

$$R'O\left[CH_2 + CH_2 + CH_2 + CH_2 + O\right]_n R''$$

$$a_n \quad b_n \quad c_n \quad d_n \ e_n$$

R' is α end group and R" is ω end group

R'O-(CH₂CH₂CH₂CH₂O)ₙ-H	$e_n"$	H-(OCH₂CH₂CH₂CH₂)ₙ₋₁-OR" $\quad z_n"$
R'O-(CH₂CH₂CH₂CH₂O)ₙ₋₁-CH₂CH₂CH=CH₂	d_n	H₂C=CHCH₂CH₂-(OCH₂CH₂CH₂CH₂)ₙ₋₁-OR" $\quad v_n$
R'O-(CH₂CH₂CH₂CH₂O)ₙ₋₁-CH₂CH₂CH₂CH=O	e_n	O=CHCH₂CH₂CH₂-(OCH₂CH₂CH₂CH₂)ₙ₋₁-OR" $\quad z_n$

Scheme 10.4. Proposed annotation for fragmentation of poly(tetramethylene glycol) (PTMEG). Structures of typically observed fragment ions are also shown.

$$M(\mathbf{v_n}) = M(R'') + nM(P) + M(Cat) - 1 \qquad \text{(Eq. 10.16)}$$

$$M(\mathbf{z_n}) = M(R'') + nM(P) + M(Cat) + 15 \qquad \text{(Eq. 10.17)}$$

$$M(\mathbf{z_n''}) = M(R'') + nM(P) + M(Cat) + 17 \qquad \text{(Eq. 10.18)}$$

where $M(\mathbf{d_n})$, $M(\mathbf{e_n})$, $M(\mathbf{e_n''})$, $M(\mathbf{v_n})$, $M(\mathbf{z_n})$, and $M(\mathbf{z_n''})$ are the mass-to-charge ratios of the fragment ions of the $\mathbf{d_n}$, $\mathbf{e_n}$, $\mathbf{e_n''}$, $\mathbf{v_n}$, $\mathbf{z_n}$, and $\mathbf{z_n''}$ series, respectively, M(R') and M(R'') are the masses of the α and ω end group, respectively, M(P) is the mass of polymer repeat unit (i.e., 72.1 Da for PTMEG), and M(Cat) is the mass of the cation.

A comparison of the typical MALDI-MS/MS spectra obtained from lithiated precursors of di-hydroxyl end-capped oligomers of PEG, PBG, and PTMEG are shown below in Figure 10.8 [5]. One of the main differences between these fragment ion spectra, and those obtained from methacrylates and styrenes (*vide infra*), is that the former contain peaks of relatively high intensity across the full mass range. This is typical of spectra from polyether polymers.

Differentiation between PBG and PTMEG polymers (which have a repeat unit with the same empirical formula, i.e., C₄H₈O) is possible from the MS/MS spectra, as minor peaks from side-chain losses are noted in the data from PBG [5]. Low-intensity peaks from the $\mathbf{c_{nb}}$ and $\mathbf{x_{nb}}$ series are noted to be present in the spectrum from PBG, which result from loss of one ethyl side chain of the polymer.

It was proposed that peaks with mass-to-charge ratios close to that of the precursor ion can be used to infer end-group information using the conditions typically used for MALDI-MS/MS experiments [5]. Sequencing of copolymers is also possible from these series of fragment ions, as has been demonstrated by means of ESI-MS/MS of alcohol-initiated ethylene oxide/propylene oxide (EO/PO) systems [6]. Experiments at higher collision energies resulted in MALDI-MS/MS spectra of PEG dominated by fragments of lower mass-to-charge ratios that did not retain the cation [9, 10]. These lower mass-to-charge ratio peaks were also proposed to be structurally significant.

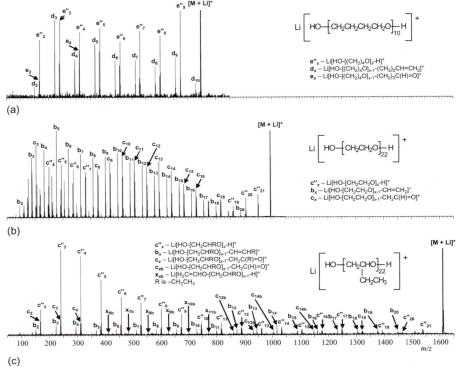

(a)

(b)

(c)

Figure 10.8.   MALDI-MS/MS spectra from lithiated di-hydroxyl end-capped polyethers. (a) Decamer of PTMEG; (b) 22-mer of PEG; and (c) 22-mer of PBG. The proposed structures of fragment ions are shown. Modified from Reference 5.

10.4.4   Polyesters and Polyesteramides

A number of polyesters and polyesteramides have been characterized by means of MALDI-MS/MS [7, 23–25], with sequence and end-group information inferred from the resulting data. An example of the MALDI-MS/MS spectrum from poly(butylene adipate) [23] is shown in Chapter 5 (Figure 5.8). The data generated enabled the sequence and end groups of this polymer to be verified. The nomenclature used for annotation of the fragment ions from polyesters and polyesteramides should be analogous to that described for the other polymer systems above. An example is given for PET [7], which is polyester derived from the reaction of terephthalic acid (or dimethyl terephthalate) with ethylene glycol. The MALDI-MS/MS spectrum from the sodiated hexamer of PET is shown in Figure 10.9. The proposed notation for annotation of this polyester is also shown in Scheme 10.5. These MALDI-MS/ MS data indicate that the end-group structure is as shown in Figure 10.9, with some indication (by the peaks at m/z 149 and 341) of the presence of an additional, alternative, structure with an acid end group and a di-ethylene glycol (DEG) unit in-chain. The presence of DEG units and acid end groups are commonly noted in PET

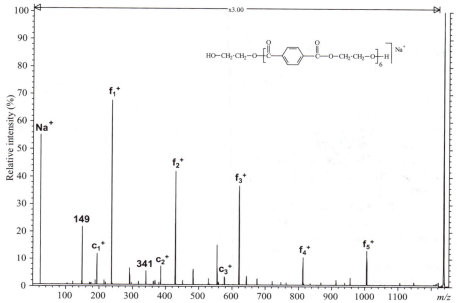

Figure 10.9. MALDI-MS/MS spectrum from sodiated hexamer of poly(ethylene terephthalate) (PET). The spectrum indicates the presence of another structure (namely oligomer with an acid end group and a di-ethylene glycol (DEG) unit in-chain). Modified from Reference 7.

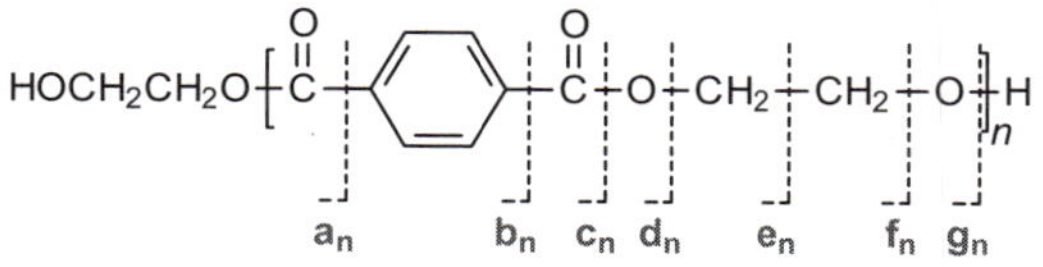

Scheme 10.5. Proposed annotation for fragmentation of poly(ethylene terephthalate) (PET).

polymers, and the presence of these functionalities were confirmed by analysis using NMR spectroscopy.

10.5 EXAMPLES OF APPLICATIONS

A number of examples are described in this section that show how both end group and sequence information may be obtained for a range of polymers and a block copolymer. Polymers generated by a number of different polymerization mechanisms have been studied. MALDI-MS/MS is an important technique in studying these different polymers, as the structures of the end groups are often an indicator of the polymerization mechanism.

10.5.1 Methacrylate Polymers and Copolymers

Industrially manufactured PMMA is typically made by free-radical polymerization. The structures of end groups can be a very useful indicator of either the initiator and/or chain transfer agent (CTA) used to make the polymer. Characterization of the end-group functionality can therefore play a very important role in the analysis of competitors' products, or when producing new polymers. MALDI-MS/MS plays an important role in this end-group characterization for PMMA, but the high average molecular weight of many polymers means that lower molecular weight extracts occasionally have to be prepared for analysis (this is also true for NMR spectroscopy and MALDI-TOF experiments). MALDI-MS/MS spectra obtained from PMMA polymers with varying initiating end-group functionality are shown in Figure 10.10 [7]. These data can be used to infer information on the masses of the initiating end group, as shown in Scheme 10.6.

A series of peaks ($z_n^\bullet$ series, m/z 110, 210, and 310) is seen at the same mass-to-charge ratios in all three spectra, indicating that the terminating end group is the same in all three samples. The mass-to-charge ratios of the $b_n^\bullet$ series for the three PMMA samples indicate that different initiators or CTAs have been used to make these polymers.

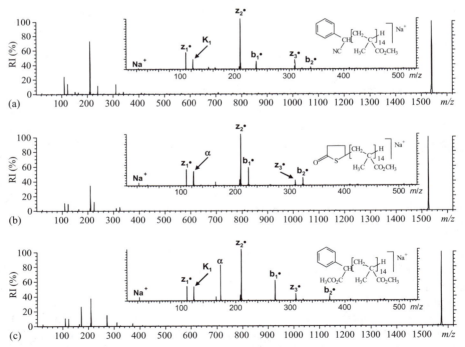

Figure 10.10. MALDI-MS/MS spectra from sodiated PMMA samples with different initiating end-group functionalities. Proposed fragment ion schemes are displayed in Scheme 10.6. Modified from Reference 7.

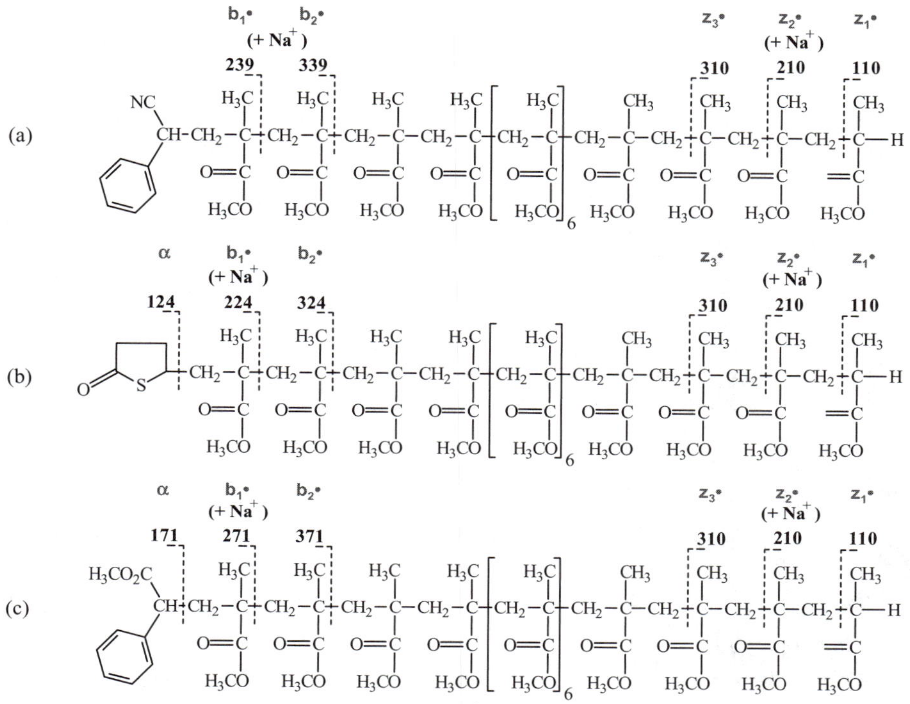

(a)

$b_1^\bullet$ $b_2^\bullet$
(+ Na$^+$)
239   339

$z_3^\bullet$ $z_2^\bullet$ $z_1^\bullet$
(+ Na$^+$)
310   210   110

(b)

α $b_1^\bullet$ $b_2^\bullet$
(+ Na$^+$)
124   224   324

$z_3^\bullet$ $z_2^\bullet$ $z_1^\bullet$
(+ Na$^+$)
310   210   110

(c)

α $b_1^\bullet$ $b_2^\bullet$
(+ Na$^+$)
171   271   371

$z_3^\bullet$ $z_2^\bullet$ $z_1^\bullet$
(+ Na$^+$)
310   210   110

Scheme 10.6. Proposed fragmentation of PMMA samples, from MALDI-MS/MS spectra shown in Figure 10.10. End-group structures can be inferred from these data. Modified from Reference 7.

An example MALDI-MS/MS spectrum for a polymer of PMMA made using atom transfer radical polymerization (ATRP) is shown below in Figure 10.11 [19]. ATRP is a living free-radical polymerization system that offers interesting possibilities for the generation of many functionalized polymers and copolymers. The end-group functionality is very important to the reactivity of the polymer and, therefore, to the possibility of further reaction of the system, such as to make a block copolymer (which is difficult by traditional free-radical techniques).

This PMMA polymer was expected to have a tosyl group at the initiating end and chloride terminating-end functionality. The mechanism of polymerization of ATRP involves a shuttling halide on the living chain end, resulting in this functionality remaining as the terminating functionality on the resulting polymer (unless the terminating end group is modified post polymerization to increase stability). One series of peaks from the $b_n^\bullet$ series (m/z 262 and 362, separated by m/z 100 which is equivalent to the mass of the repeat unit of PMMA) indicates that the initiating end group is that expected, as is shown by the fragmentation scheme in Figure 10.11. A second series ($z_n^\bullet$, m/z 128 and 228) suggests that the chloride functionality is intact on the terminating end of the polymer (see Figure 10.11).

Group transfer polymerization (GTP) is a methodology for the production of acrylic polymers and copolymers with low polydispersities. A block copolymer of

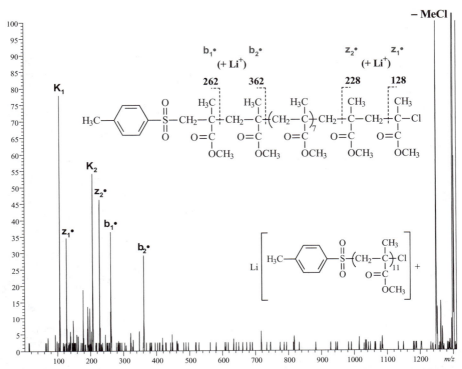

Figure 10.11.   MALDI-MS/MS spectrum from lithiated 11-mer of PMMA generated by ATRP. Proposed fragmentation scheme is shown for the determination of end groups from ions of the $\mathbf{b_n}^{\bullet}$ and $\mathbf{z_n}^{\bullet}$ series. Modified from Reference 16.

methyl methacrylate (MMA) and butyl methacrylate (BMA) has been characterized by means of MALDI-TOF and MALDI-MS/MS [20]. Information on the molecular weight distribution of the copolymer (and the individual monomers in the system) was obtained by MALDI-TOF [20]. Detailed information on the structure, including end-group structure and sequence of the copolymer, were generated using MALDI-MS/MS [20]. This technique has the capability of differentiating between block and random copolymers. The MALDI-MS/MS spectrum from a selected oligomer from the sample is displayed in Figure 10.12. The proposed fragmentation scheme for ions of low mass-to-charge ratios, used to infer end-group information, is shown in Scheme 10.7. Ions of the $\mathbf{b_n}^{\bullet}$ and $\mathbf{z_n}^{\bullet}$ series can be used to infer information on a partial sequence of the block copolymer as well as end-group structure, as the peaks from these series indicate that runs of MMA are present at the initiating end ($\mathbf{b_n}^{\bullet}$ series separated by m/z 100 for MMA repeats, i.e., m/z 108 and 208) and BMA at the terminating end ($\mathbf{z_n}^{\bullet}$ series separated by m/z 142 for BMA repeats, i.e., m/z 136 and 278). No evidence was found for the presence of BMA at the initiating end and MMA at the terminating end of the copolymer, ruling out the presence of totally random copolymer.

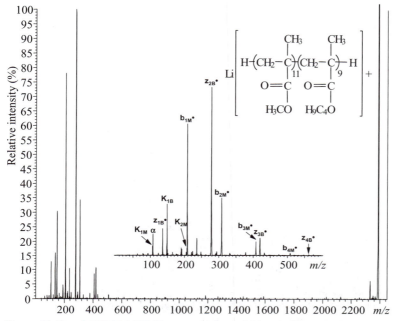

Figure 10.12. MALDI-MS/MS spectrum from lithiated oligomer of MMA/BMA block copolymer generated by GTP (inset is annotated expansion of m/z 0–600). End group and partial sequence information may be gleaned from ions of the $b_n^\bullet$ and $z_n^\bullet$ series, as is shown in Scheme 10.7. Modified from Reference 20.

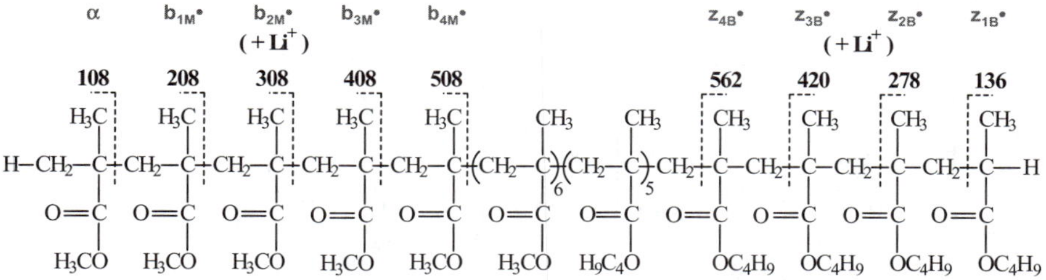

Scheme 10.7. Proposed fragmentation of MMA/BMA block copolymer, from MALDI-MS/MS spectrum shown in Figure 10.12. End-group structures and partial sequence information can be inferred from these data. Modified from Reference 20.

An expansion of the region of the MS/MS spectrum from m/z 600–2300 is shown in Figure 10.13, with peaks of very low intensity (compared to those of the $b_n^\bullet$ and $z_n^\bullet$ series) annotated. These peaks enable the full sequence of the block copolymer to be established. Peaks from four series (a_{nb}, b_n, y_{nb}, and z_n) are detected, as expected from previous data from methacrylate homopolymers (*vide infra*). The annotation has been slightly modified to account for the presence of varying amounts of two monomers in some of the fragment ions, as can be seen in the proposed fragment ion structures shown in Schemes 10.8 and 10.9.

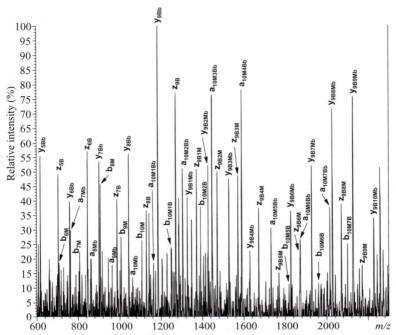

Figure 10.13. Expansion (m/z 600–2300) of MALDI-MS/MS spectrum from lithiated oligomer of MMA/BMA block copolymer generated by GTP. Sequence information may be gleaned from ions of the $\mathbf{a_{nb}}$, $\mathbf{b_n}$, $\mathbf{y_{nb}}$, and $\mathbf{z_n}$ series. Proposed structures of fragment ions, with explanation of modified annotation, are shown in Schemes 10.8 and 10.9. Modified from Reference 20.

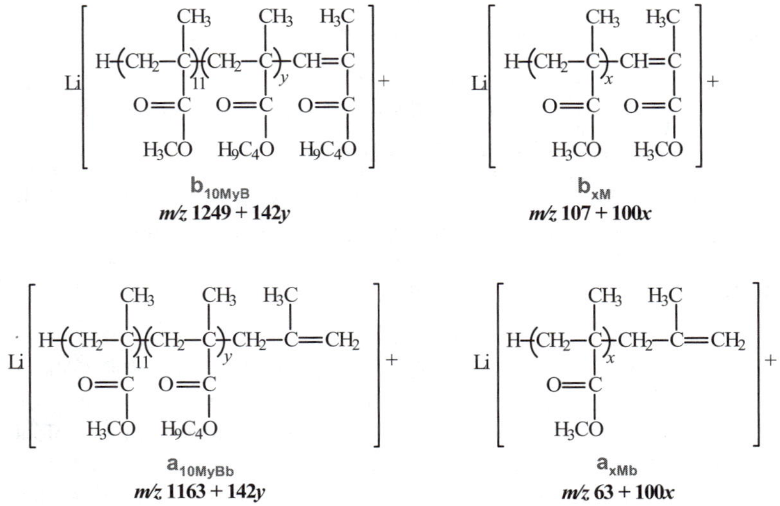

Scheme 10.8. Proposed structures of fragment ions ($\mathbf{a_{nb}}$ and $\mathbf{b_n}$) from MMA/BMA block copolymer, from expansion (m/z 600–2300) of MALDI-MS/MS spectrum shown in Figure 10.13. Sequence information can be inferred from these fragment ions. Modified from Reference 20.

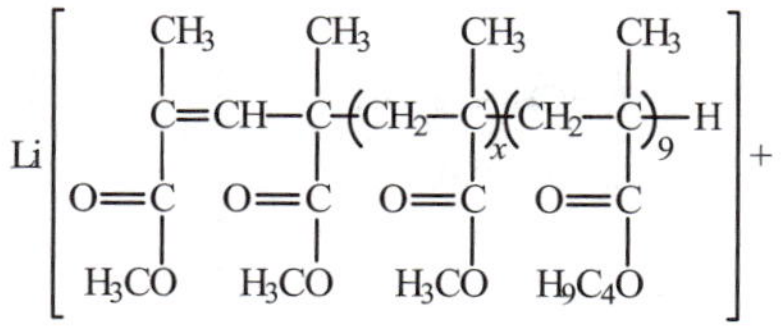

z_{9BxM}
m/z 1471 + 100*x*

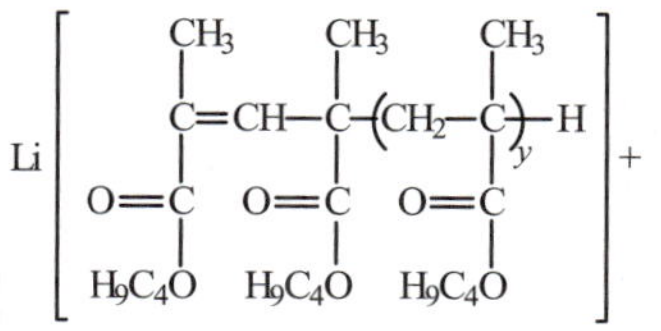

z_{yB}
m/z 277 + 142*y*

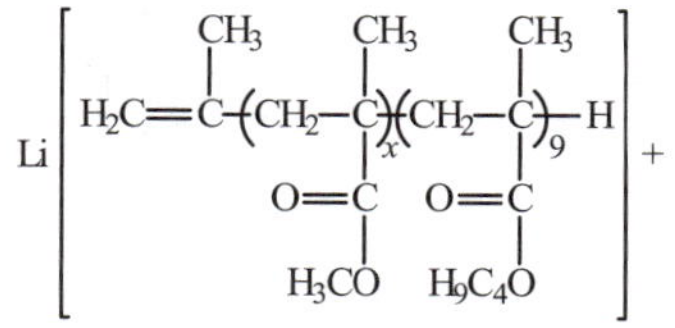

y_{9BxMb}
m/z 1327 + 100*x*

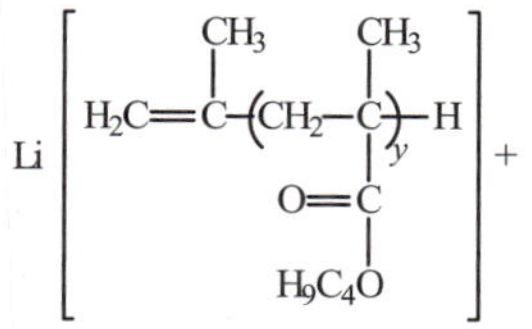

y_{yBb}
m/z 49 + 142*y*

Scheme 10.9. Proposed structures of fragment ions ($\mathbf{y}_{nb}$ and $\mathbf{z}_n$) from MMA/BMA block copolymer, from expansion (*m/z* 600–2300) of MALDI-MS/MS spectrum shown in Figure 10.13. Sequence information can be inferred from these fragment ions. Modified from Reference 20.

These data indicate that this polymer is indeed a block copolymer, with methyl methacrylate monomer added to the initiator prior to feed of butyl methacrylate monomer. Low-intensity peaks in the MS/MS spectrum (between *m/z* 1100–1400) indicate that a very low level of copolymer is present with imperfection in the block nature, probably arising due to the addition of BMA monomer just prior to total exhaustion of MMA monomer from the reaction vessel.

10.5.2 Styrene Polymers

An example is shown of how MALDI-MS/MS can be used to identify whether styrene polymers generated by anionic polymerization have been functionalized as desired by the polymer chemist. A batch of living polymer was made and split into two, with one-half having an end-group modifier added [11]. The MALDI-MS/MS spectra of oligomers from the two resulting samples are shown in Figure 10.14 (unmodified) and Figure 10.15 (modified).

Peaks from the $\mathbf{b}_n^\bullet$ and $\mathbf{z}_n^\bullet$ series can be used to clearly differentiate between the two samples and show that both have the same initiating functionality ($\mathbf{b}_n^\bullet$ series with *m/z* 253, 357, and 461), and one is modified with bulky terminating agent ($\mathbf{z}_n^\bullet$ series with *m/z* 558, 662, and 766 in Figure 10.15) and the other is unmodified ($\mathbf{z}_n^\bullet$ series with *m/z* 154, 258, 362, 466, and 570 in Figure 10.14). The precursor ions were cationized with copper ions as the basic amine initiating group led to the presence of abundant protonated ions in the MALDI-TOF spectra that overlapped with $[M + Ag]^+$ peaks when silver salts were added to cationize the polymer. Selection of protonated precursor ions for MALDI-MS/MS experiments did not yield informa-

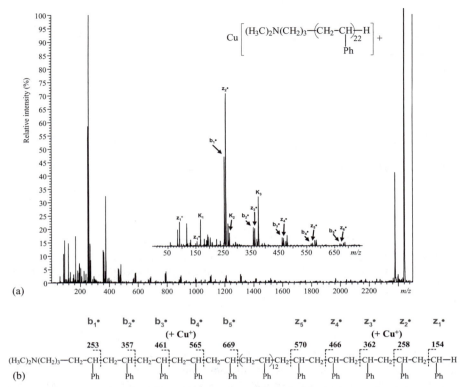

Figure 10.14.   (a) MALDI-MS/MS spectrum from [M + Cu]$^+$ of 22-mer of poly(styrene) (unmodified), (annotated in expansion of m/z 0–700 as inset) and (b) proposed fragmentation scheme indicating how the masses and structures of the end groups can be inferred from the data. Modified from Reference 11.

tion on the modification of the terminating end group as fragmentation was all directed to the initiating group (due to the presence of the basic amine functionality).

10.5.3   Polyethers

Polyethers with chain-end functionality are often used as emulsifiers, such as, for example, in personal care applications. Two examples of how MALDI-MS/MS may be employed, to confirm that the end groups have been modified and to generate information on the structure of the functionalization, are described. Generation of information on the end-group structure can aid the understanding of how emulsifiers have been produced, which can be useful in the analysis of competitors' products.

The MALDI-MS/MS spectrum from the lithiated decamer of a stearyl alcohol-initiated PPG is shown in Figure 10.16 [5]. Peaks in the higher mass-to-charge ratio region of the spectrum are useful for inferring the structure of the end groups. Loss of octadecene, plus a propylene glycol unit (sequentially), from the initiating end of oligomer gives rise to the peaks at m/z 605 and 547 (z''_{10} and z''_9) respectively,

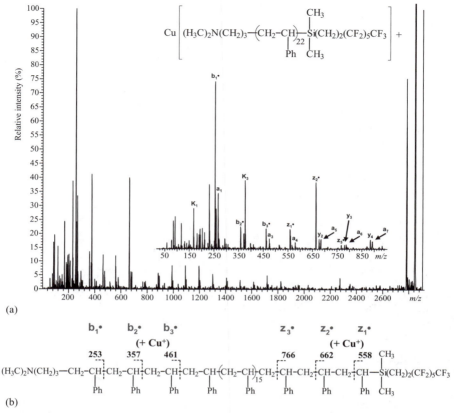

(a)

(b)

Figure 10.15. (a) MALDI-MS/MS spectrum from $[M + Cu]^+$ of 22-mer of poly(styrene) (modified), (annotated in expansion of m/z 0–900 as inset) and (b) proposed fragmentation scheme indicating how the masses and structures of the end groups can be inferred from the data. Modified from Reference 11.

whereas the ions of the c_n'' series indicate that the PPG is terminated with a hydroxyl group. A minor series of ions (from the c_{nb} series) indicate that the polymer is indeed a PPG rather than a polyoxetane (which has a repeat unit with the same empirical formula, i.e., C_3H_6O).

The MALDI-MS/MS spectra from two alkyl phenol-initiated polyethers are shown in Figure 10.17 [5]. The loss of part of, or the whole of the bulky initiated group is favored, giving rise to fragment ions that retain the terminating functionality. Losses of portions or the whole alkyl chain from the initiating groups indicate the length of the chain, plus give some information on the branching (common for industrial feedstocks used to make these materials). Loss of the whole of the alkyl phenol initiating functionality gives rise to the highest mass-to-charge ratio peak of the x_n series (x_{12} at m/z 535 for the nonylphenol ethoxylate and x_8 at m/z 583 for the dodecylphenol-initiated PBG).

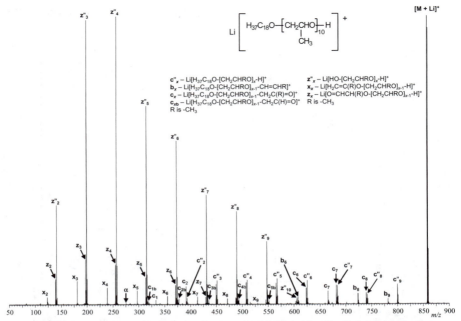

Figure 10.16. MALDI-MS/MS spectrum from the lithiated decamer of stearyl alcohol-initiated PPG. The proposed structures of the fragment ions are shown along with corresponding annotation for fragment ion peaks. Modified from Reference 5.

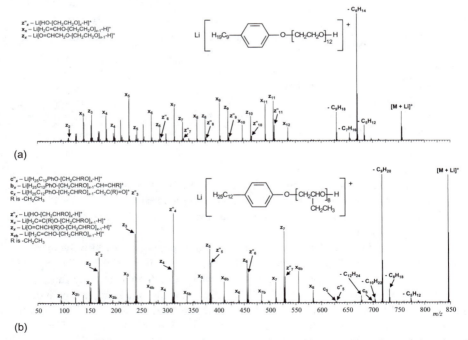

(a)

(b)

Figure 10.17. MALDI-MS/MS spectra from (a) the lithiated dodecamer of nonylphenol-initiated PEG and (b) the lithiated octamer of dodecylphenol-initiated PBG. The proposed structures of the fragment ions are shown along with corresponding annotation for fragment ion peaks. Modified from Reference 5.

Peaks from the x_{nb} series in the MALDI-MS/MS spectrum of dodecylphenol-initiated PBG indicate that this polymer was made from a 1,2-epoxide-based polyether, rather than being a PTMEG polymer (which has the same empirical formula, i.e., C_4H_8O). This was confirmed by analysis using NMR spectroscopy.

10.6   CONCLUSIONS

MALDI-MS/MS is a powerful technique for the detailed sequencing of synthetic polymers, alongside other analytical techniques such as MALDI-TOF, NMR spectroscopy, and occasionally, Fourier transform infrared (FTIR) spectroscopy. Information on end-group structure can often readily be obtained with this technique, providing useful additional data to that generated by MALDI-TOF. The coupling of other ionization techniques, such as ESI [6, 31] and desorption electrospray ionization (DESI) [32], with MS/MS can also be a powerful combination, especially as these techniques offer some advantages over MALDI. ESI can more readily be directly coupled with separation science techniques for polymers, such as GPC. Furthermore, the multiply charged ions that are often generated by ESI also offer the possibility of generating data from precursor ions of greater molecular weight, providing the selection of individual precursors is possible [31]. Ion mobility approaches may aid the selection of the required multiply charged precursor for MS/MS. DESI offers the possibility of characterizing synthetic polymers more rapidly, with little or no sample preparation [32].

Novel software has been developed to aid the interpretation of MS/MS data from synthetic polymers [45]. The software can use exact mass data to provide more confidence in the results. Further developments in this software, such as to enable the automatic interpretation of MS/MS data from synthetic polymers (as has been possible for many years for peptides), should help widen the usage of MS/MS for polymer characterization. The lengthy interpretation times that are often required for manual interpretation of MS/MS data have already been reduced with the introduction of the current software package, but a predictive version would make analysis much less time-intensive.

REFERENCES

1. Gidden, J.; Wyttenbach, T.; Jackson, A.T.; Scrivens, J.H.; Bowers, M.T., Gas-phase conformations of synthetic polymers: poly(ethylene glycol), poly(propylene glycol), and poly(tetramethylene glycol), *J. Am. Chem. Soc.*, **2000**, 122, 4692–4699.
2. Hanton, S.D.; Parees, D.A.; Owens, K.G., MALDI PSD of low molecular weight ethoxylated polymers, *Int. J. Mass Spectrom.*, **2004**, 238, 257–264.
3. Hoteling, A.J.; Kawaoka, K.; Goodberlet, M.C.; Yu, W.M.; Owens, K.G., Optimization of matrix-assisted laser desorption/ionization time-of-flight collision-induced dissociation using poly(ethylene glycol), *Rapid Commun. Mass Spectrom.*, **2003**, 17, 1671–1676.
4. Hoteling, A.J.; Owens, K.G., Improved PSD and CID on a MALDI TOFMS, *J. Am. Soc. Mass Spectrom.*, **2004**, 15, 523–535.
5. Jackson, A.T.; Green, M.R.; Bateman, R.H., Generation of end-group information from polyethers by matrix-assisted laser desorption/ionisation collision-induced dissociation mass spectrometry, *Rapid Commun. Mass Spectrom.*, **2006**, 20, 3542–3550.

6. Jackson, A.T.; Scrivens, J.H.; Williams, J.P.; Baker, E.S.; Gidden, J.; Bowers, M.T., Microstructural and conformational studies of polyether copolymers, *Int. J. Mass Spectrom.*, **2004**, 238, 287–297.

7. Jackson, A.T.; Yates, H.T.; Scrivens, J.H.; Critchley, G.; Brown, J.; Green, M.R.; Bateman, R.H., The application of matrix-assisted laser desorption/ionization combined with collision-induced dissociation to the analysis of synthetic polymers, *Rapid Commun. Mass Spectrom.*, **1996**, 10, 1668–1674.

8. Okuno, S.; Kiuchi, M.; Arakawa, R., Structural characterization of polyethers using matrix-assisted laser desorption/ionization quadrupole ion trap time-of-flight mass spectrometry, *Eur. J. Mass Spectrom.*, **2006**, 12, 181–187.

9. Botrill, A.R.; Giannakopulos, A.E.; Millichope, A.; Lee, K.S.; Derrick, P.J., Combination of time-of-flight mass analysers with magnetic-sector instruments: in-line and perpendicular arrangements: applications to poly(ethylene glycol) with long-chain end groups, *Eur. J. Mass Spectrom.*, **2000**, 6, 225–232.

10. Bottrill, A.R.; Giannakopulos, A.E.; Waterson, C.; Haddleton, D.M.; Lee, K.S.; Derrick, P.J., Determination of end groups of synthetic polymers by matrix-assisted laser desorption/ionization: high-energy collision-induced dissociation, *Anal. Chem.*, **1999**, 71, 3637–3641.

11. Jackson, A.T.; Bunn, A.; Hutchings, L.R.; Kiff, F.T.; Richards, R.W.; Williams, J.; Green, M.R.; Bateman, R.H., The generation of end group information from poly(styrene)s by means of matrix-assisted laser desorption/ionisation-collision induced dissociation, *Polymer*, **2000**, 41, 7437–7450.

12. Jackson, A.T.; Yates, H.T.; Scrivens, J.H.; Green, M.R.; Bateman, R.H., Matrix-assisted laser desorption/ionization-collision induced dissociation of poly(styrene), *J. Am. Soc. Mass Spectrom.*, **1998**, 9, 269–274.

13. Scrivens, J.H.; Jackson, A.T.; Yates, H.T.; Green, M.R.; Critchley, G.; Brown, J.; Bateman, R.H.; Bowers, M.T.; Gidden, J., The effect of the variation of cation in the matrix-assisted laser desorption/ionisation collision induced dissociation (MALDI-CID) spectra of oligomeric systems, *Int. J. Mass Spectrom.*, **1997**, 165, 363–375.

14. Gidden, J.; Bowers, M.T.; Jackson, A.T.; Scrivens, J.H. Gas-phase conformations of cationized poly(styrene) oligomers, *J. Am. Soc. Mass Spectrom.*, **2002**, 13, 499–505.

15. Gies, A.P.; Vergne, M.J.; Orndorff, R.L.; Hercules, D.M., MALDI-TOF/TOF CID study of polystyrene fragmentation reactions, *Macromolecules*, **2007**, 40, 7493–7504.

16. Polce, M.J.; Ocampo, M.; Quirk, R.P.; Leigh, A.M.; Wesdemiotis, C., Tandem mass spectrometry characteristics of silver-cationized polystyrenes: internal energy, size, and chain end versus backbone substituent effects, *Anal. Chem.*, **2008**, 80, 355–362.

17. Polce, M.J.; Ocampo, M.; Quirk, R.P.; Wesdemiotis, C. Tandem mass spectrometry characteristics of silver-cationized polystyrenes: backbone degradation via free radical chemistry, *Anal. Chem.*, **2008**, 80, 347–354.

18. Borman, C.D.; Jackson, A.T.; Bunn, A.; Cutter, A.L.; Irvine, D.J., Evidence for the low thermal stability of poly(methyl methacrylate) polymer produced by atom transfer radical polymerisation, *Polymer*, **2000**, 41, 6015–6020.

19. Jackson, A.T.; Bunn, A.; Priestnall, I.M.; Borman, C.D.; Irvine, D.J., Molecular spectroscopic characterisation of poly(methyl methacrylate) generated by means of atom transfer radical polymerisation (ATRP), *Polymer*, **2006**, 47, 1044–1054.

20. Jackson, A.T.; Scrivens, J.H.; Simonsick, W.J.; Green, M.R.; Bateman, R.H., Generation of structural information from polymers and copolymers using tandem mass spectrometry, *Polymer Preprints*, **2000**, 41, 641–642.

21. Jackson, A.T.; Yates, H.T.; Scrivens, J.H.; Green, M.R.; Bateman, R.H., Utilizing matrix-assisted laser desorption/ionization-collision induced dissociation for the generation of structural information from poly(alkyl methacrylate)s, *J. Am. Soc. Mass Spectrom.*, **1997**, 8, 1206–1213.

22. Chaicharoen, K.; Polce, M.J.; Singh, A.; Pugh, C.; Wesdemiotis, C., Characterization of linear and branched polyacrylates by tandem mass spectrometry, *Anal. Bioanal. Chem.*, **2008**, 392, 595–607.

23. Rizzarelli, P.; Puglisi, C.; Montaudo, G., Matrix-assisted laser desorption/ionization time-of-flight/time-of-flight tandem mass spectra of poly(butylene adipate), *Rapid Commun. Mass Spectrom.*, **2006**, 20, 1683–1694.

24. Rizzarelli, P.; Puglisi, C.; Montaudo, G., Sequence determination in aliphatic poly(ester amide)s by matrix-assisted laser desorption/ionization time-of-flight and time-of-flight/time-of-flight tandem mass spectrometry, *Rapid Commun. Mass Spectrom.*, **2005**, 19, 2407–2418.

25. Rizzarelli, P.; Puglisi, C., Structural characterization of synthetic poly(ester amide) from sebacic acid and 4-amino-1-butanol by matrix-assisted laser desorption ionization time-of-flight/time-of-flight tandem mass spectrometry, *Rapid Commun. Mass Spectrom.*, **2008**, 22, 739–754.

26. Chen, R.; Li, L., Lithium and transition metal ions enable low energy collision-induced dissociation of polyglycols in electrospray ionization mass spectrometry, *J. Am. Soc. Mass Spectrom.*, **2001**, 12, 832–839.

27. Chen, R.; Tseng, A.M.; Uhing, M.; Li, L. Application of an integrated matrix-assisted laser desorption/ ionization time-of-flight, electrospray ionization mass spectrometry and tandem mass spectrometry approach to characterizing complex polyol mixtures, *J. Am. Soc. Mass Spectrom.*, **2001**, 12, 55–60.

28. Craig, A.G.; Derrick, P.J., Spontaneous fragmentation of cationic polystyrene chains, *J. Am. Chem. Soc.*, **1985**, 107, 6707–6708.

29. Craig, A.G.; Derrick, P.J., Collision-induced decomposition of cationic radical polystyrene chains, *J. Chem. Soc., Chem. Commun.*, **1985**, 891–892.

30. Craig, A.G.; Derrick, P.J., Production and characterization of beams of polystyrene ions, *Aust. J. Chem.*, **1986**, 39, 1421–1434.

31. Jackson, A.T.; Slade, S.E.; Scrivens, J.H., Characterisation of poly(alkyl methacrylate)s by means of electrospray ionisation-tandem mass spectrometry (ESI-MS/MS), *Int. J. Mass Spectrom.*, **2004**, 238, 265–277.

32. Jackson, A.T.; Williams, J.P.; Scrivens, J.H., Desorption electrospray ionisation mass spectrometry and tandem mass spectrometry of low molecular weight synthetic polymers, *Rapid Commun. Mass Spectrom.*, **2006**, 20, 2717–2727.

33. Lattimer, R.P., Tandem mass-spectrometry of poly(ethylene glycol) proton-attachment and deuteron-attachment ions, *Int. J. Mass Spectrom. Ion Proc.*, **1992**, 116, 23–36.

34. Lattimer, R.P., Tandem mass-spectrometry of lithium-attachment ions from polyglycols, *J. Am. Soc. Mass Spectrom.*, **1992**, 3, 225–234.

35. Lattimer, R.P., Tandem mass-spectrometry of poly(ethylene glycol) lithium-attachment ions, *J. Am. Soc. Mass Spectrom.*, **1994**, 5, 1072–1080.

36. Selby, T.L.; Wesdemiotis, C.; Lattimer, R.P., Dissociation characteristics of [M+X](+) ions (X=H, Li, K) from linear and cyclic polyglycols, *J. Am. Soc. Mass Spectrom.*, **1994**, 5, 1081–1092.

37. Jackson, A.T.; Jennings, K.R.; Scrivens, J.H., Generation of average mass values and end group information of polymers by means of a combination of matrix-assisted laser desorption ionization-mass spectrometry and liquid secondary ion tandem mass spectrometry, *J. Am. Soc. Mass Spectrom.*, **1997**, 8, 76–85.

38. Jackson, A.T.; Jennings, R.C.K.; Scrivens, J.H.; Green, M.R.; Bateman, R.H., The characterization of complex mixtures by field desorption tandem mass spectrometry, *Rapid Commun. Mass Spectrom.*, **1998**, 12, 1914–1924.

39. Gidden, J.; Jackson, A.T.; Scrivens, J.H.; Bowers, M.T., Gas phase conformations of synthetic polymers: poly (methyl methacrylate) oligomers cationized by sodium ions, *Int. J. Mass Spectrom.*, **1999**, 188, 121–130.

40. Bateman, R.H.; Hoyes, J.B., A comparison of PSD and CID MS/MS data from MALDI-TOF and MALDI-magnetic-sector-TOF instrumentation, *Abstr. Pap. Am. Chem. Soc.*, **1995**, 210.

41. Bateman, R.H.; Green, M.R.; Scott, G.; Clayton, E., A combined magnetic sector-time-of-flight mass-spectrometer for structural determination studies by tandem mass-spectrometry, *Rapid Commun. Mass Spectrom.*, **1995**, 9, 1227–1233.

42. Medzihradszky, K.F.; Adams, G.W.; Burlingame, A.L., Peptide sequence determination by matrix-assisted laser desorption ionization employing a tandem double focusing magnetic-orthogonal acceleration time-of-flight mass spectrometer, *J. Am. Soc. Mass Spectrom.*, **1996**, 7, 1–10.

43. Laine, O.; Trimpin, S.; Rader, H.J.; Mullen, K., Changes in post-source decay fragmentation behavior of poly(methyl methacrylate) polymers with increasing molecular weight studied by matrix-assisted

laser desorption/ionization time-of-flight mass spectrometry, *Eur. J. Mass Spectrom.*, **2003**, 9, 195–201.

44. Hanton, S.D.; Parees, D.M.; Zweigenbaum, J., The fragmentation of ethoxylated surfactants by AP-MALDI-QIT, *J. Am. Soc. Mass Spectrom.*, **2006**, 17, 453–458.

45. Thalassinos, K.; Jackson, A.T.; Williams, J.P.; Hilton, G.R.; Slade, S.E.; Scrivens, J.H., Novel software for the assignment of peaks from tandem mass spectrometry spectra of synthetic polymers, *J. Am. Soc. Mass Spectrom.*, **2007**, 18, 1324–1331.

CHAPTER 11

LC-MALDI MS FOR POLYMER CHARACTERIZATION

Steffen M. Weidner and Jana Falkenhagen

Federal Institute for Materials Research and Testing (BAM), Department I,
Analytical Chemistry, Reference Materials, Berlin, Germany

11.1 INTRODUCTION

Since its earliest applications in the mid-1950s the direct coupling of chromatography to mass spectrometry has become one of the most powerful techniques for the separation and identification of chemical species. Especially for the analysis of complex mixtures of low-molar mass compounds, both methods are complementary to each other. Whereas liquid chromatography enables a separation of species depending on the retention mechanism due to differences in size, polarity, chirality, and structure isomerism, mass spectrometry is able to provide additional structural information and can be considered as a universal detector for chromatography.

Primary difficulties like the evaporation of the carrier gas in the gas chromatography-mass spectrometry (GC-MS) coupling and the more demanding evaporation of the liquid mobile phases in the high-performance liquid chromatography-mass spectrometry (HPLC-MS) coupling had been overcome by the development of new sample inlet interfaces in combination with more efficient pump stages and increasing detector sensibility.

Unfortunately, all these technical improvements of coupling of (liquid chromatography (LC) to mass spectrometry (MS) cannot be transferred one-to-one for an analysis of polymers using a matrix-assisted laser desorption/ionization-time-of-flight (MALDI-TOF) mass spectrometer for one main reason: the *sample preparation.*

No matter what principle of mass spectrometric ionization (chemical ionization [CI], electron impact [EI], electro spray ionization [ESI], atmospheric pressure chemical ionization [APCI], field desorption [FD], etc.) is applied in MS, each compound being investigated finally has to be transferred into the gas phase of the mass spectrometer. Polymers exhibit a very high molar mass (up to millions of Daltons), and therefore, show a very low volatility. The principle of matrix-assisted laser desorption/ionization-time-of-flight mass spectrometry (MALDI-TOF MS) was developed in order to "evaporate" the intact polymer molecules from solid state

MALDI Mass Spectrometry for Synthetic Polymer Analysis, Edited by Liang Li
Copyright © 2010 John Wiley & Sons, Inc.

into the gas phase. This method requires a relatively complex sample preparation procedure. Sample molecules have to be embedded in an excess of suitable matrix molecules. The more homogeneous the mixture is, the better the ionization process will be, and the more reproducible data can be obtained. Finally, this *solid-state mixture* is fired at by a laser followed by a fast desorption of intact molecules and their ionization.

Another important limitation of combining these methods is set by the principle of chromatographic separation.

The majority of LC investigations of polymers is based on the principle of size exclusion. Due to conformational changes, polymer molecules are separated from each other according to their hydrodynamic volume by means of specific stationary phases. This LC technique is called size exclusion chromatography (SEC) or, based on the colloquial name for stationary phases, gel permeation chromatography (GPC). Ideally, no interaction between stationary phase and dissolved polymer should occur. The determination of molar masses and mass distributions is performed by a previous calibration with suitable polymer standards.

Beside a molar mass distribution (MMD) nearly all synthetic polymers exhibit an additional chemical heterogeneity distribution (CHD, see Figure 11.1).

Another chromatographic method based on interactions between sample molecules and mobile phase, on the one hand, and the stationary phase, on the other hand, is the liquid adsorption chromatography (LAC). In this method, every liquid chromatogram of polymers consists in a superposition of chromatograms caused by the two above described different separation mechanisms.

Such a principle of LC analysis of polymers was introduced by Gorshkov and colleagues [1–4]. Molecules are separated under so-called "critical conditions of

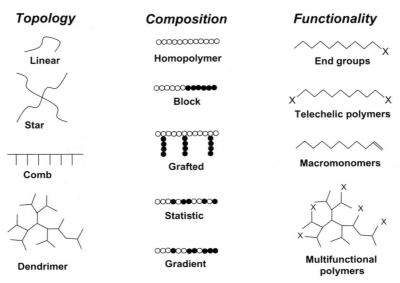

Figure 11.1. Chemical heterogeneity distribution of polymer.

adsorption" according to their chemical heterogeneity (functionalities, chemical distribution composition, chirality, tacticity, etc.). At these chromatographic conditions, the separation mechanism is characterized by a compensation of entropic and enthalpic contributions. The thermodynamic interpretation is given by the Gibbs–Helmholtz equation for the free enthalpy

$$\Delta G = \Delta H - T \Delta S \qquad \text{(Eq. 11.1)}$$

In contrast to the previously described size exclusion mode of chromatography with $\Delta S < 0$ and $\Delta H = 0$, the adsorption mode (LAC) is characterized by enthalpic interactions between the stationary phase and polymer molecules. The LAC method can be expressed through $\Delta H < 0$ and $T \Delta S \ll \Delta H$. The compensation of enthalpic and entropic interactions of the repeating unit of the polymer chain at the "critical point of adsorption" leads to $\Delta G = 0$. At this specific condition, identical polymer molecules with the same repeat unit eluate independent of their molar mass. Typical calibration curves obtained in the various modes of chromatography are presented in Figure 11.2. Critical conditions can be adjusted, for example, by changing solvent composition and/or temperature [5].

A more detailed description of the principle of liquid adsorption chromatography at critical conditions (LACCC) can be found in References 6–9.

A coupling of both of the chromatographic modes (SEC and LACCC) seemed to be obvious. This can be done in the two-dimensional "orthogonal" liquid adsorption chromatography (2D-LC). In the first dimension of 2D-LC, the molecules are separated under "critical conditions of adsorption" according to their chemical heterogeneity distribution. By the use of a switching valve, the collected samples can

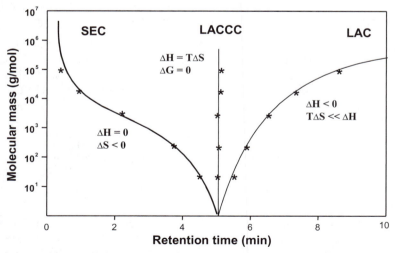

Figure 11.2. Polymer calibration curves obtained in three different modes of chromatography (size exclusion chromatography [SEC], liquid adsorption chromatography at critical conditions [LACCC], liquid adsorption chromatography [LAC]) and contribution of enthalpic and entropic terms of Gibbs-Helmholtz equation.

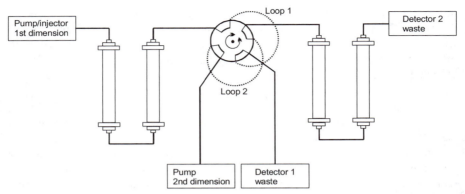

Figure 11.3.   Principle of two-dimensional chromatography (loop 1 is filled with eluate from the first dimension; previously filled loop 2 is simultaneously emptied into the second chromatographic system).

be transferred online from the first chromatographic dimension (LACCC) into the second chromatographic dimension, in which the polymers are separated according to their molar mass by means of conventional SEC. This is schematically shown in Figure 11.3.

However, the coupling of two different chromatographic systems is an expensive method, is time-consuming in the case of quantitative transfer, and considerable efforts to determine exactly the critical conditions of adsorption as well as to calibrate the SEC are needed. Therefore, attempts have been made to substitute one dimension by mass spectrometry.

The coupling of MALDI-TOF mass spectrometry to liquid chromatography seems to be a promising tool to approach some burning issues.

1. Many industrial polymers are characterized by a broad molar mass distribution (polydispersity). MALDI-TOF MS, however, only pictures the "true" mass distribution if the polydispersity is below 1.2. By means of SEC, a separation into fractions having narrower polydispersity can be achieved. The MALDI-TOF MS results of these fractions in turn can be used for an accurate calibration of the SEC.

2. Due to the lack of appropriate standards for many industrial polymers, so-called absolute methods for the mass determination have to be applied. Besides viscometry or light scattering methods, MALDI-TOF MS could be used as a detector for the calculation of the number-averaged molar mass (M_n) of polymers.

3. As mentioned above, nearly all industrial polymers exhibit a chemical heterogeneity distribution. Since the separation principle of the SEC is according to the hydrodynamic volume, any SEC analysis of polymers with identical molar masses but different shape leads to wrong results. MALDI-TOF MS can provide mass and perhaps structural information simultaneously.

4. A conventional 2D-HPLC analysis (LACCC-SEC, see Figure 11.3) is based on a transfer of fractions of the first dimension into the second dimension of the chromatography. Naturally, the solvent composition in both of the dimensions is quite different. Whereas SEC requires a thermodynamically "good" solvent, "critical conditions" of adsorption in LACCC can be adjusted by adding a thermodynamically "bad" solvent to the SEC solvent. Transferring fractions from LACCC to SEC therefore causes solvent interactions; that is, polymer signals in the low-molar mass region of the SEC can be overlaid by strong solvent signal from LACCC fractions. This can be avoided by means of an evaporative light scattering detector (ELSD) in which the solvent is completely vaporized. Many polymer LC methods use high boiling solvents so that a complete vaporization cannot be entirely achieved. A substitution of the second dimension (SEC) by MALDI-TOF MS can avoid these problems because the MALDI-TOF sample preparation nevertheless requires a complete removing of solvent.

5. Since in the MALDI-TOF process the ionization probability of polymers with identical monomer units and various end groups or structures (e.g., cycles vs. linear chains) is quite different and signal intensities are not deducible, any comparison of mass spectra of those species is absolute inadmissibly. In contrast to that, applying a suitable detector with nearly linear response chromatography could be a quantitative method. Thus, the coupling of chromatography to MALDI-TOF can result in more reliable (but not necessarily quantitative) results.

11.2   PRINCIPLES OF COUPLING

As previously mentioned, ordinary online coupling techniques (e.g., HPLC-MS) cannot be simply adapted to a coupling of LC to MALDI-TOF MS, because the sample preparation procedure in MALDI-TOF MS differs from other MS preparation procedures. Therefore, mostly offline methods have been developed and applied for that purpose. Nevertheless, a number of applications dealing with online coupling devices have been described.

Online Coupling Methods

Besides quaint appearing interfaces, like rotating belts, discs, and balls, which were used for the deposition of the polymer eluate after separation in chromatography [10–12], some more applicable approaches have been developed. According to the general principle, they are falling into two main categories: (I) spray (aerosol) methods and (II) continuous flow methods.

(**I**) Applying the online spray method, schematically shown in Figure 11.4, the solvent of the polymer solution has to be evaporated by spraying into a heated tube. The aerosol, formed by nebulization with N_2, enters the mass spectrometer and will be ionized by a laser. It becomes clear that the matrix has to be added before

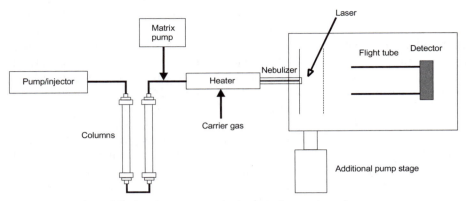

Figure 11.4.   Principle of online spray method generating an aerosol.

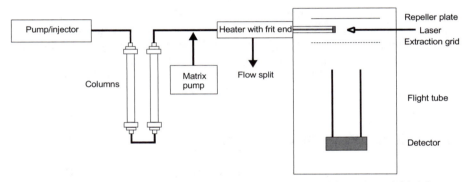

Figure 11.5.   Principle of continuous flow coupling method applying a heated frit inlet system.

nebulization simply by mixing the eluate with matrix solution via a T-fitting. Obtained MALDI-TOF mass spectra are mostly characterized by a poor peak resolution and a low reproducibility [13–17].

(II) Continuous flow coupling methods do not involve a separate evaporation of the solvent prior to ionization. The dissolved sample is introduced into the mass spectrometer by means of a small capillary, frit or filter (paper). The laser is focussed at the tip of the inlet system. In order to maintain a sufficiently low vacuum, only small flow rates can be used (typically 1–5 µL/min), which drastically reduce the overall amount of sample to be analyzed [18–21]. The principle is shown in Figure 11.5.

Recently, a new modification was presented by Laiko et al. [22] and Daniel et al. [23] that describes an atmospheric pressure (AP)-MALDI-TOF mass spectrometer for the analysis of synthetic and biopolymers. In that particular case, a small droplet is formed at the end of the capillary. A pulsed laser beam is focussed onto the surface of the droplet. Generated ions are guided by an electric field into the mass spectrometer, which is orthogonally arranged to the capillary. In contrast to the other online methods described in (I) and (II) polyethylene glycol PEG 1000 was detected in a surprisingly high resolution. Meanwhile, a considerable range of commercial AP-MALDI-TOF interfaces are available [24, 25].

Offline Coupling Methods

The manual taking of fractions can be regarded as one of the simplest offline coupling methods. Due to the high sensitivity of modern MALDI-TOF mass spectrometers, these fractions often can be used without further effort for sample spot preparation, for example, just by dropping a small volume onto a dry matrix spot (dried droplet method) [26–31]. Depending on the chromatographic problem (low concentration, poor peak resolution, shoulders, etc.) it might be advantageous to repeat fractionation several times. In those cases, a very high reproducibility of the chromatographic system is required. This drawback can be overcome by means of a fraction collector. Modern devices enable a simultaneous collecting and sample preparation (e.g., Probot™ Microfraction Collector, LC Packings.; DiNa Map MALDI Spotter, Kromatek, UK; MALDILC™ System, Gilson; SunChrom, Germany) [32–35]. Eluates can be collected in vials and, after finishing fractionation, small volumes can be premixed with matrix solution. Afterward, a small amount of this mixture (normally only a few microliters) will be deposited on the MALDI-TOF target plate [36–38].

Another method of offline coupling represents the use of a spray interface (e.g., LC-Transform series, LabConnections) [39]. A general scheme is shown in Figure 11.6. The sample solution passes a heated capillary. The solvent will be evaporated either without additional carrier gases (e.g., by ultrasonic nebulization) or by means of a carrier gas, simply by spraying at elevated temperatures onto the MALDI-TOF target [40–45]. The matrix can be added via a T-fitting. Alternatively, precoated targets can be used too. The target is fixed on a holder, which can be moved in any direction. As a result, a complete chromatographic run is mapped.

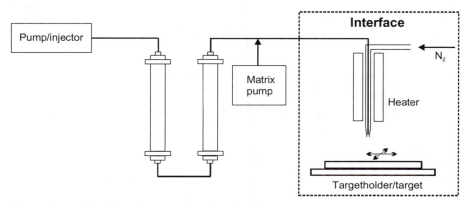

Figure 11.6. Principle of a spray deposition interface.

Depending on the target velocity during deposition, any retention time recorded in a chromatogram can be attributed to the corresponding target position, that is, to a MALDI-TOF mass spectrum.

Another approach to transfer liquid fractions on a MALDI probe represents the "heated droplet-interface" [46]. This interface consists of a transfer tube having an inlet and an outlet, the inlet being adapted to accept the LC effluents and the outlet being adapted to form continuously replaced, hanging droplets of the liquid stream. A heated MALDI sample plate is mounted below the outlet of the transfer tube for collecting the droplets. The liquid stream in the transfer tube is heated to a temperature (partial evaporation of the solvent from the hanging droplets). This heated droplet interface does not introduce sample loss, and the detection sensitivity of LC/MALDI is similar to that of standard MALDI. A further development of that principle represents the "impulse-driven heated droplet" deposition in which droplets are actively dislodged from the exit capillary onto the MALDI plate by means of a solenoid plunger [47].

11.3 EXAMPLES

Many applications for coupling of MALDI-TOF MS to chromatography for the analysis of polymers can be found in the literature. A comprehensive overview can be found in References 48 and 49. Most of them comprise the coupling of SEC [28, 45, 50–56] and LAC [30, 31, 57–61] to MALDI-TOF MS. Some other coupling principles using capillary electrophoresis (CE) or supercritical fluid chromatography (SFC) shall be not discussed in more detail because these applications are mainly focussed on the analysis of biopolymers and low-mass additives.

Often, an unambiguous definition of the separation principle in chromatography cannot be done. Chemical modifications (e.g., of end groups, polymer backbone) or structural variations of polymers (e.g., linear chains, cycles) result in a change of the chemical properties of the polymer, and unavoidably, in its behavior in a chromatographic analysis. For instance, the "ideal" SEC behavior is frequently overlaid by an additional interaction mechanism (e.g., LAC), and vice versa. Therefore, the following examples are not describing applications with regard to a certain chromatographic principle. They are focussed on coupling methods with regard to changes of the chemical heterogeneity of polymers (according to Figure 11.1).

Functionalization of the Polymer Backbone (Chemical Composition Distribution, [CCD])

In Figure 11.7a, a SEC chromatogram of a polybutadiene 5000 before (A) and after epoxidation (B) of its double bonds is shown. At first view, due to higher retention times, the molar masses after epoxidation seem to be lower than before, which is in contrast to the experimental data. The polarity of the polymer after epoxidation increases drastically. Hence, additional interactions between the stationary phase (column) and polymer molecules have to be expected. In this particular case, a previously performed calibration of the SEC system with polybutadiene standards

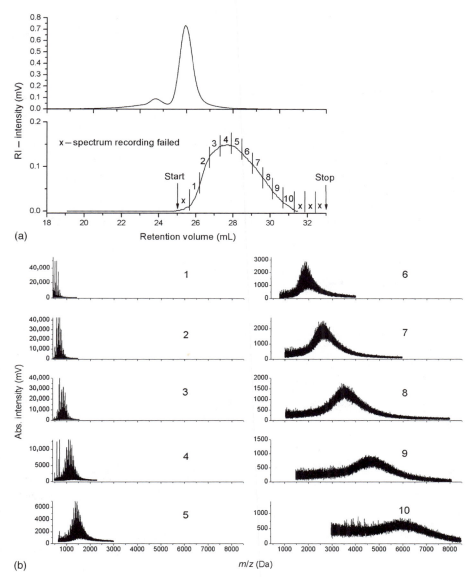

Figure 11.7. (a) GPC plots of PB 5000 before (top) and after epoxidation (bottom) (fraction marks are indicated in figure (b); reprinted with permission from Reference 45). (b) MALDI-TOF mass spectra of PB 5000 after epoxidation (number of spectrum corresponds to fraction number in Figure 11.7a, reprinted with permission from Reference 45).

is absolutely useless. The use of MALDI-TOF MS can be very helpful, because this method provides absolute molar masses and, in addition, the degree of epoxidation can be easily determined. Using the coupling interface as shown in Figure 11.6, the epoxidized sample was transferred in a MALDI-TOF mass spectrometer. The corresponding mass spectra are shown in Figure 11.7b.

The obtained masses of single fractions were used to create a MALDI calibration curve, which, directly compared to the original SEC calibration curve, clearly shows strong differences (see Figure 11.8).

Low-Molar Mass Block Copolymers (CCD)

The mass spectrum of an original ethylene oxid-propylene oxide (EO-PO) block copolymer is shown in Figure 11.9 [42]. The enlargement of the spectrum shows complex peaks, which can be attributed to various EO-PO series. Due to the short peak-to-peak distances of single series, it was impossible to determine the complete copolymer composition by MALDI solely. For that purpose, the principle of two-dimensional chromatography (see Figure 11.3) can be applied. In the first dimension, "critical" conditions of separation for one block are adjusted. Ideally, in the investigated mass range, the separation occurs according to the molar mass of the second

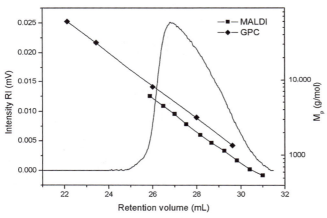

Figure 11.8. SEC chromatogram and calibration curves of PB 5000 after functionalization, calibration curves were obtained by SEC using PI standards as well as by using Mp data of SEC fractions determined by MALDI MS (from Figure 11.7b) (reprinted with permission from Reference 45).

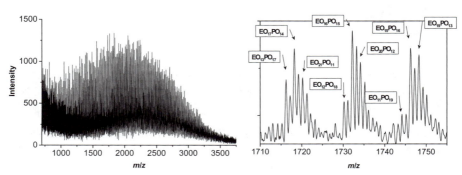

Figure 11.9. MALDI-TOF mass spectrum of EO-PO block copolymer (left) and enlargement of a region (right).

block. Additionally, fractions were taken and transferred into the second dimension (SEC), which provides molar masses. However, this principle is very time-consuming and, for some polymers, standards for a calibration of SEC are not available. Hence, MALDI-TOF MS was used to substitute the second dimension. For better resolution, the chromatographic conditions of the first dimension (reverse phase system) were slightly changed; that is, the separation mode was shifted from "critical" conditions of adsorption of PO toward the adsorption mode by changing the composition of the mobile phase. Due to the interaction between polymers and stationary phase, the peak recorded in "critical" mode (Figure 11.10a) became broader (Figure 11.10b) and could be fractionated (14 fractions). Finally, well-resolved MALDI-TOF mass spectra could be obtained (Figure 11.11) and were combined with chromatographic data in a two-dimensional plot (Figure 11.12), which simultaneously gives the number of EO and PO units.

Mixtures of Polymers with Different Topology (Topology Type Distribution [TTD])

A number of industrial polymers, especially products synthesised via polycondensation reactions, contain species of various topology, for example, cycles or stars beside linear products [62, 63]. Typically, the relative amount of these species is very low (1–3%), and their molar masses are below 10,000 Da. A routinely investi-

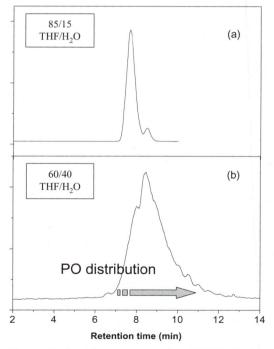

Figure 11.10. Chromatogram of an EO-PO block copolymer recorded in "critical" mode of chromatography (a) and in adsorption mode (b).

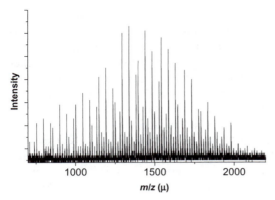

Figure 11.11.   MALDI-TOF mass spectrum of a fraction of an EO-PO block copolymer taken from adsorption chromatography.

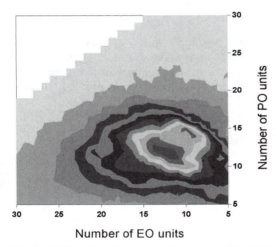

Figure 11.12.   Two-dimensional plot of an EO-PO block copolymer combining the information of adsorption chromatography and of 14 MALDI-TOF spectra. See color insert.

gation by means of SEC often gives no clue for their existence. Even though these species can be detected, the applied calibration is not exact. Typical examples are polyamides (PA) and polyethylene terephthalates (PET). The coupling of chromatography with MALDI-TOF MS can provide supplementary results. The SEC chromatogram of a PA 6 (Figure 11.13a) does not indicate an inhomogeneous sample, whereas MALDI-TOF mass spectra (Figure 11.13b) give a clear hint for the existence of additional cyclic structures. The chromatographic data, which are quantitative, and MALDI-TOF data continuously recorded from chromatographic fractions were combined in a two-dimensional plot, shown in Figure 11.14. In this plot the existence of cycles among linear polymers was clearly demonstrated.

For some specific applications (especially for polymers with comparatively low molar masses of <15,000 Da), this coupling method can be used to substitute the two-dimensional (orthogonal) analysis (shown in Figure 11.3).

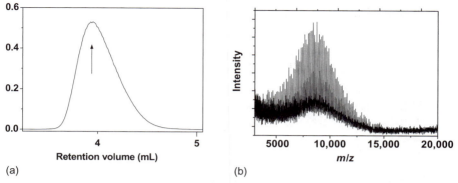

(a)   (b)

Figure 11.13.   (a) SEC elugram of a PA 6 (arrow indicates the elution volume where MALDI-TOF mass spectrum (b) was recorded.

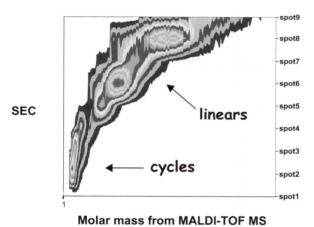

Figure 11.14.   Two-dimensional plot created using SEC data (Figure 11.13a) and nine MALDI-TOF mass spectra (mass spectra of spot 8 is exemplarily shown in Figure 11.13b). See color insert.

Mixtures of Polymers with Different End Groups (Functionality Type Distribution [FTD])

In Figure 11.15 the molar mass distribution curve of OH-terminated polydimethylsiloxane (OH-PDMS) after separation in SEC mode is shown [64]. Due to the relatively high resolution of the SEC columns in the lower mass region, the chromatogram shows single resolved peaks.

In the so-called "critical" mode of chromatography, a separation of polymers having identical structures but different end groups can be established. Polymers having identical monomer units *and identical* end groups elute at one retention time independently of their molar mass.

Polymers with identical monomer units *and different* end groups elute at different retention times. Adjusting conditions near the critical mode, one can observe various separation mechanisms in one chromatogram. This is exemplarily shown in

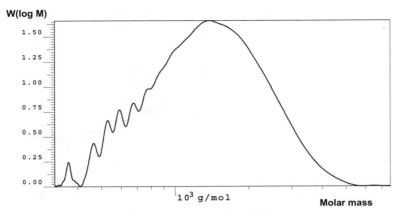

Figure 11.15. Molar mass distribution of OH-PDMS observed in SEC mode.

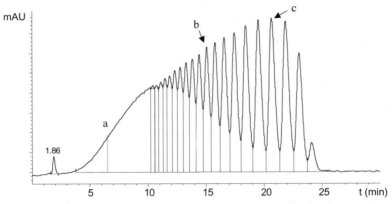

Figure 11.16. Chromatogram of OH-terminated polydimethylsiloxane observed at critical conditions for polydimethylsiloxane.

Figure 11.16. At critical conditions for PDMS, a peak at 1.86 min can be obtained, which can be attributed to PDMS with methyl end groups. According to the elution order of the OH-terminated species, the separation seems to occur by the size of the molecules, in SEC mode. Actually, these species elute in adsorption mode (LAC), showing a very good peak resolution of low-mass oligomers. This is due to the fact that the impact of interaction of the terminal OH groups with the stationary phase (silicagel) becomes much higher the shorter the OH-PDMS chain is.

The chromatographic run (Figure 11.16) was sprayed on a matrix-precoated MALDI-TOF target, and a set of mass spectra (Figure 11.17) was continuously recorded. In the low-molar mass region (corresponding to 15–25 min retention time), single mass peaks could be obtained that could easily be attributed to corresponding structures. But even in the higher mass range showing absolutely no resolution in chromatography, very good mass spectra were obtained. These MALDI data were used to create a calibration curve shown in Figure 11.18.

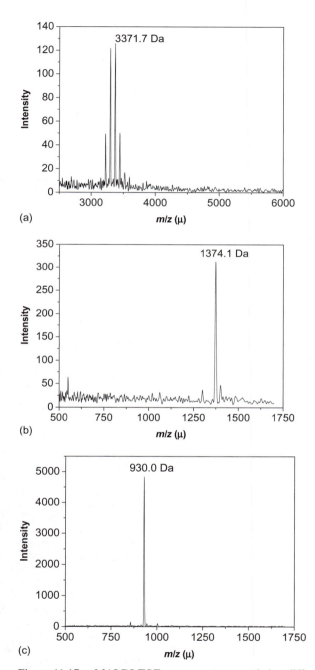

Figure 11.17.   MALDI-TOF mass spectra recorded at different positions (see Figure 11.16).

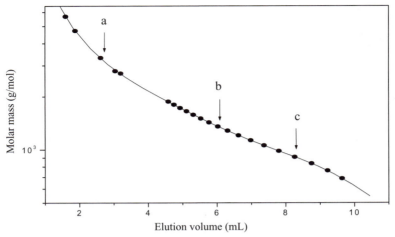

Figure 11.18.   Calibration curve of OH-PDMS by means of MALDI-TOF mass spectrometry.

11.4   CONCLUSIONS

The coupling of chromatography with MALDI-TOF MS represents an excellent tool for the investigation of heterogeneous polymers. Both methods can be regarded as complementary to each other. Limits of MALDI-TOF MS can be overcome by a previous chromatographic separation according to molar mass and/or heterogeneity of polymers. Simultaneously, MALDI-TOF MS can be used as a valuable detector for chromatography.

REFERENCES

1. Gorshkov, A.V.; Much, H.; Becker, H. et al., Chromatographic investigations of macromolecules in the critical range of liquid-chromatography: 1. functionality type and composition distribution in polyethylene oxide and polypropylene oxide copolymers, *Journal of Chromatography*, **1990**, 523, 91–102.
2. Schulz, G.; Much, H.; Krueger, H. et al., Determination of functionality and molecular-weight distribution by orthogonal chromatography, *Journal of Liquid Chromatography*, **1990**, 13(9), 1745–1763.
3. Pasch, H.; Much, H.; Schulz, G., Determination of polymer heterogeneity by two-dimensional orthogonal liquid chromatography, *Trends in Polymer Science*, **1993**, 3, 643.
4. Krüger, R.P.; Much, H.; Schulz, G., *International Journal of Polymer Analysis and Characterization*, **1996**, 2, 221.
5. Chang, T.Y.; Lee, H.; Lee, W. et al., Polymer characterization by temperature gradient interaction chromatography, *Macromolecular Chemistry and Physics*, **1999**, 200(10), 2188–2204.
6. Belenki, B.G.; Gankina, E.; Tennikov, N., et al. The chromatographic study of the basic regularities of macromolecules adsorption on the porous sorbents, *Doklady Akademii Nauk SSSR*, **1976**, 231, 1147.
7. Belenki, B.G.; Gankina, E.; Tennikov, N. et al., Fundamental aspects of adsorption chromatography of polymers and their experimental verification by thin layer chromatography, *Journal of Chromatography*, **1978**, 147, 99.

8. Falkenhagen, J., Kopplung von chromatographischen und spektroskopischen Methoden zur Bestimmung der Heterogenitäten von Polymeren. PhD Thesis, Berlin: Mensch & Buch Verlag, **1998**.

9. Pasch, H.; Trathnigg, B., *HPLC of Polymers*. Berlin: Springer, **1997**.

10. Ling, H.; Liang, L.H.; Lubman, D.M., Continuous-flow MALDI mass spectrometry using an ion trap/reflectron time-of-flight detector, *Analytical Chemistry*, **1995**, 67, 4127.

11. MacFadden, W.H.; Schwarz, H.L.; Evans, S.J., Direct analysis of liquid chromatographic effluents, *Journal of Chromatography*, **1976**(122), 389.

12. Murray, K.K.; He, L., 337 nm matrix-assisted laser desorption/ionization of single aerosol particles, *Analytical Chemistry News & Features*, **1999**, 518.

13. Murray, K.K.; Russell, D.H., Liquid sample introduction for matrix-assisted laser-desorption ionization, *Analytical Chemistry*, **1993**, 65(18), 2534–2537.

14. Beeson, M.D.; Murray, K.K.; Russell, D.H., Aerosol matrix-assisted laser-desorption ionization—effects of analyte concentration and matrix-to-analyte ratio, *Analytical Chemistry*, **1995**, 67(13), 1981–1986.

15. Murray, K.K.; Lewis, T.M.; Beeson, M.D. et al., Aerosol matrix-assisted laser-desorption ionization for liquid-chromatography time-of-flight mass-spectrometry, *Analytical Chemistry*, **1994**, 66(10), 1601–1609.

16. Fei, X.; Murray, K.K., On-line coupling of gel permeation chromatography with MALDI mass spectrometry, *Analytical Chemistry*, **1996**, 68(20), 3555–3560.

17. Fei, X.; Wei, G.; Murray, K.K., Aerosol MALDI with a reflectron time-of-flight mass spectrometer, *Analytical Chemistry*, **1996**, 68(7), 1143–1147.

18. Tsuda, T.; Yamamoto, K., Device to transform the liquid-chromatographic effluent to a continuous solid plug—a new approach to direct liquid introduction for liquid-chromatography mass-spectrometry, *Journal of Chromatography*, **1988**, 456(2), 363–369.

19. Williams, E.R.; Jones, G.C.; Fang, L.; Zare, R.N., Using tandem time-of-flight mass spectrometry, *Proceedings of the ASMS Conference on Mass Spectrometry*, **1991**, 39, 1273.

20. Li, L.; Wang, A.P.L.; Coulson, L.D., Continuous-flow matrix-assisted laser desorption ionization mass-spectrometry, *Analytical Chemistry*, **1993**, 65(4), 493–495.

21. Nagra, D.S.; Li, L., Liquid chromatography-time-of-flight mass-spectrometry with continuous-flow matrix-assisted laser-desorption ionization, *Journal of Chromatography A*, **1995**, 711(2), 235–245.

22. Laiko, V.V.; Baldwin, M.A.; Burlingame, A.L., Atmospheric pressure matrix-assisted laser desorption/ionization mass spectrometry, *Analytical Chemistry*, **2000**, 72(4): 652–657.

23. Daniel, J.M.; Ehala, S.; Fries, S.D. et al., On-line atmospheric pressure matrix-assisted laser desorption/ionization mass spectrometry, *Analyst*, **2004**, 129(7), 574–578.

24. Available at http://www.apmaldi.com/

25. Available at http://www.kranalytical.co.uk/mass-spec.html

26. Montaudo, G.; Montaudo, M.; Puglisi, C. et al., Molecular-weight distribution of poly(dimethyl-siloxane) by combining matrix-assisted laser-desorption ionization time-of-flight mass-spectrometry with gel-permeation chromatography fractionation, *Rapid Communications in Mass Spectrometry*, **1995**, 9(12), 1158–1163.

27. Pasch, H.; Rode, K., Use of matrix-assisted laser-desorption/ionization mass-spectrometry for molar mass-sensitive detection in liquid-chromatography of polymers, *Journal of Chromatography A*, **1995**, 699(1–2), 21–29.

28. Danis, P.O.; Saucy, A.; Huby, F.J., Applications of MALDI/TOF mass spectrometry coupled to gel permeation chromatography, *Abstracts of Papers of the American Chemical Society*, **1996**, 211, 368–POLY.

29. Nielen, M.W.F.; Malucha, S., Characterization of polydisperse synthetic polymers by size-exclusion chromatography matrix-assisted laser desorption/ionization time-of-flight mass spectrometry, *Rapid Communications in Mass Spectrometry*, **1997**, 11(11), 1194–1204.

30. Wachsen, O.; Reichert, K.H.; Krueger, R.P. et al., Thermal decomposition of biodegradable polyesters: 3. studies on the mechanisms of thermal degradation of oligo-L-lactide using SEC, LACCC and MALDI-TOF-MS, *Polymer Degradation and Stability*, **1997**, 55(2), 225–231.

31. Kruger, R.P.; Much, H.; Schulz, G. et al., New aspects of determination of polymer heterogeneity by 2-dimensional orthogonal liquid chromatography and MALDI-TOF-MS, *Macromolecular Symposia*, **1996**, 110, 155–176.

32. Available at http://www.lcpackings.com/products/Instruments/Probot01.html

33. Available at http://www.kromatek.co.uk/MALDI_Fraction_Collector_40.asp

34. Available at http://www.gilson.com/Products/product.asp?pID=116

35. Available at http://www.sunchrom.de/pdf/SunChrom%20SunCollect.pdf

36. Keil, O.; LaRiche, T.; Deppe, H. et al., Hyphenation of capillary high-performance liquid chromatography with matrix-assisted laser desorption/ionization time-of-flight mass spectrometry for nano-scale screening of single-bead combinatorial libraries, *Rapid Communications in Mass Spectrometry*, **2002**, 16(8), 814–820.

37. Nielen, M.W.F., Polymer analysis by micro-scale size exclusion chromatography MALDI time-of-flight mass spectrometry with a robotic interface, *Analytical Chemistry*, **1998**, 70(8), 1563–1568.

38. Grimm, R., Capillary LC with automated on-line microfraction collection onto MALDI/TOF MS targets, *International Biotechnology Laboratory*, **1997**, 58, 17–21.

39. Available from: http://www.labconnections.com/products.htm

40. Krüger, R.-P.; Falkenhagen, J.; Schulz, G.; Gloede, J., Charakterisierung von Calixarenen durch Kopplung von Flüssigkeitschromatographie und Maldi-TOF-MS, *GIT Fachzeitschrift fuer Labor*, **2001**, 45, 380.

41. Montag, P.; Falkenhagen, J.; Krueger, R.P., Strukturanalyse in komplexen Polymersystemen, *CLB Chemie in Labor und Biotechnik*, **1999**, 50, 253–257.

42. Falkenhagen, J.; Schulz, G.; Weidner, S. et al., Liquid adsorption chromatography near critical conditions of adsorption coupled with matrix-assisted laser desorption/ionization mass spectrometry, *International Journal of Polymer Analysis and Characterization*, **2000**, 5(4–6), 549–562.

43. Falkenhagen, J.; Weidner, S., Detection limits of matrix-assisted laser desorption/ionisation mass spectrometry coupled to chromatography—a new application of solvent-free sample preparation, *Rapid Communication in Mass Spectrometry*, **2005**, 19, 3724–3730.

44. Maziarz, E.P.; Liu, M., A modified thermal deposition unit for gel-permeation chromatography with matrix-assisted laser desorption/ionization and electrospray ionization time-of-flight analysis, *European Journal of Mass Spectrometry*, **2002**, 8(5), 397–401.

45. Kona, B.; Weidner, S.; and J.F. Friedrich, Epoxidation of polydienes investigated by MALDI-TOF mass spectrometry and GPC-MALDI coupling, *International Journal of Polymer Analysis and Characterization*, **2005**, 10(1–2), 85–108.

46. Zhang, B.Y.; McDonald, C.; Li, L., Combining liquid chromatography with MALDI mass spectrometry using a heated droplet interface, *Analytical Chemistry*, **2004**, 76(4), 992–1001.

47. Young, J.B.; Li, L., Impulse-driven heated-droplet deposition interface for capillary and microbore LC-MALDI MS and MS/MS, *Analytical Chemistry*, **2007**, 79(15), 5927–5934.

48. Pasch, H.; Schrepp, W., *MALDI-TOF Mass Spectrometry of Polymers*. Berlin, Heidelberg and New York: Springer, **2003**.

49. Montaudo, G.; Lattimer, R.P., *Mass Spectrometry of Polymers*. Boca Raton, FL, London, New York, and Washington, D.C.: CRC Press, **2002**.

50. Rader, H.J.; Spikermann, J.; Kreyenschmidt, M. et al., MALDI-TOF mass spectrometry in polymer analytics: 2. molecular weight analysis of rigid-rod polymers, *Macromolecular Chemistry and Physics*, **1996**, 197(10), 3285–3296.

51. Belu, A.M.; DeSimone, J.M.; Linton, R.W. et al., Evaluation of matrix-assisted laser desorption ionization mass spectrometry for polymer characterization, *Journal of the American Society for Mass Spectrometry*, **1996**, 7(1), 11–24.

52. Tatro, S.R.; Baker, G.R.; Fleming, R. et al., Matrix-assisted laser desorption/ionization (MALDI) mass spectrometry: determining Mark-Houwink-Sakurada parameters and analyzing the breadth of polymer molecular weight distributions, *Polymer*, **2002**, 43(8), 2329–2335.

53. Montaudo, M.S., Full copolymer characterization by SEC-NMR combined with SEC-MALDI, *Polymer*, **2002**, 43(5), 1587–1597.

54. Maziarz, E.P.; Liu, X.M.; Quinn E.T. et al., Detailed analysis of alpha,omega-bis(4-hydroxybutyl) poly(dimethylsiloxane) using GPC-MALDI TOF mass spectrometry, *Journal of the American Society for Mass Spectrometry*, **2002**, 13(2), 170–176.

55. Esser, E.; Keil, C.; Braun, D. et al., Matrix-assisted laser desorption/ionization mass spectrometry of synthetic polymers: 4. coupling of size exclusion chromatography and MALDI-TOF using a spray-deposition interface, *Polymer*, **2000**, 41(11), 4039–4046.

56. Montaudo, M.S., Puglisi, C.; Samperi, F. et al., Application of size exclusion chromatography matrix-assisted laser desorption/ionization time-of-flight to the determination of molecular masses in polydisperse polymers, *Rapid Communications in Mass Spectrometry*, **1998**, 12(9), 519–528.

57. Kruger, R.P.; Much, H.; Schulz, G., Determination of functionality and molar-mass distribution of aliphatic polyesters by orthogonal liquid-chromatography: 1. off-line investigation of poly(1,6-hexanediol adipates), *Journal of Liquid Chromatography*, **1994**, 17(14–15), 3069–3090.

58. Keil, C.; Esser, E.; Pasch, H., Matrix-assisted laser desorption/ionization mass spectrometry of synthetic polymers: 5. analysis of poly(propylene oxide)s by coupled liquid chromatography at the critical point of adsorption and MALDI-TOF mass spectrometry, *Macromolecular Materials and Engineering*, **2001**, 286(3), 161–167.

59. Trathnigg, B., Maier, B.; Schulz, G. et al., Characterization of polyethers using different methods: SFC, LAC and SEC with different detectors, and MALDI-TOF-MS, *Macromolecular Symposia*, **1996**, 110, 231–240.

60. Mengerink, Y.; Peters, R.; deKoster, C.G. et al., Separation and quantification of the linear and cyclic structures of polyamide-6 at the critical point of adsorption, *Journal of Chromatography A*, **2001**, 914(1–2), 131–145.

61. Pasch, H.; Deffieux, A.; Ghahary, R. et al., Analysis of macrocyclic polystyrenes: 2. mass spectrometric investigations, *Macromolecules*, **1997**, 30(1), 98–104.

62. Weidner, S.; Kuhn, G.; Just, U., Characterization of oligomers in poly(ethylene-terephthalate) by matrix-assisted laser-desorption ionization mass-spectrometry, *Rapid Communications in Mass Spectrometry*, **1995**, 9(8), 697–702.

63. Weidner, S.M.; Just, U.; Wittke, W. et al., Analysis of modified polyamide 6.6 using coupled liquid chromatography and MALDI-TOF-mass spectrometry, *International Journal of Mass Spectrometry*, **2004**, 238(3), 235–244.

64. Falkenhagen, J.; Kruger, R.-P.; Schulz, G., Characterization of silicon-containing polymers by coupling of HPLC-separation methods with MALDI-TOF mass spectrometry. In *Silicon Chemistry— From Small Molecules to Extended Systems* (Jutzi, P.; Schubert, U., editors). Wiley-VCH: Weinheim, **2003**, pp. 406–418.

CHAPTER 12

MALDI MS APPLICATIONS FOR INDUSTRIAL POLYMERS

Scott D. Hanton[1] and Kevin G. Owens[2]

[1]Air Products and Chemicals, Inc., Allentown, PA, and [2]Department of Chemistry, Drexel University, Philadelphia, PA

12.1 INTRODUCTION

With the development of "soft" ionization sources, mass spectrometry techniques could be applied to characterizing the chemical structures of large, nonvolatile molecules such as synthetic polymers. Each successive development from plasma desorption, field desorption (FD), fast atom bombardment (FAB), secondary ion mass spectrometry (SIMS), to electrospray ionization (ESI) and matrix-assisted laser desorption/ionization (MALDI) enabled larger molecules to be analyzed and more complex problems to be solved. In this chapter we will concentrate on the application of MALDI to solve a variety of problems related to synthetic polymers.

Very early in the development of the MALDI technique, Tanaka et al. recognized the applications to the oligomer distributions that compose polymer samples [1]. After the introduction of MALDI, early researchers explored how to apply the new technique to address a number of polymer chemical structure questions [2–8]. While MALDI is most often used today to explore the chemical structures of biomolecules, there exists a rich literature of applications of MALDI to synthetic polymers. A few recent review articles and books provide an effective gateway to this literature [9–13].

12.2 PRIMER ON INDUSTRIAL POLYMERS

Synthetic polymers are produced by the chemical industry in many varieties and in huge volumes. Polymers are used commercially to make many everyday items, including things like plastics, synthetic fibers, surfactants, paints, glues, and coatings. Consumers purchase these commercial materials for their performance in specific applications—what the material can do (Figure 12.1). The performance is

MALDI Mass Spectrometry for Synthetic Polymer Analysis, Edited by Liang Li

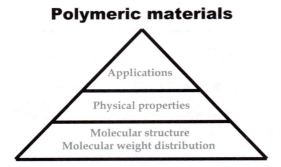

Polymeric materials

End₁ – (Monomer1)ₙ – (Monomer2)ₘ – End₂

Figure 12.1.   Schematic showing the relationship of the applications depending on the physical properties which in turn depend on the underlying molecular structure of the polymer. Also showing a schematic of a simple polymer molecule.

delivered from the physical properties of the polymer. The physical properties are dependent on the underlying chemical structure of the polymer.

Synthetic polymers are a class of materials that are created by sequential reaction of one or more small molecule(s), called monomer(s), to form a chain of repeating functions (Figure 12.1). This chain is often terminated at both ends with different atoms or molecules called the end groups. If there is a single monomer, the material is known as a homopolymer. If there are two or more monomers forming the chain, the material is known as a copolymer because the different monomers are co-polymerized. Due to the sequential reaction synthesis, all polymers are composed of a distribution of different chain lengths. Each individual chain length is called an oligomer, or for low-mass polymers, a telomer.

The chemical structure of a polymer is defined by the chemical species of the end groups and the repeat group(s), and the molecular weight distribution of the different chain lengths. Polymer chemists need the molecular structure and molecular weight distribution of the polymer to design and synthesize new polymers, and troubleshoot polymer problems. MALDI has proven to be an effective analytical tool contributing to the determination of polymer chemical structure, especially when combined with more traditional analytical tools for polymer characterization, such as nuclear magnetic resonance (NMR) or infrared (IR) spectroscopy for molecular structure information, and size exclusion chromatography (SEC) for molecular weight distribution data.

12.3   PRIMER ON MALDI

The MALDI experiment is a rather easy experiment to complete successfully, but one that has rather complex mechanisms occurring beyond the direct observation of the analyst. Many important details of the MALDI experiment are discussed in other chapters of this book [14]. The key elements of the MALDI experiment are

- choice of an effective matrix for the analyte [15–18],
- choice of an effective cationization agent for the analyte [18],
- effective mixing of the matrix and the analyte (including choice of solvent, if solvent is used),
- effective deposition of the matrix and analyte on the target,
- absorption of laser energy by the matrix,
- desorption of the analyte from the target surface,
- cationization of the analyte; and
- mass separation and detection of the analyte ions.

Each of these elements must be successful to obtain an analytically useful MALDI mass spectrum. Most of the time in MALDI, the instrument works well, and the problems in obtaining good mass spectra can be solved with improved sample preparation techniques at the laboratory bench.

While a variety of different mass spectrometers are used today for MALDI experiments, most practitioners use time-of-flight mass spectrometers (TOFMS). The TOFMS provides the performance required for the analysis of a wide variety of synthetic polymers, such as high mass range, sufficient mass resolution and mass accuracy, and the multiplex advantage. TOFMS instruments are also widely available and affordable. The key disadvantages of the TOFMS are a limited intensity dynamic range (generally only 256) and limited tandem mass spectrometry capability (via post-source decay [PSD]). An increase in the dynamic range of the data system of the instrument would yield significant new advantages for polymer MALDI [19].

12.4 MALDI SAMPLE PREPARATION FOR POLYMERS

There are two main methods of MALDI sample preparation, liquid phase (with solvent) and solid phase (solvent-free). These sample preparation methods are discussed in much more detail in Chapters 6 and 7 of this book, respectively. In the solvent sample preparation methods, a single, good solvent is chosen for the analyte, matrix, and any cationization agent. A typical solvent method for a relatively low mass (~5000 Da) polymer would be to generate a 5 mg/mL-solution of the analyte and a 0.25-M solution of the matrix. The solutions are then combined 10:100 by volume. For many low-mass polymer samples, a molar matrix to analyte ratio (M/A) of between 1000–5000 is often effective. If the polymer will effectively cationize with H^+, almost all MALDI matrices will provide it. If the polymer will effectively cationize with Na^+, many matrices (such as 2,5-dihydroxy benzoic acid [DHB]) also provide this cation directly, as they are contaminated with Na from the synthetic process. Alternatively, the sample can be prepared in a soft glass vial. Sufficient Na for the experiment will be leached out of the soft glass (not borosilicate glass) by most solvents. If the polymer requires Ag^+ cationization, then a 5-mg/mL solution of a soluble silver salt, for example, silver trifluoroacetate (AgTFA), can be added at the same volume as the analyte.

Once the combined solution is prepared, the sample is deposited on the target plate. The simplest deposition method is to deposit about 0.2 μL of solution and allow it to air dry. To achieve improved quantitation, the solution can be electro-sprayed on the target [20–21].

In the solvent-free MALDI sample preparation method, the matrix, analyte, and any necessary cationization agent are combined through grinding with no solvent present (see Chapter 7 of this book) [22]. Solvent-free methods have been demonstrated for a variety of different polymers. While the mechanisms of the mixing are still not well understood, the method performs very well. Solvent-free sample preparation methods have been shown to provide mass spectra with improved resolution, increased signal-to-noise, and reduced background than comparable solvent sample preparations [22]. One simple, solvent-free method involves combining about 0.2 mg of analyte and about 40 mg of matrix in a small glass vial, adding two standard air rifle BBs, and vortexing for about 60 s [23]. A small amount of the resulting powder is applied to a target plate with a small metal spatula to form a thin film. This method has been shown to be effective with a wide variety of materials, including some crystalline polymers, waxes, liquid samples, emulsified samples, and even solid residues of polymer samples on a laboratory wipe [23].

Solvent-free methods are particularly useful in solving problems of two different types. The most important use of solvent-free methods is with insoluble analytes. Prior to the development of solvent-free sample preparation methods, MALDI was limited to soluble analytes. The development of the solvent-free methods has enabled MALDI analysis of many insoluble polymers. Figure 12.2 shows a MALDI mass spectrum obtained from an insoluble polymer that was created as a 1% solids emulsion in water. The emulsified materials consisted of the low-mass insoluble polymer of interest and a much higher-mass polymer used as diluent.

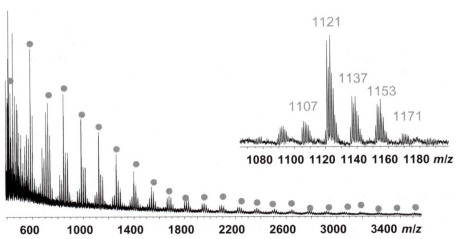

Figure 12.2. MALDI mass spectrum obtained on an insoluble polymer that was one part of a 1% solids blend emulsified in water. This component is insoluble in all solvents attempted. The sample was prepared with the BB solvent-free method.

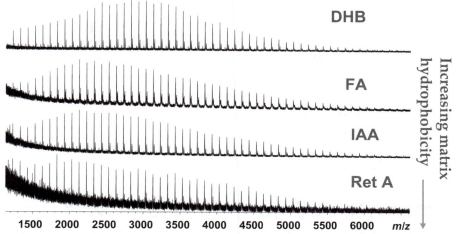

Figure 12.3.   MALDI mass spectra of polymethylmethacrylate 4k obtained from solvent-free sample preparation using different matrices having very different liquid solubilities.

In Figure 12.2 we can observe a very interesting repeat structure. We see a single monomer unit with several different end groups. The MALDI analysis was critical in determining the chemical structure of this novel advanced material. Insoluble materials provide a significant challenge for most analytical labs. Most of the typical analytical tools require a soluble sample (liquid chromatography [LC], LC/MS, and liquids NMR), or a significant amount of analyte (solids NMR). MALDI achieved this analysis using only about 0.1 mg of material in the sample preparation. While we did not measure the amount of sample consumed in the actual analysis, MALDI is extremely sensitive, and only a very small fraction of the analyte used in the sample preparation created the observed ions, probably on the order of femtomoles.

Another important use of the solvent-free sample preparation methods is to use matrices that are not a good solubility match for the analyte (for more information on this topic, see Chapter 6 of this book). Even though the analyte and the matrix may be soluble, they do not share a good solvent. Figure 12.3 shows example mass spectra of polymethylmethacrylate (PMMA) 4k prepared with different matrices spanning a range of liquid solubilities.

The data shown in Figure 12.3 demonstrates the utility of the solvent-free sample preparation to solve problems encountered in trying to match a specific analyte with a specific matrix.

12.5   MALDI OF A SIMPLE HOMOPOLYMER

Figure 12.4 shows the MALDI mass spectrum obtained from a solvent sample preparation of polystyrene (PS) 4k sample prepared with retinoic acid as the matrix, tetrahydrofuran (THF) as the solvent, and silver trifluoroacetate (AgTFA) as the cationization reagent.

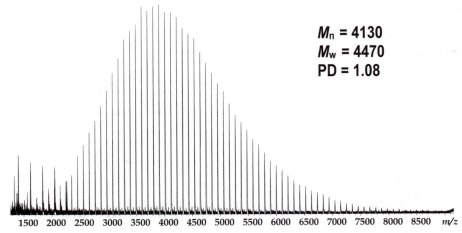

M_n = 4130
M_w = 4470
PD = 1.08

Figure 12.4.   MALDI mass spectrum of polystyrene 4k obtained using retinoic acid as the matrix, tetrahydrofuran as the solvent, and silver trifluoroacetate as the cationization agent.

In the mass spectrum we can clearly see the distribution of oligomers that define this polymer sample. From the masses of adjacent ion peaks, we can measure the mass of the repeat unit as 104 Da, the correct mass for the PS monomer unit. From the residual mass, we can determine that the mass of the cation + the mass of the end groups is 165 Da, which corresponds to one $(CH_3)C-$ and one $-H$ end group, plus a Ag^+ cation.

Interestingly, not all of the repeating masses in Figure 12.4 have spacing of 104 Da. The peaks at the low-mass end of the PS distribution that are observed most clearly, between 1400 and 2300 Da, have a spacing of about 108 Da. These ions are not PS, but are silver clusters from the cationization agent [24]. The intensity of these clusters is related to the laser fluence used in the analysis and the amount of cationization agent used in the sample preparation. These ions illustrate the necessity of taking the time to understand the details of the mass spectrum. Simply observing a repeating set of ions does not ensure that they represent the polymer analyte.

12.6   MOLECULAR WEIGHT DISTRIBUTION MEASUREMENTS

Since MALDI produces almost all intact molecular ions, the distribution of observed ions is a good measure of the polymer molecular weight distribution. If the MALDI mass spectrum contains any significant intensity of fragment ions, then it is no longer assured that the distribution of observed ions represents the actual polymer molecular weight distribution. For most typical synthetic polymers, fragmentation is not a significant issue.

The average molecular weights of the polymer samples can be determined directly from the peak areas observed in the MALDI mass spectra. Chapter 8 of this book provides much more detailed information on this topic. The standard average molecular weight measures used by polymer chemists are the first two moments of the distribution, the number-average molecular weight, M_n, and the weight-average molecular weight, M_w. The polydispersity [PD] is the ratio of M_w/M_n. These equations are used to calculate the average molecular weights:

$$M_n = \Sigma M_i N_i / \Sigma N_i \qquad \text{(Eq. 12.1)}$$

$$M_w = \Sigma (M_i)^2 N_i / \Sigma M_i N_i \qquad \text{(Eq. 12.2)}$$

$$\text{PD} = M_w/M_n \qquad \text{(Eq. 12.3)}$$

where M_i is the mass of oligomer i, and N_i is the peak area of oligomer i in the mass spectrum.

Using these equations, we can measure the average molecular weights of the PS sample analyzed in Figure 12.4. The average molecular weights are $M_n = 4130\,Da$ and $M_w = 4470\,Da$ with PD = 1.08. Excellent precision can be obtained in measurements of average molecular weights by MALDI, often less than 1%. This precision can be further improved with the use of electrospray deposition in the sample preparation [21].

For many relatively low-mass polymers with narrow PD, MALDI average molecular weight measurements have been shown to be accurate compared to the traditional SEC measurements [25–27]. The capability of MALDI to measure accurate average molecular weights for polymer samples is inversely related to the PD of the sample. For most polymer samples with a PD $\leq$ 1.25, MALDI will provide highly accurate average molecular weights. For samples with PD between 1.25–1.50, MALDI average molecular weights will often be analytically useful, but there will be some error in the results. For samples with PD between 1.50–1.75, MALDI will provide useful measures of M_n, but the M_w values will contain significant errors. For samples with PD greater than 1.75, MALDI may provide some interesting mass spectra, but the average molecular weight values will both have significant errors.

Many commercial synthetic polymers are part of related product families. Sometimes the relationship involves the same end groups and repeat units, but changes in the degree of polymerization or molecular weight of the materials. By determining the average molecular weight, the different family members can be identified and distinguished. Surfynol® (Air Products and Chemicals, Inc., Allentown, PA) surfactants have differing amounts of ethoxylation on a backbone of 2,4,7,9-tetramethyl-5-decyne-4,7-diol (Surfynol 104). The performance of these materials as surfactants depend on the degree of ethoxylation. Surfynol surfactants offer several advantages as surfactants: rapid migration, low dynamic surface tension, defoaming capability, FDA compliance, and stability in high electrolyte brine systems. Figure 12.5 shows MALDI mass spectra obtained for some Surfynol surfactants [28].

The mass spectra in Figure 12.5 show increasing average molecular mass for this series of materials. While these data were collected on a low-mass resolution instrument (prior to the development of delayed extraction), the average molecular

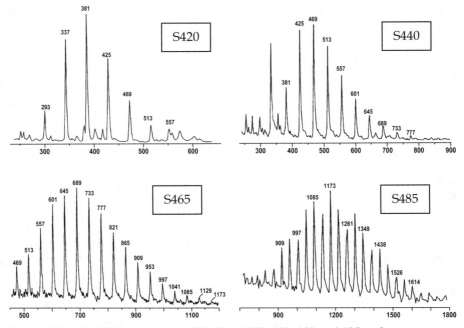

Figure 12.5. MALDI mass spectra of Surfynol 420, 440, 465, and 485 surfactants showing the average molecular mass increase in this product family.

TABLE 12.1. Average Molecular Weights of Four Surfynol Surfactants as Measured by MALDI-MS

Sample	M_N	M_W	PD
S420	370	380	1.03
S440	490	510	1.04
S465	680	720	1.04
S485	1200	1240	1.03

weights are still easily calculated from the mass spectra. Table 12.1 shows the average molecular weights calculated from the MALDI mass spectra:

These data also demonstrate a limitation of MALDI in determining average molecular weights for low-mass polymer samples. Short chain length telomers of oxygen-functional repeat units may not have sufficient cation stability to be detected quantitatively by MALDI [29]. We find that quantitative detection by metal cationization requires about four repeat units. Although telomers with fewer than four repeat units may be observed, the average molecular weights will err high. Using multiple methods, including gas chromatography mass spectrometry (GC/MS), the

S420 average molecular weights were shown to be $M_n = 280\,Da$ and $M_w = 290\,Da$ [28].

One additional challenge in characterizing low-mass polymer samples is interference from low-mass background ions. A variety of low-mass background ions are observed from different matrix ions and matrix cluster ions. By taking a background spectrum, the interfering ions can be documented and excluded from calculations on the low-mass polymer analytes.

12.7 MOLECULAR STRUCTURE—REPEAT UNITS AND END GROUPS

As observed above in Figure 12.4, the MALDI data can readily provide the mass of the repeat unit and the end groups. The mass of the repeat unit is often sufficient to characterize the monomer used to synthesize the material. The list of typical monomers is finite, and there are few monomers with the same nominal mass. The end groups, however, can have significantly more chemical diversity, and the mass alone is not usually sufficient to identify the end groups. Usually more information, either from the polymer chemistry, application, or analytical data from techniques like NMR or IR combined with the mass data can provide definitive characterization of the end groups.

If some chemical information is known about the polymer, for example, that it is an alkyl phenol ethoxylate, then the mass data will quickly identify which alkyl phenol is the end group. Figure 12.6 shows expansions of two MALDI mass spectra of common alkyl phenol ethoxylates [27].

In the top mass spectrum of Figure 12.6, we observe ions assigned as octyl phenol ethoxylates cationized with Na^+ at 757 and 801 Da. For each of those telomers we also see ions 16 Da higher in mass which are assigned as the same telomer cationized with K^+. In the bottom mass spectrum, we observe ions assigned as nonyl phenol ethoxylates cationized with Na^+ at 771 and 815 Da. The ion at 787 is a K^+ cationized telomer. From these mass spectra, the mass difference between an octylphenol and a nonylphenol end group is readily apparent.

Accurate mass information can be obtained from MALDI experiments. Some of the more powerful TOFMS instruments are capable of mass accuracy of less than 5 ppm. There are times when accurate mass measurements are useful in distinguishing different repeat units or end-group possibilities. Figure 12.7 shows an expansion of a MALDI mass spectrum obtained from a blend of two polyethers, a polytetramethylene glycol (PTMEG) and an ethoxylated surfactant.

Using the PTMEG ions as an internal calibrant yields an average mass error in the ethoxylate ions of 5 ppm [30]. Using the ethoxylate ions as an external mass calibration yields an average mass error in the PTMEG ions of 14 ppm. While it is possible to measure the individual oligomers with high mass accuracy, once we establish the assignment of the repeat unit, its mass is then known. As we calculate the residual mass for the polymer, all of the mass error accumulates in the residual

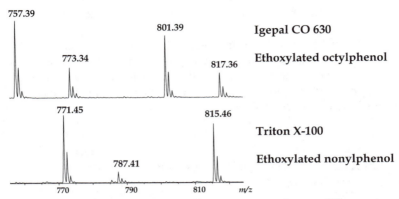

Figure 12.6. Small segments of MALDI mass spectra for two different ethoxylated surfactants. Mass calculations show that the upper spectrum is an ethoxylated octylphenol and the lower spectrum is an ethoxylated nonylphenol. MALDI can easily differentiate these materials. Each spectrum shows both Na- and K-cationization which results in ions spaced by 16 Da. Used with permission from *JCT Coatings Technology* from Reference 27.

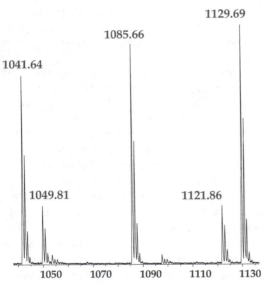

Figure 12.7. Small segment of a MALDI mass spectrum for a blend of two polyethers, polytetramethylene glycol (PTMEG), and an ethoxylated surfactant. An internal calibration on the PTMEG yields an average 5 ppm mass error and an external calibration on the ethoxylate yields an average 14 ppm mass error.

mass. The residual mass is usually much smaller than the mass of an individual oligomer. This often leads to mass error of hundreds of parts per million in the residual mass. This level of mass error can still be useful to reduce the number of possibilities for the identity of the end groups, but unfortunately is not sufficient to uniquely determine the identity of the end groups.

12.8 MOLECULAR STRUCTURE—NOVEL SYNTHESIS

MALDI can effectively contribute to the chemical characterization of novel materials. Despite the chemical diversity of existing commercial synthetic polymers, polymer scientists still synthesize novel polymers. Figure 12.8 shows the MALDI mass spectrum of a novel polyol polymer developed for use as a coatings material [27].

This polymer is made using a novel synthesis involving cationic polymerization and an emulsified epoxy resin [31]. The MALDI analysis was part of an integrated characterization involving IR, NMR, and SEC. In Figure 12.8, we see clusters of ions separated by 358 Da, the expected repeat mass corresponding to a bisphenol—A diglycidyl ether (BADGE) resin + water. The average molecular weights calculated from this spectrum are M_n = 2100 Da and M_w = 2900 Da with a PD = 1.4. The expansion region of Figure 12.8 shows one repeat region with peak labels for many of the significant ions. Figure 12.9 shows the chemical structures assigned to the main series of peaks observed in Figure 12.8.

The four main series correspond to three different end groups (A–C) and an extra, partial repeat unit (D). The main product corresponds to a copolymerization of bisphenol A diglycidyl ether (BADGE) + water with glycol end groups (A). The series labeled B and C are assigned with defects in the end groups. The B series has

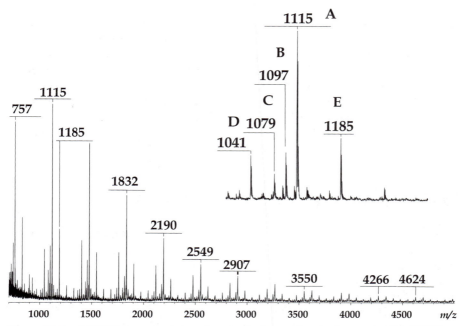

Figure 12.8. MALDI mass spectrum of a novel polymer showing a repeat unit of 358 Da. The expansion shows the distribution of different species observed. Most of the labeled ions have been assigned. See Figure 12.9 for the chemical structures of the assignments.

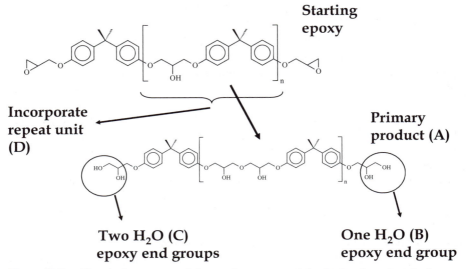

Figure 12.9.   Chemical structures of the starting epoxy and the final polymeric polyol, including the various end groups observed in the MALDI mass spectrum. Common impurities in the polyol are species retaining one or two epoxy end groups. Used with permission from *JCT Coatings Technology* from Reference 27.

one epoxy end group, and the C series has two epoxy end groups. The series labeled D in Figure 12.8 has an extra epoxy unit from the starting BADGE material. These additional partial repeat units are common in epoxy chemistry. While the MALDI data cannot produce accurate absolute quantitation of the amounts of the different polymers in mixtures, the relative amounts of these species can be important in identifying different commercial products (please see discussion of blend quantitation issues below). The MALDI data provided important chemical structure information for this novel material. The repeat units and end groups were verified and the average molecular weights were measured.

12.9   MOLECULAR STRUCTURE—COPOLYMERS

Copolymers are polymers composed of more than one type of repeat unit. Copolymers present additional challenges for characterization by MALDI. First, the mass spectra can become substantially more complex due to the different oligomers that can be constructed from the different repeat units. Figure 12.10 shows a MALDI mass spectrum of a novel surfactant that is composed of two repeat units and three end groups. Overall we can assign seven different series in this spectrum. This is an example of the complexity that can regularly challenge MALDI, even at a relatively low molecular mass.

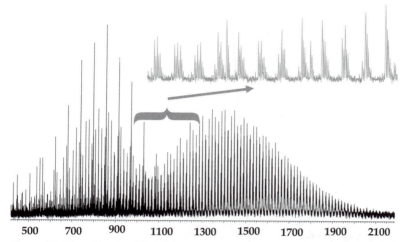

500   700   900   1100   1300   1500   1700   1900   2100

Figure 12.10.   Complex MALDI mass spectrum obtained from a novel surfactant containing two repeat units and three end groups.

While the chemical characterization of copolymers can be decidedly more complex than the analysis of homopolymers, MALDI analysis can contribute significantly to the initial chemical characterization of novel copolymer syntheses. Figure 12.11 shows the MALDI mass spectrum of a novel polymer resulting from the copolymerization of N-ethylaziridine and carbon monoxide [32]. These polymers are being developed for biomimetic applications. A full chemical characterization was needed to verify the synthesis and optimize the catalyst.

The mass spectrum in Figure 12.11 was obtained with a solvent sample preparation using methanol as the solvent and DHB as the matrix [16]. The mass spectrum confirms the expected repeat units and end groups in the polymer. When combined with NMR and SEC data, a full chemical characterization of this material was realized.

12.10   DETERMINATION OF MINOR COMPONENTS

Most of the time, MALDI experiments are used to characterize the majority species in a material. MALDI excels as a bulk analysis tool. In some circumstances, however, MALDI can take advantage of its inherently high sensitivity to provide needed data about a minority species in a particular material. One specific example is the analysis of defoamer species in commercial polyvinyl alcohol (PVOH) products. One of the applications of PVOH is in the manufacture of cloth. In this application the formation of bubbles and foam results in negative performance. To reduce foam, different kinds of molecular defoamers are added at low levels to the PVOH product. Many of these defoamers contain low- to medium-mass polyols which are quite

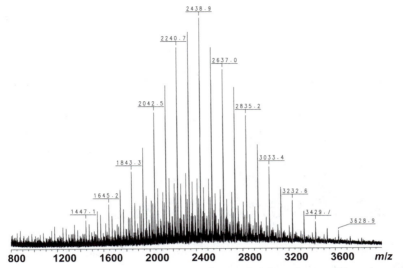

Figure 12.11. MALDI mass spectrum of a novel polymeric material produced from the copolymerization of N-ethylaziridine with CO. The sample was prepared using methanol and DHB. Used with permission from the *Journal of the American Society for Mass Spectrometry* from Reference 16.

amenable to MALDI analysis. The key to these experiments is the difficulty of obtaining MALDI mass spectra for the bulk component—PVOH [33]. Since PVOH responds very poorly in the MALDI experiment, using the sample preparation designed for the analysis of polyols, the molecular defoamers can be easily detected and characterized despite being minor components. Figure 12.12 shows the MALDI mass spectrum of a typical molecular defoamer used with PVOH.

In this example, the defoamer is composed of two different polypropylene glycol (PPG) polymers, one with a molecular mass around 2000 Da and another with a molecular mass around 5000 Da. Once the chemical structure of the defoamer is characterized, the structure/property relationships of various defoamers can be determined.

The combination of chromatography techniques with MALDI was discussed in detail in Chapter 11 of this book. One of the key advantages of using chromatography to generate samples for interrogation by MALDI is to separate minor components from major components to enable the characterization of the minor components. Here are two examples using SEC to size-separate a minor component from a major component. In both cases, detection and/or characterization of the minor component was severely compromised by the intensity of the ions from the major component. For these SEC experiments, a liquid chromatography transform (LCT) was used to capture the eluant from the SEC experiment, preserving the chromatographic resolution of the SEC experiment [34]. In these LCT experiments, the eluant was pneumatically sprayed from a heated nozzle onto a plate prepared with a specific MALDI matrix.

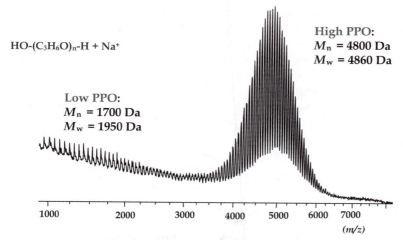

HO-(C₃H₆O)ₙ-H + Na⁺

High PPO:
M_n = 4800 Da
M_w = 4860 Da

Low PPO:
M_n = 1700 Da
M_w = 1950 Da

(*m/z*)

Figure 12.12.   MALDI mass spectrum of a molecular defoamer composed of a blend of two different polypropylene glycol polymers.

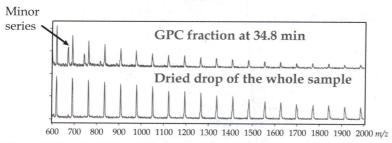

Minor series

GPC fraction at 34.8 min

Dried drop of the whole sample

Figure 12.13.   Partial MALDI mass spectra obtained from PTMEG 1000 samples. Top: Analysis of the position on the LCT track corresponding to the SEC elution time of 34.8 min. Bottom: Analysis of the whole sample without chromatographic separation. Used with permission from *Analytical Chemistry* from Reference 35.

SEC can readily separate the oligomers of PTMEG 1000 and those separated samples can be readily mass analyzed by MALDI [35]. These experiments can easily show the breadth of oligomer chain lengths present in PTMEG 1000, enable the accurate calibration of the SEC for PTMEG, and enable the characterization of a minor species that was not detected in the MALDI analysis of the whole sample without chromatographic separation. Figure 12.13 shows partial MALDI mass spectra of an SEC separated sample (top), and the whole sample (bottom).

The minor species is clearly visible in the size-separated sample, while it is not visible in the whole sample analysis. This minor species can be assigned to a cyclic polymer that has lost water during the cyclization reaction. Since these oligomers will not react without the –OH end groups, characterizing the presence of these species can be important to some applications. Mapping these species also show that they have a narrower molecular mass distribution than the rest of the sample.

Size separation can also be used to refine an analysis of a higher molecular mass component of a new material in the presence of a highly responsive lower mass component. Figure 12.14 shows the mass spectrum of a blended material.

Without using the TOFMS low-mass blanking capability, a minor component is not detected. With mass blanking, the presence of the higher mass component of the blend is detected, but the analysis is not sufficient for chemical characterization. Figure 12.15 shows the higher mass MALDI mass spectra obtained after separating the sample with SEC and capturing the eluant with the LCT device.

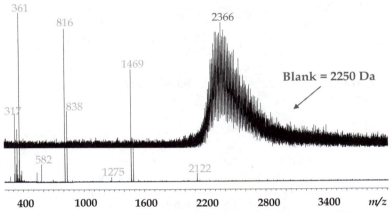

Figure 12.14.   MALDI mass spectra of a novel blended surfactant. Without blanking (bottom), the high mass component is not detected. With blanking (top), the high mass component is detected but not well characterized.

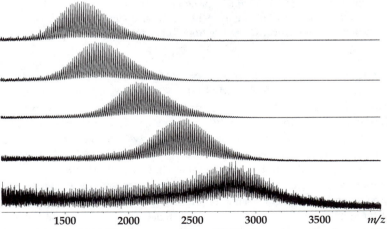

Figure 12.15.   MALDI mass spectra from SEC separated sample of a novel blended surfactant. Each mass spectrum was obtained from the material eluted by the SEC at a different time.

By analyzing different spots on the LCT plate, which correspond to different elution volumes, different regions of the oligomer distribution of the minor component are characterized.

12.11   BLEND QUANTITATION ISSUES

Hopefully, through all of the chapters of this book, the many capabilities of MALDI to analyze synthetic polymers have been illustrated. There are many different experiments that work exceedingly well, even if the underlying MALDI mechanisms are not well understood. Despite all of that success, there is one experiment in which MALDI methods have struggled to succeed. Many synthetic polymers are used commercially in products that contain blends or mixtures of different materials. While MALDI can provide detailed chemical characterization information about the components of the blend, MALDI experiments generally struggle to quantify the relative amounts of the different components in the blend.

MALDI mass spectra can be generated of the materials that compose a blend of different polymers, as shown in the molecular defoamer in Figure 12.12. While MALDI will often provide analytically useful quantitative results for the average molecular weights of the individual components, MALDI is usually challenged to quantitatively measure the relative amounts of the components of the blend. One of the difficulties, as discussed earlier in Chapter 11, is the accurate assignment of the spectrum baseline and calculation of the area of the mass spectral peaks. A more significant scientific problem, however, is that under many sample preparation conditions, the relative responses of different materials in a single MALDI mass spectrum have little correlation to the concentrations of those species in the sample vial. Different materials appear to have very different response factors for MALDI. Response factors can vary by more than two orders of magnitude for a cationic surfactant compared to a nonionic surfactant. These types of response factor differences have also been observed by SIMS [36]. So far, there is very little in the literature measuring MALDI response factors for synthetic polymers. A few recent works have been published which either concentrate on the quantitation of biomolecules [37] or show specialized results for one class of polymer [38]. A recent paper has applied the concept of using an internal standard to use MALDI to obtain quantitative data [39]. Many other analytical techniques use internal standards, and this idea will need to be explored more fully for use in MALDI.

Figures 12.16 and 12.17 show MALDI mass spectra obtained for an equimolar blend of PEG 1500 and PEG 4600 with different instrument conditions.

Figure 12.16 was collected on a home-built TOFMS without delayed extraction. Figure 12.17 was collected on a modern commercial instrument with delayed extraction. It is clear that the details of the instrumental conditions make a significant impact on the relative responses measured for these polymer samples. Additional experiments (data not shown) show that the issue is not simply mass resolution, but is a complex mixture of issues including instrument and sample preparation. In a detailed study of these issues, Goldschmidt completed a series of factorial-design experiments to probe the impact of experimental conditions on quantitation of blends

1:1 Molar blend of PEG 1500:PEG 4600

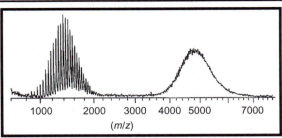

Relative area = PEG 1500/PEG 4600 = 0.30

Figure 12.16.   Low-resolution MALDI mass spectrum of an equimolar blend of PEG 1500 and PEG 4600 collected without delayed extraction. The relative areas of the two species are 1500/4600 = 0.30.

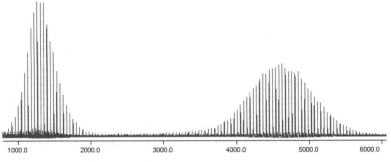

Relative area = PEG 1500/PEG 4600 = 0.88

Figure 12.17.   High-resolution MALDI mass spectrum of an equimolar blend of PEG 1500 and PEG 4600 collected with delayed extraction. The relative areas of the two species are 1500/4600 = 0.88.

of the same repeat unit and different molecular masses [40]. Figure 12.18 shows the results of one factorial experiment on PEG blends probing the relative concentrations of the polymers and the matrix.

These experiments show that while blended samples produce MALDI spectra, the quantitative relative areas require great care in their interpretation.

The experiments described above have all considered blends of materials of the same chemistry but different molecular mass. If molecular mass is the key variable, can MALDI successfully quantitate blends of polymers with different repeat units but the same molecular mass? Figure 12.19 shows a MALDI mass spectrum of equal concentrations of PEG 1900, monomethyl PEG 2000 (PEGOMe), PPG 2025, and Igepal Co890 (an ethoxylated nonylphenol with molecular mass of about 2000 Da).

This experiment was done with THF as the solvent, DHB as the matrix, and sodium trifluoroacetate (NaTFA) as the cationization agent. We clearly see signifi-

PEG 1500/PEG 4600 (area/mole)

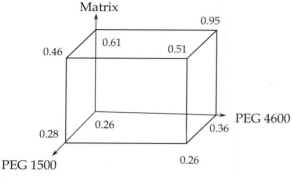

Figure 12.18. Results from a factorial experiment examining the relative intensity of PEG 1500/PEG 4600 with respect to the amount of PEG 1500, amount of PEG 4600, and the amount of the DHB matrix. The values at the cube vertices show how dependent the quantitation is upon the details of the experiment.

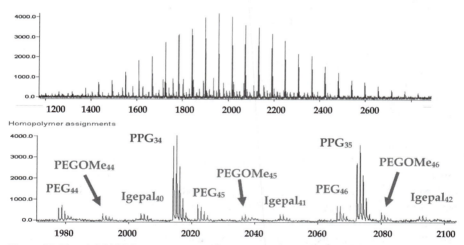

Figure 12.19. A MALDI mass spectrum from an equal mass blend of four common polymers with molecular mass of about 2000 Da: PEG, monomethyl PEG (PEGOMe), PPG, and an ethoxylated nonylphenol (Igepal Co890). The relative peak intensities show very different relative response factors for these polymers in this experiment in DHB. See color insert.

cant differences in the relative response factors for these four common polymers, despite their chemical similarity. Table 12.2 below shows the results of analysis of these data.

These data reinforce the results from the earlier, same chemistry, different molecular mass experiments. There is still a significant lack of understanding about quantitating polymer blends.

TABLE 12.2.   Relative Response of Four Polymers from an Equimolar Blend

Polymer	% Relative Intensity	Response Factor[a]
PEG	18	1.00
PEGOMe	6	0.33
PPG	68	3.78
Igepal	8	0.44

[a]The response factor in this table is relative to the PEG peak intensity.

12.12   SUMMARY

All of the chapters in this book show numerous experiments of solving interesting problems in polymer chemistry with MALDI mass spectrometry. In this chapter several examples have been shown on how to solve problems relevant to industrial polymer chemists. MALDI data have been shown to be valuable in determining the chemical structure of a variety of materials. For more complex problems, MALDI experiments often provide complementary data to other analytical measurements such as NMR, IR, GC/MS, and SEC.

To utilize all of the power of MALDI to solve important problems, care must be taken with the sample preparation. Choice of the appropriate matrix and cationization agent and use with either solvent or solvent-free sample preparation methods are critical to obtaining analytically useful data. The literature continues to grow with innovative additions to our knowledge of matrices and sample preparation methods.

Despite the power of current MALDI technology, there are two key challenges to further improve MALDI to solve synthetic polymer problems. First, improve the dynamic range of the TOFMS instruments. Second, additional work is required to understand the issues preventing quantitative analysis of polymer blends, as discussed above. The characterization of the individual components of the blend is very useful, but once the question of, *what is it* is answered, it is quickly followed by *how much is there?* Improvements in blend quantitation and dynamic range would enable MALDI to provide even more important data.

ACKNOWLEDGMENTS

The authors would like to acknowledge Air Products and Chemicals, Inc., Drexel University, John Sadowski for critical review of the manuscript, the editor for the opportunity to contribute this chapter, and several of our key colleagues over more than a decade of joint research and collaboration, including Dave Parees, Rick King, Bob Goldschmidt, Michael Liu, Yansan Xiong, Fritz Walker, Renee Keller, Li Jia, Andrew Hoteling, and Bill Erb.

REFERENCES

1. Tanaka, K.; Waki, H.; Ido, Y.; Akita, S.; Yoshido, Y.; Yoshido, T., Protein and polymer analyses up to m/z 100,000 by laser ionization time-of-flight mass spectrometry, *Rapid Commun. Mass Spectrom.*, **1988**, 2, 151.
2. Bürger, M.; Müller, H.; Seebach, D.; Börnsen, O.; Schär, M.; Widmer, M., Matrix-assisted laser desorption and ionization as a mass spectrometric tool for the analysis of poly[(r)-3-hydroxybutanoates], comparison with gel permeation chromatography, *Macromolecules*, **1993**, 26, 4783.
3. Blais, J.; Tessier, M.; Bolbach, G.; Remaud, B.; Rozes, L.; Guittard, J.; Brunot, A.; Maréchal, E.; Tabet, J., Matrix-assisted laser desorption ionization time-of-flight mass spectrometry of synthetic polyesters, *Int. J. Mass Spectrom. Ion Proc.*, **1995**, 144, 131.
4. Montaudo, G.; Montaudo, M.S.; Puglisi, C.; Samperi, F., Characterization of polymers by matrix-assisted laser desorption ionization-time of flight mass spectrometry: end group determination and molecular weight estimates in poly(ethylene glycols), *Macromolecules*, **1995**, 28, 4562.
5. Danis, P.; Karr, D.; Mayer, F.; Holle, A.; Watson, C., The analysis of water-soluble polymers by matrix-assisted laser desorption time-of-flight mass spectrometry, *Org. Mass Spectrom.*, **1992**, 27, 843.
6. Danis, P.; Karr, D., A facile sample preparation for the analysis of synthetic organic polymers by matrix-assisted laser desorption/ionization, *Org. Mass Spectrom.*, **1993**, 28, 923.
7. Lloyd, P.; Suddaby, K.; Varney, J.; Scrivener, E.; Derrick, P.; Haddleton, D., A comparison between matrix-assisted laser desorption/ionization time-of-flight mass spectrometry and size exclusion chromatography in the mass characterization of synthetic polymers with narrow molecular-mass distributions: poly(methyl methacrylate) and poly(styrene), *Euro. J. Mass Spectrom.*, **1995**, 1, 293.
8. Belu, A.; DeSimone, J.; Linton, R.; Lange, G.; Friedman, R., Evaluation of matrix-assisted laser desorption ionization mass spectrometry for polymer characterization, *J. Am. Soc. Mass Spectrom.*, **1996**, 7, 11.
9. Anton, S.D., Mass spectrometry of polymers and polymer surfaces, *Chem. Rev.*, **2001**, 101, 527.
10. Nielen, M.W.F., MALDI time-of-flight mass spectrometry of synthetic polymers, *Mass Spectrom. Rev.*, **1999**, 18, 309.
11. Pash, H.; Schrepp, W (editor), *MALDI-TOF Mass Spectrometry of Synthetic Polymers*. Berlin: Springer-Verlag, **2003**.
12. Gross, M.L. (editor), *Encyclopedia of Mass Spectrometry Vol. 6 (Molecular Ionization Methods)*. Amsterdam: Elsevier, **2007**.
13. Montaudo, G.; Lattimer, R.P. (editor), *Mass Spectrometry of Polymers*. Boca Raton, FL: CRC Press, **2002**.
14. See Chapter 2 of this book.
15. Hanton, S.D.; Clark, P.A.C.; Owens, K.G., Investigations of matrix-assisted laser desorption/ionization sample preparation by time-of-flight secondary ion mass spectrometry, *J. Am. Soc. Mass Spectrom.*, **1998**, 9, 282.
16. Hanton, S.D.; Owens, K.G., Using MESIMS to analyze polymer MALDI matrix solubility, *J. Am. Soc. Mass Spectrom.*, **2005**, 16, 1172–1180.
17. Hoteling, A.J.; Erb, W.J.; Tyson, R.J.; Owens, K.G., Exploring the importance of the relative solubility of matrix and analyte in MALDI sample preparation using HPLC, *Anal. Chem.*, **2004**, 76, 5157–5164.
18. See Chapter 6 of this book.
19. Mourey, T.H.; Hoteling, A.J.; Balke, S.T.; Owens, K.G., Molar mass distributions of polymers from SEC and MALDI-TOF MS: methods for comparison, *J. Appl. Polym. Sci.*, **2005**, 97, 627–639.
20. Hensel, R.R.; King, R.C.; Owens, K.G., Electrospray sample preparation for improved quantitation in matrix-assisted laser desorption/ionization time-of-flight mass spectrometry, *Rapid Commun. Mass Spectrom.*, **1997**, 11, 1785.
21. Hanton, S.D.; Hyder, I.Z.; Stets, J.R.; Owens, K.G.; Blair, W.R.; Guttman, C.M.; Giuseppetti, A.A., Investigations of electrospray sample deposition for polymer MALDI mass spectrometry, *J. Am. Soc. Mass Spectrom.*, **2004**, 15, 168–179.

22. Trimpin, S., Rouhanipour, A., Räder, H.J.; Müllen, K., New aspects in matrix-assisted laser desorption/ionization time-of-flight mass spectrometry: a universal solvent-free sample preparation, *Rapid Commun. Mass Spectrom.*, **2001**, 15, 1364–1373.

23. Hanton, S.D.; Parees, D.M., Extending the solvent-free MALDI sample preparation method, *J. Am. Soc. Mass Spectrom.*, **2005**, 16, 90–93.

24. Macha, S.F.; Limbach, P.A.; Hanton, S.D.; Owens, K.G., Silver cluster interferences in MALDI mass spectrometry of nonpolar polymers, *J. Am. Soc. Mass Spectrom.*, **2001**, 12, 732–743.

25. Montaudo, G.; Montaudo, M.S.; Puglisi, C.; Samperi, F., Characterization of polymers by matrix-assisted laser desorption/ionization time-of-flight mass spectrometry: molecular weight estimates in samples of varying polydispersity, *Rapid Commun. Mass Spectrom.*, **1995**, 9, 453–460.

26. Zhu, H.; Yalcin, T.; Li, L. Analysis of the accuracy of determining average molecular weights of narrow polydispersity polymers by MALDI TOFMS, *J. Am. Soc. Mass Spectrom.*, **1998**, 9, 275–281.

27. Hanton, S.D., New mass spectrometry techniques for the analysis of polymers for coatings applications: MALDI and ESI, *JCT Coatings Tech.*, **2004**, 1, 62–68.

28. Parees, D.M.; Hanton, S.D.; Cornelio Clark, P.A.; Willcox, D.A., Comparison of mass spectrometric techniques for generating molecular weight information on a class of ethoxylated oligomers, *J. Am. Soc. Mass Spectrom.*, **1998**, 9, 282–291.

29. Barry, J.P.; Caron, W.J.; Pesci, K.M.; Anselmo, R.T.; Radke, D.R.; Evans, J.V.; Derivatization of low molecular weight polymers for characterization by MALDI TOFMS, *Rapid Commun. Mass Spectrom.*, **1997**, 11, 437–442.

30. See Chapter 3 of this book.

31. Walker, F.H.; Dickenson, J.B.; Hegedus, C.R.; Pepe, F.R.; Keller, R.J., New polymeric polyol for thermoset coatings:superacid-catalyzed copolymerization of water and epoxy resins, *JCT Coatings Tech.*, **2002**, 74, 928.

32. Jia, L.; Sun, H.; Ding, E.; Allegeier, A.M.; Hanton, S.D., Living alternating copolymerization of N-alkylaziridines and carbon monoxide as a route for synthesis of poly-β-peptoids, *J. Am. Chem. Soc.*, **2002**, 124(25), 7282.

33. Danis, P.O.; Karr, D.E., Applications of MALDI in polymer chemistry. Presented at the ASMS Fall workshop in Boston, MA, November 12–13, **1993**.

34. Dwyer, J.; Botten, D., A novel sample preparation device for MALDI-MS, *Am. Lab.*, **1996**, 28, 51–54.

35. Hanton, S.D.; Liu, X.M., GPC separation of polymer samples for MALDI analysis, *Anal. Chem.*, **2000**, 72, 4550–4554.

36. Benninghoven, A.; Ruedenauer, F.G.; Werner, H.W., *Secondary ion mass spectrometry*, New York: John Wiley and Sons, **1987**.

37. Chavez-Eng, C., Quantitative aspects of matrix-assisted laser desorption/ionization using electrospray deposition. PhD Thesis, Drexel University, **2002**.

38. Yan, W.; Joseph, A.; Gardella, J.; Wood, T.D., Quantitative analysis of technical polymer mixtures by MALDI TOFMS, *J. Am. Soc. Mass Spectrom.*, **2002**, 13, 914–920.

39. Chen, H.; Me, M., Quantitation of synthetic polymers using an internal standard by MALDI TOFMS, *J. Am. Soc. Mass Spectrom.*, **2005**, 16, 100–106.

40. Goldschmidt, R.J., Some quantitative aspects of the analysis of synthetic polymers by MALDI TOFMS. PhD Thesis, Drexel University, **1998**.

INDEX

Note: Page numbers in *italics* refer to Figures; those in **bold** to Tables.

MALDI Mass Spectrometry for Synthetic Polymer Analysis, Edited by Liang Li
Copyright © 2010 John Wiley & Sons, Inc.

CHEMICAL ANALYSIS

A SERIES OF MONOGRAPHS ON ANALYTICAL CHEMISTRY
AND ITS APPLICATIONS

Series Editor
J. D. WINEFORDNER